ISODYNE STRESS ANALYSIS

ENGINEERING APPLICATION OF FRACTURE MECHANICS

Editor-in-Chief: George C. Sih

VOLUME 8

Isodyne Stress Analysis

J. T. Pindera

Department of Civil Engineering
University of Waterloo
Waterloo, Ontario, Canada

and

M.-J. Pindera

Department of Civil Engineering
University of Virginia
Charlottesville, Virginia, U.S.A.

Kluwer Academic Publishers

Dordrecht / Boston / London

Library of Congress Cataloging-in-Publication Data

Pindera, Jerzy-Taduesz.
    Isodyne stress analysis / J.T. Pindera and M.-J. Pindera.
        p.   cm. -- (Engineering application of fracture mechanics)
    Includes bibliographies and index.
    ISBN-13:978-94-010-6927-4   e-ISBN-13.978-94-009-0973-1
    DOI: 10.1007/978-94-009-0973-1

    1. Strains and stresses. 2. Photoelasticity.   I. Pindera, M.-J.
(Marek-Jerzy), 1951- II. Title.   III. Title: Isodyne stress.
IV. Series.
TA417.6.P46   1989
620.1'124--dc20   89-8037

Published by Kluwer Academic Publishers,
P.O. Box 17, 3300 AA Dordrecht, The Netherlands.

Kluwer Academic Publishers incorporates the publishing programmes of
D. Reidel, Martinus Nijhoff, Dr W. Junk, and MTP Press.

Sold and distributed in the U.S.A. and Canada
by Kluwer Academic Publishers,
101 Philip Drive, Norwell, MA 02061, U.S.A.

In all other countries, sold and distributed
by Kluwer Academic Publishers Group,
P.O. Box 322, 3300 AH Dordrecht, The Netherlands.

printed on acid free paper

Dedicated to
Aleksandra Anna Pindera

Contents

Contents IX

Series on engineering application of fracture mechanics

Fracture mechanics technology has received considerable attention in recent years and has advanced to the stage where it can be employed in engineering design to prevent against the brittle fracture of high-strength materials and highly constrained structures. While research continued in an attempt to extend the basic concept to the lower strength and higher toughness materials, the technology advanced rapidly to establish material specifications, design rules, quality control and inspection standards, code requirements, and regulations of safe operation. Among these are the fracture toughness testing procedures of the American Society of Testing Materials (ASTM), the American Society of Mechanical Engineers (ASME) Boiler and Pressure Vessel Codes for the design of nuclear reactor components, etc. Step-by-step fracture detection and prevention procedures are also being developed by the industry, government and university to guide and regulate the design of engineering products. This involves the interaction of individuals from the different sectors of the society that often presents a problem in communication. The transfer of new research findings to the users is now becoming a slow, tedious and costly process.

One of the practical objectives of this series on *Engineering Application of Fracture Mechanics* is to provide a vehicle for presenting the experience of real situations by those who have been involved in applying the basic knowledge of fracture mechanics in practice. It is time that the subject should be presented in a systematic way to the practising engineers as well as to the students in universities at least to all those who are likely to bear a responsibility for safe and economic design. Even though the current theory of linear elastic fracture mechanics (LEFM) is limited to brittle fracture behavior, it has already provided a remarkable improvement over the conventional methods not accounting for initial defects that are inevitably present in all materials and structures. The potential of the fracture mechanics technology, however, has not been fully recognized. There remains much to be done in constructing a quantitative theory of material damage that can reliably translate small specimen data to the design of large size structural components. The work of the physical metallurgists and the fracture mechanicians should also be brought together by reconciling the details of the material microstructure with the assumed continua of the computational methods. It is with the aim of developing a wider appreciation of the fracture mechanics technology applied to the design of engineering structures such as aircrafts, ships, bridges, pavements, pressure vessels, off-shore structures, pipelines, etc. that this series is being developed.

Undoubtedly, the successful application of any technology must rely on the soundness of the underlying basic concepts and mathematical models and how they reconcile with each other. This goal has been accomplished to a large extent by the book series on *Mechanics of Fracture* started in 1972. The seven published volumes offer a wealth of information on the effects of defects or cracks in cylindrical bars, thin and thick plates, shells, composites and solids in three dimensions. Both static and dynamic loads are considered. Each volume contains an introductory chapter that illustrates how the strain energy criterion can be used to analyze the combined influence of defect size, component geometry and size, loading, material properties, etc. The criterion is particularly effective for treating mixed mode fracture where the crack propagates in a non-self similar fashion. One of the major difficulties that continuously perplex the practitioners in fracture mechanics is the selection of an appropriate fracture criterion without which no reliable prediction of failure could be made. This requires much discernment, judgement and experience. General conclusion based on the agreement of theory and experiment for a limited number of physical phenomena should be avoided.

Looking into the future the rapid advancement of modern technology will require more sophisticated concepts in design. The micro-chips used widely in electronics and advanced composites developed for aerospace applications are just some of the more well-known examples. The more efficient use of materials in previously unexperienced environments is no doubt needed. Fracture mechanics should be extended beyond the range of LEFM. To be better understood is the entire process of material damage that includes crack initiation, slow growth and eventual termination by fast crack propagation. Material behavior characterized from the uniaxial tensile tests must be related to more complicated stress states. These difficulties should be overcome by unifying metallurgical and fracture mechanics studies, particularly in assessing the results with consistency.

This series is therefore offered to emphasize the applications of fracture mechanics technology that could be employed to assure the safe behavior of engineering products and structures. Unexpected failures may or may not be critical in themselves but they can often be annoying, time-wasting and discrediting of the technical community.

G. C. Sih
Editor-in-Chief

Bethlehem, Pennsylvania

Preface

"It is true that
"Nothing is more practical than a theory"
Provided — however —
That the assumptions on which the theory is founded
Are well understood.
— But, indeed, engineering experience shows that
"Nothing can be more disastrous than a theory
When applied to a real problem
Outside of the practical limits of the assumptions made",
Because of an homonymous identity
With the problem under consideration."

(J.T.P.)

The primary objective of this work is to present the theories of analytical and optical isodynes and the related measurement procedures in a manner compatible with the modern scientific methodology and with the requirements of modern technology pertaining to the usefulness of the stress analysis procedures. The selected examples illustrate some major theses of this work and demonstrate the particular efficiency of the isodyne methods in solving the technologically important problems in fracture mechanics and mechanics of composite structures including new materials.

To satisfy this objective it was necessary to depart from the common practice of presenting theories and techniques of experimental methods as a compatible system of equations and procedures without mentioning the tacitly accepted assumptions and their influence on the theoretical admissibility of analytical expressions and the reliability of the experimental or analytical results. It was necessary to design a more general frame of reference which could allow to assess the scientific correctness of isodyne methods and the reliability of experimental results.

The key issue of this work is the reliability and accuracy of experimental and analytical procedures of experimental mechanics, and particularly of analytical and experimental procedures of stress and strength analysis, which are used in structural design and optimization. This involves reliability of analytical models of the load-stress-strain-deformation relations, of the constitutive materials relations, and of the materials strength hypotheses pertaining to fatigue, fracture mechanics, and inelastic failures. Thus the format and scope of this work is determined by the stated and perceived requirements of modern industry with regard to such issues as the damage tolerance studies related to reliability of the product, the assurance of the quality and acceptable cost of the product, risk analysis, the product liability, and the liability of researchers and designers resulting from the concept of global knowledge.

The title of the book, "Isodyne Stress Analysis" is too concise — out of necessity — to satisfactorily represent its content and chosen approach. We use the preface to present the reasons for the chosen approach and particularly the reason why the theory, techniques, and applications of the recently developed method of the isodyne stress analysis are presented within a more general scientific framework. This framework comprises theories of modeling reality, perception, observation, measurement, and experimentation, or shortly, the theory of scientific methods and the influence of the prevailing, or established, paradigm. To do this, we refer to the present situation in science and technology.

The present period is characterized by a fantastic rate of development in science and by resulting thereof achievements of modern technology. The cause of this progress is socio-economic pressure related to societal needs and demands, competition, defense, safety, or ecology. As a result of this development, for the first time in history the rapidly changing technologies profoundly influence all aspects of societal life. They influence the entire emerging knowledge and practice, including the whole educational and training systems. It is recognized that technological progress influences directly the life of the present generation and decides about the future of future generations. The related technological risk becomes rapidly a major field of study; the developing procedures of risk analysis and the notions of product liability already directly influence various design and manufacturing procedures. With regard to applied mechanics this leads to reassessment of procedures of analytical and experimental stress and fracture analysis with regard to their reliability, that is their ability to correctly simulate and predict the behaviour of real engineering structures. This is especially true when the optimization of composite structures and structures made of new materials is considered.

In addition, the progress in applied physics, mathematics, electronics, information theory and electronic signal processing has led to the development of measurement instruments and systems, including commercially available instruments, whose performance and accuracy surpasses typical requirements of engineering research. In principle, the problems of sufficiently accurate instruments in experimental mechanics ceased to exist.

The achievements of modern technology are based on results of pertinent scientific and engineering research, which is performed according to requirements of modern industry including the relevant conditions and constraints, and which utilizes results obtained in various fields of sciences. This wing of engineering research, subjected to strong influence of scientific methodology through direct contacts with scientists and mathematicians and to the demanding requirements of the aeronautical and aerospace industries including defense, develops rapidly, is insignificantly influenced by the traditional paradigms, and is very innovative. However, numerous published papers and books show that this advanced scientific approach in experimental and analytical mechanics is not generally accepted, and that various analytical and experimental procedures are developed, or used, whose theoretical foundations are weak or incorrect since they often are theoretically inadmissible, either mathematically, or physically, or both. In this traditional wing of experimental and analytical research the influence of obsolete paradigms established long ago is prevailing.

As a result, a profound dichotomy has developed in experimental mechanics between two wings of the engineering research: the research which is modern technology-oriented, which is rooted in the traditions of technologically leading aeronautical engineering and which complies with requirements of scientific methodology; and the traditional, general engineering research which utilizes analytical and experimental procedures whose theoretical background contains mathematical and physical inadmissibilities accepted for various reasons. Contribution of the prevailing paradigms to the development and growth of this dichotomy between the physical and phenomenological methodologies in engineering research is intrinsic. This dichotomy makes it impossible or difficult to compare published research results and to assess their reliability with regard to the actual physical processes. Even the terminology is influenced as the terms such as theory, property, rigorousness, response, material constants, stress, strain, often are used to denote incompatible notions. Thus the demarcation line between the traditional and the advanced experimental mechanics is determined by the rank of the accepted theoretical foundation: either related to nontestable phenomenological theories or to testable physical theories.

A corollary problem has developed. Presently, hundreds of technical papers in particular fields of engineering research are published each year. The problem is how to choose the criteria for selecting papers which should be read in order to keep abreast with the rapidly advancing frontiers of knowledge. Thus far, the only benchmark is the physical inadmissibility caused by violation of the second law of thermodynamics which separates acceptable papers from unacceptable ones. There is a need for sufficiently comprehensive demarcation lines based on the notions of the theory of scientific methods.

In summary, the consequences of the dichotomy that presently exists in experimental mechanics are serious with regard to the reliability of published results and their applicability to the very demanding tasks of modern technology. In the modern technological world, the research, development, design, manufacturing and servicing are considered to be components of a system whose aim is to economically produce reliable and competitive products. The problem of risk analysis and product liability is becoming very acute. This leads to the notion that personal responsibility of researcher and designer is a component of the product liability, in accordance with the developing principle of global knowledge. Rapid development of new materials and new composite structures not amenable to traditional procedures of stress analysis and material characterization compounds the problem. The general socio-technological framework outlined above is of importance to the theories and techniques of experimental and analytical stress analysis. The fact that at the present time the term "Experimental Mechanics" is used to denote two incompatible systems of theories and procedures is detrimental to technological progress.

We believe that this new situation requires that contemporary books on stress analysis emphasize two interrelated issues: the issue of the scientific correctness of the presented theories, and the issue of the reliability of obtained results with respect to the actual physical processes occurring in real bodies. In accordance with this axiom, the content of this book is divided into two series of chapters. Chapters 1—8 which constitute Part 1 of the book provide the theoretical background needed to correctly analyse and assess the experimental results presented in Chapters 9—12 which constitute Part 2. In this way the efficacy and efficiency of isodyne methods can be assessed, particularly with regard to the local three-dimensional effects in regions of cracks and notches and the mechanics of composite structures.

Particular attention is given in Chapters 1 and 2 to the basic scientific and methodological problems of importance to experimental mechanics, including the theory of modeling reality, limits of applicability of notions and solutions of infinitesimal elasticity, and discussions of local, three-dimensional effects.

Theories of plane and differential analytical isodynes are presented in Chapter 3 within the framework of linear elasticity. The derived relations, together with the models of interaction between radiation and matter presented in Chapter 4, are utilized to develop theory of optical isodynes presented in Chapter 5.

Pertinent constitutive material relations are discussed in Chapter 6 where examples illustrating actual, multi-mechanism, inelastic mechanical and optical responses of several polymeric materials are presented.

A set of basic experimental techniques of isodynes, including the basic techniques of the amplitude modulation and spatial frequency modulation is presented in Chapter 7.

The selected examples given in Part 2 encompass semi-plane stress states, contact problems, three-dimensional local effects, and stresses in composite structures. These examples show that the isodyne methods are suitable for performing the required testing of analytical solutions and the quantification of ranges of applicability of singular solutions.

It is our pleasant duty to acknowledge the conceptual and factual contributions of various individuals to the format and contents of this work. We acknowledge and appreciate contributions of our co-workers, whose names are listed in references. In particular, the contributions of Prof. Zhang Yuanpeng and Ms. Xinhua Ji deserve special recognition. Discussions of basic theoretical issues with Prof. H. H. E. Leipholz and Prof. George C. Sih significantly enhanced the format of this work. We are particularly obliged to Prof. George C. Sih for his suggestion to write a book on isodyne methods in fracture mechanics of homogeneous and composite structure and for his patience and encouragement. The format and approach chosen in this book is strongly influenced by the format of courses taken by the first author at Warsaw

Technical University from professor of physics, Mieczysław Wolfke, and professor of mathematics, Witold Pogorzelski, as well as from professors of aircraft design. The support and understanding given by Aleksandra Anna Pindera were invaluable for completion of this work.

The research work whose results are utilized in this book has been supported by the Natural Sciences and Engineering Research Council of Canada under Grant A-2939. The available space does not permit to list all the persons who contributed to the production of the manuscript and whose faultless contributions are sincerely appreciated; thus individual acknowledgements are limited to Mrs. Nadia Bahar who efficiently drafted all the figures, and Mr. Derek W. Hitchens who rendered invaluable photographic services. It should be mentioned that the concept of isodynes was developed during the research on behaviour of bolted flanged connections which had been supported by the Pressure Vessels Research Committee of the American Society of Mechanical Engineers. Authors are also indebted to the technical staff of the Machine Shop in the Faculty of Engineering at the University of Waterloo, who manufactured measurement instruments utilized in the presented research.

Waterloo, Ontario, Canada   Jerzy Tadeusz Pindera
October 1988   Marek-Jerzy Pindera

List of major symbols and definitions

a_i	Principal components of index tensor ($i = 1, 2, 3$)
a_{ij}	Components of index tensor ($i, j = x, y, z$)
b, b_0	Thickness
c	Velocity of the light in vacuum or in air
C_ε	Photoelastic strain coefficient
$C_{1\varepsilon}, C_{2\varepsilon}$	Absolute photoelastic strain coefficients
$C_\varepsilon(t)$	Photoelastic stress relaxation modulus function
C_σ	Photoelastic stress coefficient
$C_{1\sigma}, C_{2\sigma}$	Absolute photoelastic stress coefficients
$C_\sigma(t)$	Photoelastic creep compliance function
D	Electric displacement vector
$D(t)$	Creep compliance in tension
E	Modulus of elasticity
$\overline{E}$	Electric vector
$E(t)$	Relaxation modulus in tension
F	Transfer function
f	Frequency
$f_x(y), f_y(x)$	Integration boundary functions for normal stress components
$g_x(y), g_y(x)$	Integration boundary functions for shear stress components
G	Shear modulus
I, I_0	Intensity of electromagnetic radiation (radiant power per unit area)
k	Absorption index; factor of proportionality
K	Coefficient
ℓ_{sx}, ℓ_{sy}	Order of lines of constant normal stress components
m	Isochromatic order
m_i	Isopachic order
m_{sx}, m_{sy}	Isodyne order
M	Integer order of isodynes
n_i	Principal components of index of refraction ($i = 1, 2, 3$)
n	Index of refraction
$\bar{n}$	Unit vector, normal to the boundary directed outwards

p_ε	Relative strain-optic constant
p_{ij}	Elasto-optic or strain-optic constant $(i, j = 1, 2, \ldots, 6)$
p_x, p_y	Intensities of normal forces [force/length]
p_{nx}, p_{ny}	Normalized intensities of normal forces
P_x, P_y	Intensities of total normal forces
q_σ	Relative stress optic constant
q_{ij}	Piezo-optic or stress-optic constant $(i, j = 1, 2, \ldots, 6)$
$\bar{q}$	Load intensity at the boundary
q_x, q_y	Components of load intensity at the boundary
r	Distance from the point of scattering to the point of observation
R	Relative retardation
$\overline{R}$	Reflected electric vector
s_r	Output signal
s_e	Input signal (entrance signal)
s_σ	Model stress-optic coefficient [force/area]
$s(x, y)$	Boundary of the plate
S_x, S_y	Characteristic directions related to x- or y-directions
S_s	Arbitrary chosen elastic isodyne coefficient [force/length]
S_ε	Strain-optic material coefficient
S_σ	Stress-optic material coefficient [force/length]
t	Time
T	Temperature
$\overline{T}$	Transmitted electric vector
t_{xy}, t_{yx}	Intensities of shear forces
T_{xy}, T_{yx}	Total shear forces
v	velocity of light in the medium
V	Volume of the particle
W, W_0	Radiant power per unit cross-sectional area per unit bandwidth
α	Angle between axis of polarization and one of the principal directions
β	Angle between analyzer axis and normal to polarizer axis
δ	Angle of incidence of the light beam with the outer normal to the boundary
ε_i	Principal components of strain tensor $(i = 1, 2, 3)$
ε_i^s	Secondary principal strain components in the plane perpendicular to the characteristic direction $(i = 1, 2, 3; s = x, y, z)$
ε_{ij}	Components of strain tensor; components of dielectric tensor $(i, j = x, y, z)$
$\varepsilon_{\ell\ell}$	Linear limit strain
θ	Angle between the direction of propagation of incident radiation and direction of observation (observation angle)
λ	Wavelength of radiation
μ	Magnetic permeability
ν	Poisson ratio
ρ	Radius of curvature
σ_i	Principal components of stress tensor $(i = 1, 2, 3)$

σ_i^s	Secondary principal stress components in the plane perpendicular to the characteristic directions ($i = 1, 2, 3; s = x, y, z$)
σ_{ij}	Components of stress tensor ($i, j = x, y, z$)
σ_{ij}^n, σ_{nij}	Normalized components of stress tensor ($i, j = x, y, z$)
$\sigma_{\ell\ell}$	Linear limit stress
ϕ	Relative phase retardation
$\Phi(x, y)$	Airy stress function
Φ	Angle between the direction of polarization of incident radiation and the scattering plane (azimuthal angle)

Purpose. Approach. Methodology

1.1 Concept of isodynes

1.1.1 Treatment of the subject

The general aim of this work is to present a unified and consistent picture of the theory and techniques of isodynes developed hitherto in a manner consistent with the requirements of scientific approach. The manner and scope of presentation of the topic has been chosen according to the requirements of the testability of theories and procedures. This encompasses discussion and assessment of assumptions, derivations, and procedures of collecting experimental data and evaluating of results.

The spontaneous and rapid development of various methods of isodynes, in particular of the methods related to certain industrial problems, has resulted in some incongruities regarding the terminology, assumptions, or interpretation of the results. For instance, according to the obtained empirical evidence, in several well defined cases it is necessary to consider the influence of the mirage effect which has been used as a foundation of the strain gradient methods [29, 30, 34, 37] and which limits the applicability of the integrated photoelasticity and of the caustic methods [1, 2]. Also, the resolution of the optical isodyne measurements can be so high that the influence of the geometrically nonlinear effects (e.g. in the finite deformation problems) becomes noticeable. The rapidly developing concepts of the differential isodynes, dealing with three dimensional problems, yield requirements which must be satisfied by a theory more general than the particular theory of the plane analytical isodynes presented so far. The requirements discussed in the sequel, in particular the requirements following from Bohr's correspondence principle [10], must be satisfied to assure that the theories of isodynes and related measurements are presented in a manner consistent with the theoretical framework of other advanced optical methods of experimental mechanics which are presently developed for advanced industrial purposes. In this context it is interesting to note that the theoretical background of many optical methods developed under the pressure of industry is rather sophisticated in comparison with the theoretical background accepted in the average engineering research. In other words, the modern high technology industry presents requirements with regard to the format of methods and techniques under development which are often the most demanding requirements concerning the theoretical background of the experimental engineering methods acceptable in modern engineering practice.

Summarizing, the rapid development of the theory, techniques, and practical applications of isodynes made it impossible to systematically develop the pertinent terminology before the major perspectives of development were perceived and identified. It appears that now the time has arrived to unify the theory and various techniques and to introduce a more consistent terminology in order to establish a sufficiently general theoretical framework. In this process it was unavoidable to critically assess, and to revise when necessary, several conceptions, definitions, formulations and derivations which had been presented in the earlier papers of the first of the authors and his co-workers, in order to present a consistent theory, not contradictory with itself and the correspondence principle. The theory and related techniques presented in this work are sufficiently general to be used as a basis for further generalization encompassing arbitrary three-dimensional stress states.

It must be noted that the treatment of the subject is basically limited to the physically and geometrically linear relations between stresses, strains, and patterns of interaction between radiation and deformed solid bodies. In other words, all the accepted physical and mathematical models of involved phenomena are linear: the values of all major experimental physical parameters of the experimental techniques are chosen accordingly. The actually occurring local nonlinear effects, both physical and geometric, are discussed and their influence on the predictions of linear models is presented and assessed.

For the sake of simplicity, the patterns of propagation of radiant energy and its interaction with deformed solids are discussed in terms of the wave theory, with some unavoidable exceptions.

When developing a new analytical-experimental method which should satisfy the minimal acceptable requirements of scientific rigorousness and methodology on one hand, and the requirements of engineering usefulness and reliability on the other hand, it is not possible to avoid inconsistencies in the presentation of the topic, in the interpretation of the developed relations and in the terminology. This work removes inconsistencies in treatment of the subject and in presentation of analytical-experimental results which occured in former papers on the theory and techniques of isodynes, the representative samples of which are listed in pertinent references [18, 30, 32, 34, 35].

1.1.2 Development of the concept of isodynes

It has been shown that the integrated photoelastic method developed by J. T. Pindera and P. Straka in 1971 [17], yields photographic recordings of the transmitted and scattered light intensities, each of which contains enough information to determine easily and reliably all stress components of a plane stress field along a principal stress trajectory when this trajectory is a straight line. A possibility of generalization of this method and application to contact problems has been suggested by Pindera and Sze in 1972 [18].

The term "photoelastic isodyne" was proposed by J. T. Pindera and S. Mazurkiewicz in 1977 [20] to denote a new family of characteristic lines of plane stress fields obtained by using the theory and technique presented in [17] and [18]. In 1981 J. T. Pindera introduced the term "elastic isodynes" [23] to denote a family of characteristic lines of plane stress field, related to first derivatives of Airy stress function, and has shown that the elastic isodynes and plane photoelastic isodynes are identical when some particular conditions are satisfied. The capacity of isodyne methods in stress analysis of contact problems was demonstrated by Pindera and Mazurkiewicz in 1981 [24].

In 1983 J. T. Pindera and B. R. Krasnowski [32] introduced the concepts of generalized plane elastic isodynes related directly to first derivatives of Airy stress functions and of isodyne surfaces. The notion of differential isodynes has been introduced in 1985 by J. T. Pindera, B. R. Krasnowski and M.-J. Pindera [35].

Basic theory of isodynes has been presented by J. T. Pindera and B. R. Krasnowski in 1983 [32], by J. T. Pindera in 1984 [34] and by J. T. Pindera, B. R. Krasnowski and M.-J. Pindera in 1985 [35].

1.1.3 Characteristic features of isodynes

Isodynes are always related to a particular direction in a particular stress field in a stressed body; that direction is called "characteristic direction". Isodynes are particular lines on the isodyne surfaces which are spanned over a particular plane located at the surface of a stressed body, or inside of this body. This plane is called "characteristic plane". Specifically, isodynes are lines of constant elevation of the isodyne surface with respect to the characteristic plane. Projection of those lines on the characteristic plane represents "field of isodynes".

Plane elastic analytical isodynes carry information on the total normal forces acting on the cross-sections which are parallel to chosen characteristic directions or, more rigorously, on the intensities of the total normal forces existing in characteristic planes. Specifically, the cross-sections through an isodyne surface, parallel and normal to the characteristic directions, yield the parallel and normal isodyne distribution functions, the values of which are proportional to the values of the normal and shear force intensities. The slopes of these functions are proportional to the corresponding normal and shear stress components, respectively.

The term "experimental isodynes" denotes isodyne fields obtained experimentally, for instance by using optical methods of the optical isodyne techniques.

It has been shown during the last ten years that the methods of plane isodynes can be successfully applied to solve or experimentally study various classes of stress analysis problems, such as [24—28, 31, 34—35]:

— determination of stress components of stress fields in plates which are typically considered plane stress fields but actually are weakly three-dimensional;
— determination of stresses caused by concentrated loads (contact problems);
— determination of particular functions of stress components, e.g. the geometric loci of points at which one stress component is equal to zero;
— verification of analytical solutions for stress states and analysis of their reliability;
— determination of regions where the assumed plane stress field is becoming a clearly three dimensional stress field, which occurs in contact, notch or crack problems;
— determination of major components of local, three-dimensional stress states in plates which are often denoted by the term "local effects";
— determination of surface stresses in actual structures using isodyne coatings;
— determination of stress states in composite structures;
— determination of pertinent Airy stress functions.

It has also been shown that it is feasible to generalize the concept of isodynes to solve certain general three-dimensional stress problems.

Such a rapid development of various versatile methods and techniques based on the concept of isodynes, leads to the introduction of the term "isodyne methods" which denotes a new group of the analytical-experimental methods of stress analysis. At the present time, the methods of optical isodynes are sufficiently well developed to satisfy more demanding requirements related to fracture mechanics and mechanics of laminated connections and composite structures.

1.2 Scientific framework

1.2.1 Preamble

The final objective of this work is to present a particular system of new methods and techniques of stress analysis and its major components, in a manner consistent with the requirements of modern design engineers regarding the reliability and accuracy of stress analysis data, both analytical and experimental. This system is denoted by the term "isodynes", and has been developed to satisfy stress analysis requirements of fracture mechanics and mechanics of composite materials.

To satisfy modern scientific and utilitarian requirements, the theories and techniques of isodynes are presented in a manner which allows rational assessment and analysis of their theoretical foundations, and permits qualitative and quantitative testing of procedures and results. Such a format of presentation of the main topic required the choice of a sufficiently comprehensive theoretical framework which influenced the scope of this work.

The choice of a more general frame of reference was needed because contemporary engineering research exhibits various incompatibilities and contradictions at all theoretical and utilitarian levels. For example, there still exists the comfortable belief — which is common within the engineering community and is reflected in the structure of the majority of engineering texts and reference books — that the development of science in general, and engineering mechanics in particular, is a continuous intellectual process based on inductive logic. This belief is no longer tenable and useful. Thus in this work it was accepted as an axiom that a more critical insight into the foundations and meaning of the utilitarian engineering formulas is needed to assure that the particular engineering procedures of stress analysis are not used outside of their ranges of applicability.

In the remainder of this Chapter we outline a sufficiently general frame of reference which satisfies the related scientific, intellectual, and utilitarian requirements.

1.2.2 Introduction

The traditional division between science and technology, according to which scientists produce knowledge and expand our understanding of nature, and engineers apply this knowledge in technological problem-solving activities which are oriented towards practical, limited missions, is no longer clearly evident in various fields of modern engineering. The merging of frontiers in science and some selected fields of technology began almost a century ago when research and development engineers commenced to develop rational foundations for aeronautical engineering. The fantastic growth of aeronautical engineering and its transformation into aerospace engineering is a result of dreams and visions transformed into actions based on adapted scientific methodology and influenced by induced and growing societal needs. The patterns of development and growth of high technology, the objective of which is to satisfy the existing and anticipated needs of modern societies, are very similar. In both cases, the necessary research and development activities have been performed by research groups consisting of physicists, mathematicians, and of engineers having strong scientific background and appreciation of scientific methodology.

With respect to engineering mechanics, encompassing applied mechanics, and mechanics of solids and fluids, this above outlined development has led to the loss of internal coherence regarding the intellectual, scientific, and professional aspects. This loss of internal coherence, which has been apparent for a long time, has been drastically exposed by the recent progress in the theory and techniques of measurement; the modern instruments, easily available, allow the testing of analytical predictions and the assessment of the quality of engineering experiments with a practically unlimited accuracy [8].

Research in engineering mechanics is presently split profoundly along underlying theoretical lines into a basically phenomenological-hypothetical or speculative research, and a basically physical-theoretical (or physical-scientific) research, regardless of whether or not this is analytical or experimental research. It appears that the underlying reasons of this unsatisfactory situation are of epistemological and ontological character. Differences in the understanding of basic notions of science lead to a different understanding of reality, and different notions of theory [3, 10, 12, 14, 19, 22, 42]. In particular, this dichotomy leads to a different understanding of the term "speculative-phenomenological" in scientific research and in engineering research. In contrast with the approach in science, in the engineering speculative-phenomenological approach it is not customary to present and justify the complete system of assumptions and to quantify simplifications, neither in analytical nor in experimental mechanics. It is often tacitly accepted that some of the assumptions and the conclusion may be contradictory and mutually exclusive. Nevertheless, very often such analytical relations and experimental procedures are considered "rigorous", despite the fact that they may be based on assumptions, or lead to the conclusions, which are physically, that is theoretically, inadmissible.

Parallel to the still-existing belief that no theory is needed to understand the obtained experimental data and results because "nature always gives correct answers", there exists a belief that the correctly transformed basic analytical relations, for instance, the phenomenological constitutive material relations, must rigorously portray phenomena of interest, regardless of their theoretical foundations. Very often the predictions of such phenomenological relations are taken as measures for the assessment of some experimental procedures, and it is quite easy to find conclusions in published papers that "the agreement between the theory and the experiment is excellent". This leads to a dangerous overestimation of the validity, predictive power, and reliability of analytical relations of engineering mechanics, which results in the well-known attitude which is either stated explicitly or hidden in the structure of various engineering courses: "As an engineer, I do not care where my formula came from; all I want to know is how to use it." Obviously, the roots of such an odd understanding of the meaning and reliability of analytical relations and experimental data, and the lack of knowledge that both the analytical and experimental procedures must be based on reliable theories to be meaningful, are very deep and are embedded in the educational system. Two examples illustrate this issue.

The Tacoma Narrows Bridge, the design of which was extensively tested in a university laboratory using reduced models, collapsed in 1940 because of an unanticipated aeroelastic interaction. However, the same aeroelastic interaction, which may induce so-called "flutter" of airplane wings, was included in core courses in aeronautical engineering as early as 1930's, that is, about 10 years before the collapse of the Tacoma Narrows Bridge.

The first passenger jet airplane, the "Comet", was withdrawn from service in the 1950's after a series of tragic failures which were caused by metal fatigue in stress concentration regions. However, both phenomena, the metal fatigue and the stress concentration, were known already in the 1930's, as was the knowledge of three-dimensional stress states in the region of notches in plates. The actual failures were not predicted analytically, nor were they observed during the tests, because the theoretical model of the airplane taken as the basis of experiments was at variance with the actual working conditions. In both cases — the Tacoma Narrows Bridge failure and the failure of the airplane Comet — as in many others, the accumulated and available knowledge, which could have prevented failures, was not utilized.

A very particular issue of contemporary applied mechanics is the unwillingness or reluctance of researchers in the field of experimental mechanics to recognize and adapt recent advances in the field of analytical mechanics, and vice versa, as demonstrated in numerous published research papers. This leads to development of analytical relationships which are often based on physically inadmissible assumptions, and to the development of experimental procedures which are based on physically inadmissible assumptions and on impertinent or even incommensurate analytical relations. The resulting errors very often exceed values of the pertinent factors of safety.

The consequences of the above outlined theoretical and practical dichotomy in basic and applied research in engineering mechanics are often disastrous. They result in numerous critical assessments of the reliability of experimental procedures and obtained results, such as "half or more of the numerical data published by (engineering) scientists in their journal articles are unusable because there is no evidence that the researcher actually measured what he thought he was measuring, and no evidence that the possible sources of error were eliminated or accounted for" [4].

An analogous situation exists regarding the common application of various simplified analytical solutions of the plane theory of elasticity to real engineering problems. For instance, the rapid progress in fracture mechanics results in development of new concepts and new analytical and experimental procedures which are based on the acknowledgement of the actually existing three-dimensional stress states [22, 31] and local nonlinearities [15, 43, 44]. Nevertheless, a major part of the extensive and expensive research work is still being conducted within the framework of plane and linear fracture mechanics and it is often concluded in the published papers that the analytical predictions, called "theoretical results" are in excellent agreement with the obtained experimental evidence.

It appears that it should be possible to minimize the existing theoretical and practical confusion and contradictions in research in engineering mechanics by stressing the most pragmatic issue of the actual reliability of the analytical, analytical-experimental, and numerical methods being developed and applied to specific engineering tasks. In this approach the term "reliability" denotes the degree of correlation between the results obtained by means of various procedures of applied mechanics and applied materials sciences, and the actual responses of real physical bodies and systems to the actual energy inputs caused by actual loads.

Looking for the roots of the problem of an unsatisfactory correspondence between the results of the applied mechanics methods and the actual behaviour of physical bodies and systems, one can conclude that they are related to specific deficiencies of our educational system.

1.2.3 *Major driving forces in contemporary applied mechanics*

It has been common to believe, or to assume, that research in applied mechanics is basically a pure intellectual activity and, as such, is not influenced by the societal, economic or psychological factors. Intellectual curiosity was supposed to be the main driving force, not particularly influenced by pragmatic reasons; the use of mathematical logic was supposed to eliminate or suppress psychological factors. This belief is no longer tenable. The rapid development and application of physical, scientific methodology in applied mechanics is occurring under the influence of a system of factors briefly outlined below. These factors impose the basic requirement that the methods and techniques of applied mechanics which are meant to be applied for practical purposes must be related to the real universe and deal with real physical problems.

These requirements have, in effect, altered essentially the profile of experimental mechanics. Particular measurement procedures and their application to determine responses of materials, bodies and systems no longer represent the main tasks of experimental mechanics. Firstly, physicists are better equipped than engineers to develop advanced measurement systems and procedures, as proven extensively during the last decade. Secondly, the major problem of applied mechanics is at present the reliability of analytical and experimental procedures, which is becoming the major task of modern experimental mechanics. As a result, two distinct groups have emerged in the field of experimental mechanics: the phenomenological group which uses traditional concepts and approaches, and the physical group which applies concepts and methods of modern science. The following considerations illustrate this issue.

1.2.3.1 Requirements of modern societies. For the first time in history the rapidly developing new technologies influence profoundly all aspects of societal life. They influence the entire emerging knowledge and practice, including general university curricula and all other aspects of societal life.

Three emerging issues deserve particular attention from the point of view of the already developed socioeconomic framework, with regard to activities of theorists, analytists and experimentalists in the field of applied mechanics: recognition of the fact that engineering is related to social science, and does not only represent technology; ethical and legal aspects of the quality of the product of work of researchers and designers, such as the legal responsibility for results, the liability for defects and failures, and the underlying new principle of global knowledge; and risk (hazard) analysis — a new field of applied research [45].

The last issue, risk analysis, is of special interest to the researchers in mechanics. It allows a rational assessment of the degree (or level) of the reliability of analytical, experimental and numerical procedures used in engineering design by including human health and life as major components of criteria of designing, and by quantifying the notion of the acceptable human life and economic losses. This, in turn, allows the establishment of rational criteria for the formulation of quantified answers to questions such as "how small is small", "how big is big", and "how wrong is wrong". Such questions arise when the influence of the physical and mathematical simplifications on the reliability of analytical and experimental procedures adapted in engineering practice is considered.

1.2.3.2 Requirements of modern technology. The rapidly growing system of theories, processes, materials and technologies denoted by the term "high technology" requires advanced analytical and experimental procedures and methods, the development and testing of which require a more rigorous and comprehensive theoretical basis than that presently accepted as sufficient or satisfactory. It is worthwhile to note that the revolutionary changes in the advanced sector of technology denoted by the term "high technology" have been made possible by: utilization of the incredible progress in related fields of science such as physics, chemistry, electronics, information theory, mathematics; consideration of real physical systems such as structures, processes, materials; consideration of the actual interaction of components of the system in terms of energy flow and material and system parameters, presented analytically by overall transfer functions, impedances, responses and signal/noise ratios; development of advanced design based on comprehensive criteria of optimization and comprehensive analysis of structural reliability in actual working conditions.

The pattern, scope and depth of this development influences directly and indirectly basic and applied research in applied mechanics, and introduces a new major, physical trend.

1.2.3.3 Progress in the reliability and accuracy of measurements. Development of theories and techniques of engineering measurements has been traditionally considered an integral, or sometimes major, part of experimental mechanics; not much attention has been paid to the reliability of analytical relations relating the measured quantities and the static and dynamic responses of engineering bodies and systems to various energy inputs.

Requirements regarding the accuracy of engineering measurements have been much lower than those accepted in metrology and natural sciences because of theoretical and technical limitations — errors within the range of a few, or several, percent have been considered small. These limitations have been practically eliminated during the last decade. Instruments are commercially available, the resolution of which was considered theoretically impossible only a few years ago [8]. Methods and techniques have been developed which allow the collection, processing and retrieval of numerical and analogous information on particular parameters of the actual responses of materials, bodies and systems with a practically unlimited degree of accuracy and reliability. As a result, the resolution of instruments, the accuracy of measurements, and the reliability of recorded data ceases to be a problem in experimental mechanics when the theoretical background of the observer or experimentalist is adequate. It should be noted that this progress has occurred under strong socioeconomic pressure related to safety, ecology, competition, defense, etc.

A few examples illustrate the state-of-the-art: measurable time periods: 6—10 femtosecond $(10^{-15}\,\mathrm{s})$; molecular movements: picosecond spectroscopy $(10^{-12}\,\mathrm{s})$; photographic recordings: 300 picosecond time resolutions; electro-optic sampling: less than 300 femtosecond temporal resolution; recent features of hand-held calculators: symbolic calculations; holography: free-standing, three-dimensional images in space; storing capacity of new, dynamic random-access memory chips: 16 million binary bits of information; resolution of commercial image-processing systems: $2048 \times 2048 \times 8$ bits frame memory; transition temperature of new superconductors: 94 K or higher; spatial resolution of commercial micropositioners: 20—50 nanometer at 25 mm travel; spatial resolution of recordings and measurements: 1 angstrom.

The last development is remarkable from the point of view of solid state physics and of constitutive relations for engineering materials: the positions of atoms within crystalline and molecular networks and their displacements during the dislocation process and during elastic and inelastic deformation can be reliably observed.

Progress in designing superconducting materials has opened entirely new horizons in experimental mechanics, the consequences of which cannot be foreseen at the present time.

1.2.4 Interdisciplinary cooperation

As mentioned earlier, in the field of aeronautical engineering the close coopera-tion between engineers, physicists, chemists, physiologists and mathematicians has resulted in development of a scientific-technological basis which was suffi-ciently comprehensive and reliable to allow a successful transformation of aeronautical engineering into aerospace engineering, and made it possible to reach planets of our solar system.

Analogous forms of cooperation between engineering researchers and scien-tists are being developed extensively in various fields of technology. The results of such cooperation are remarkable regarding the development of the physical approach in the field of applied mechanics. Modern scientific methodology is being introduced, in particular, the methodology of modelling reality; mathe-matical rigorousness is more widely required; new mathematical theories are being utilized, such as catastrophe theory, theory of strange attractors, mathe-matics of chaos, or fractal mathematics [41].

1.2.5 Development in materials science

Three aspects of progress in this field can be considered:

- a new body of knowledge about the actual structure of materials, and about actual structural changes occurring in materials during deformation at the atomic and molecular levels.
- progress in analytical micromechanics related to simplified models of discrete structure of materials which may lead to relations representing a

bridge between the actual structure of material and the analytical representation of material responses, based on the concepts of continuum mechanics.

— development of constitutive relations for materials, characterized by two weakly related trends of development: a speculative-phenomenological description based on the concept of a single mechanism of an isothermal deformation of a hypothetical material continuum; a physical-theoretical description based on the concepts of averaged responses actually occurring at the atomic, molecular, and microscopic levels, which are related to various, simultaneously occurring mechanisms of deformation, inherently coupled with thermal effects.

It should be noted that the constitutive relations for materials are the basis of analytical and experimental procedures applied in stress analysis and in design. However, the state-of-the-art is not satisfactory at the present time with regard to the requirements of design, of fracture mechanics, and of mechanics of composite materials. Intensive team research work is being conducted with the objective to construct reliably not only the explanatory and descriptive but also predictive physical models of materials. So far the electronic theory for materials science is in the first stage of development and the theoretical-physical materials science is not yet invented [5].

The practicality of a more theoretical approach is demonstrated in [11]. The authors of that paper utilized effects at the quantum mechanics level to develop a new diagnostic procedure. Similarly, a new stress analysis procedure was developed on the basis of the strain gradient optical effect which has been commonly neglected as an unpleasant noise in the system [37] and a new stress analysis procedure is available using the thermoelastic effect [21].

1.2.6 Interface problems of experimental and analytical mechanics

The classical analytical solutions of infinitesimal, two-dimensional elasticity are based on the concept of a globally and locally homogeneous and isotropic material continuum (characterized by a single mechanism of deformation), and on the concept of generalized plane stress state when applied to plate problems. It has been known for over 50 years that the analytical solutions based on the concept of a generalized plane stress state yield incorrect predictions of stress values in plates in regions of notches, cracks, edges, and local loads, usually on the unsafe side [46]. The influence of the accepted paradigm in engineering mechanics has been so strong, however, that the applicability of two-dimensional analytical and numerical solutions to the determination of stresses in real plates of a finite thickness was not questioned until recently, despite the accumulated empirical evidence to the contrary. The psychological breakthrough was apparently caused by industrial needs related to the development and application of composite structures. At the present time one may notice the dynamic development of analytical and numerical methods of applied mechanics for the determination of the actual, local, three-dimensional stress states in real plates made of homogeneous or composite materials. Such stress states are called "local effects", "edge effects", "boundary effects", etc. [13, 39].

A more difficult situation exists in the realm of finite deformations. It has been shown that uniqueness of solutions of finite elasticity is not to be expected, in a general case [7]. This creates certain serious theoretical and practical problems, particularly in model techniques of the experimental stress analysis.

1.2.7 Scientific methodology. Sources of dichotomy in experimental mechanics.
       Modeling reality

It is interesting to note that in engineering research papers it is not customary to present the theoretical framework of the developed analytical relations, or of the experimental procedures, with only some exceptions. Indeed, it is considered inappropriate, or even in poor taste, to present or to discuss the pertinent theoretical background. This leads to a situation where the physical reality is confused with a hypothetical one; such a situation occurs often in research whose objective is to verify experimentally the predictions of some phenomenological analytical solutions developed on the basis of a hypothetical reality, which are actually physically inadmissible.

The problem of the speculative versus the physical methodology in engineering is compounded by the fact that several opposite views of development of science presently co-exist: the classical view influenced by logical empiricism, the deductive method of testing and refuting theories which develop continuously, the notion of development through scientific revolutions, etc.

In the classical view, science is a strictly logical process based on inductive logic. Scientific theories should yield quantifiable predictions which are confirmed or refuted by experiments. The successive theories progress closer to the truth, in a manner implied by typical textbooks. In engineering research, this view leads to belief in formulas; it also leads to neglecting of limitations imposed by underlying assumptions, and to unwillingness to quantify the influence of simplifications.

With regard to theories and procedures of analytical and experimental mechanics, three particular methodological issues deserve attention:

— the principle of correspondence formulated by Niels Bohr;
— the theory of patterns and conditions of scientific discovery developed by Karl R. Popper;
— the theory of revolutionary character of scientific development and the related notion of paradigm in science presented by Thomas S. Kuhn.

The principle of correspondence presented by Bohr in various papers, and being an object of extensive discussions [3, 10] requires that a new, more comprehensive theory encompasses as a special case the theory being replaced provided that the old theory satisfies the scientific requirements. One of the consequences of the correspondence principle is that the new theory or procedure must be noncontradictory with all basic, nonrefuted theories. The practical consequence of this principle is that the assumptions underlying a physical theory (components and parameters of the assumed physical model) and the solutions must be physically admissible — for instance, energy and power must be finite, an interpenetration of two bodies is not acceptable, and solutions containing singularities are theoretically incorrect, particularly within the linear framework.

The theory of scientific methods developed by Karl R. Popper [40] has essentially influenced modern science and modern research. Popper introduced notions of testing a theory based on the theory of the deductive method of testing as opposed to inductive methods. The view that the logic of scientific discovery is based on inductive logic is rejected by Popper and is replaced by the view that a hypothesis (theory, model) can only be empirically tested, after it has been advanced. According to this framework, a theory can never be empirically verified, but can be empirically refuted or falsified. A system can be accepted as empirical or scientific only if it is capable of being tested by experience.

A criterion of demarcation is the falsifiability of a system, not its verifiability, where the notion of demarcation denotes the barrier which separates science from metaphysical speculation. Thus, a theory is accepted by a process of conjecture and refutation rather than by confirmation of its prediction by an observation or a specially designed experiment.

According to the concept and theory of paradigm in science developed by Thomas S. Kuhn [12], no coherent direction of ontological development can be observed in successive theories in science. The scientists share, in fact, a paradigm or a set of paradigms. The development of prevailing paradigms is related to some psychological features of the human mind, and to the influence (pressure) of the scientific community. Consequently, acceptance of a more advanced theory occurs often with a significant delay because of the influence of the prevailing paradigm, rooted in certain sociological and psychological factors, which claims to be "the true" theory. Thus, the evolutionary stages of development of a given field of science or a given theory when scientists work sharing the same paradigm (basic theory, law, model) are separated by revolutionary stages of development when the old, refuted paradigm is replaced by a new paradigm providing a better insight which is compatible with a new piece of empirical evidence. The influence of the established paradigms is particularly strong in traditional experimental mechanics.

The modern scientific methodology, as outlined above, is generally accepted as the theoretical foundations of research in science, in modern technology, and in advanced experimental mechanics. It is also accepted in advanced undergraduate text books in engineering [14]. It is not accepted in traditional experimental mechanics.

1.2.8 Methodology of modeling in mechanics

The majority of modern scientific theories are based on axiomatic (and optimistic) assumptions that the world is real and recognizable, and that the

human mind is able to develop commensurate models of particular features of
the real, complex world; this world can be understood as a system characterized
by an infinite number of coupled responses. However, reality is not directly
accessible to the human mind. Thus, the basic theoretical problems are the
methodology of modeling, of formulating the criteria of modeling, and of
formulating the so-called criteria of truth, that is, the criteria for testing the
reliability of the developed models with respect to the real world. Consequently,
any scientific theory represents a model (a picture, a caricature) of some
selected features of the physical reality [3, 9, 12, 22, 40, 42—44]. This model is
formulated in conceptual form, verbally, and is finally presented in symbolic
terms of mathematical logic and/or mathematical analysis. The verbal formulation specifies and defines the underlying concepts and outlines the perceived
and accepted mechanisms of phenomena. Two existing incompatible methodologies of the development of models of real processes in mechanics, physical and
phenomenological, are presented in Figure 1.1.

It occurs often in applied mechanics that a theory is developed on the basis
of a speculative, assumed reality, because of a perceived practical necessity.
However, the requirements of scientific rigorousness are usually not observed:
quite often either the assumed reality is not compatible with the physical reality,
or the predictions violate the assumptions; this is not acceptable in a scientific-
speculative theory. The resulting problem is that the basic theoretical issue,
namely, the correlation between the real event and the assumed event, is not
demonstrated, and the limitations of the derived relations are not quantified. As
a result, it occurs often that either the assumptions, or the predictions, or both,
are physically (theoretically) not admissible.

Summarizing, a scientific theory is characterized by

— explanatory power, compatible with unrefuted, scientific notions and
    theories;
— descriptive power, related to the variation of all involved parameters;
— predictive power, allowing extrapolation of empirical results.

Thus, the notions of theory in the natural and general sciences, and in
engineering science, are often different despite the identical terminology, and
the speculative-phenomenological approaches in science and in engineering
mechanics are often incompatible. This is the basis of the dichotomy between
the more theoretical (explanatory, descriptive and predictive in purpose) and
more phenomenological (narrowly descriptive in essence) trends in engineering
research.

This dichotomy in understanding reality and models of reality which lead to
different notions of theory results in the development of two classes of
analytical and experimental procedures for the analysis of behaviour of
materials and structures; both classes often yield incompatible predictions.

The process of model development in the second, physical-theoretical
approach can be presented by the following scheme:

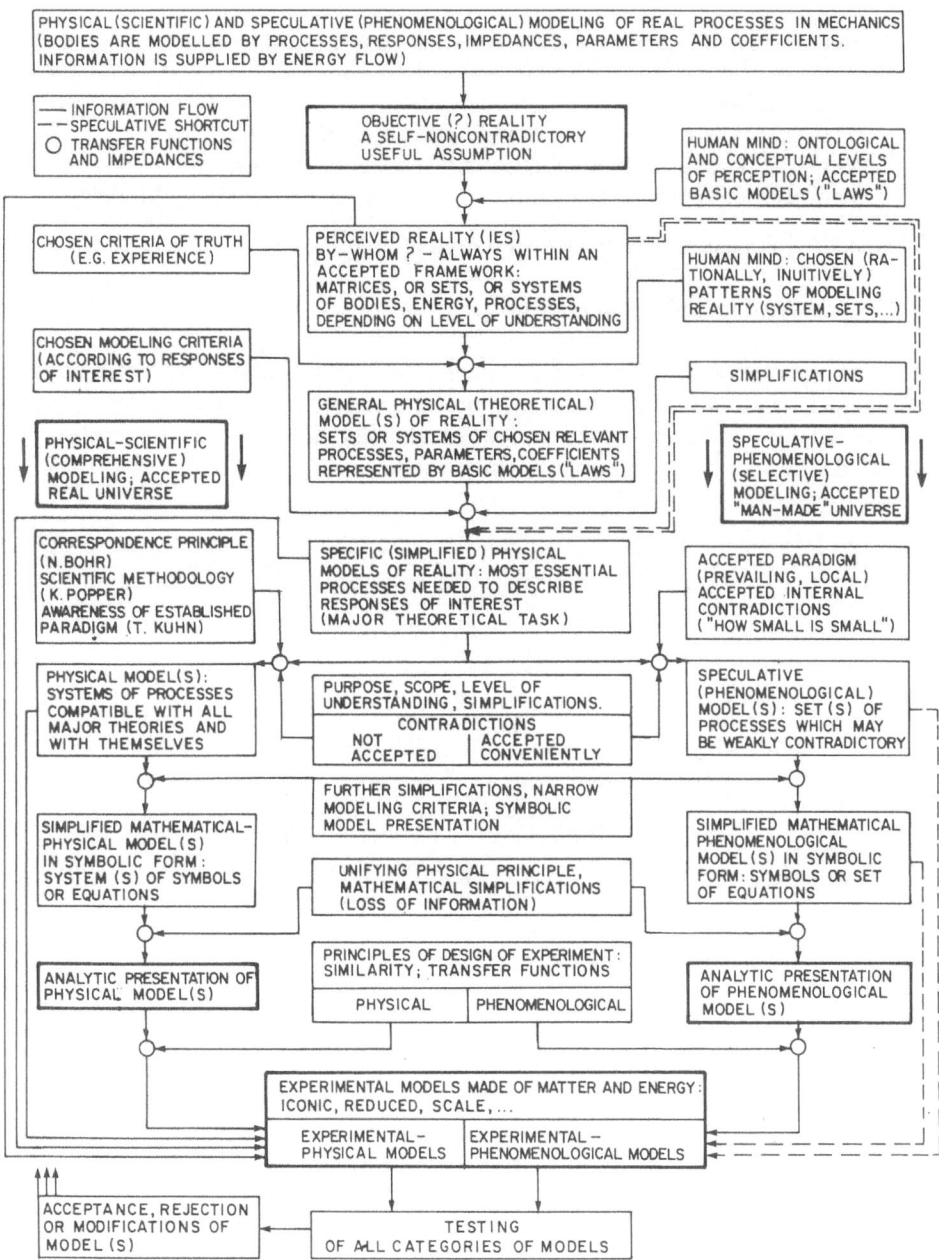

Fig. 1.1. Speculative-phenomenological and physical-scientific methodology of modeling of structures and responses of real bodies and of real processes.

Reality.
Various Physical Models Related to Chosen Modeling Criteria.
Various Mathematical Models Related to Conceptual and Mathematical Simplifications.
Various Analytical, Numerical and Experimental Models.
Various Procedures for Evaluation of Responses of Interest.
Various Methods and Procedures for Testing All Categories of Models.

This methodology is schematically presented in Figure 1.2.

1.2.9 Advanced experimental mechanics

Modern experimental mechanics is not a set of measurement procedures and techniques, and the major issue in modern experimental mechanics is not measurements. Modern experimental mechanics is a system of critically selected, or developed, components which encompass:

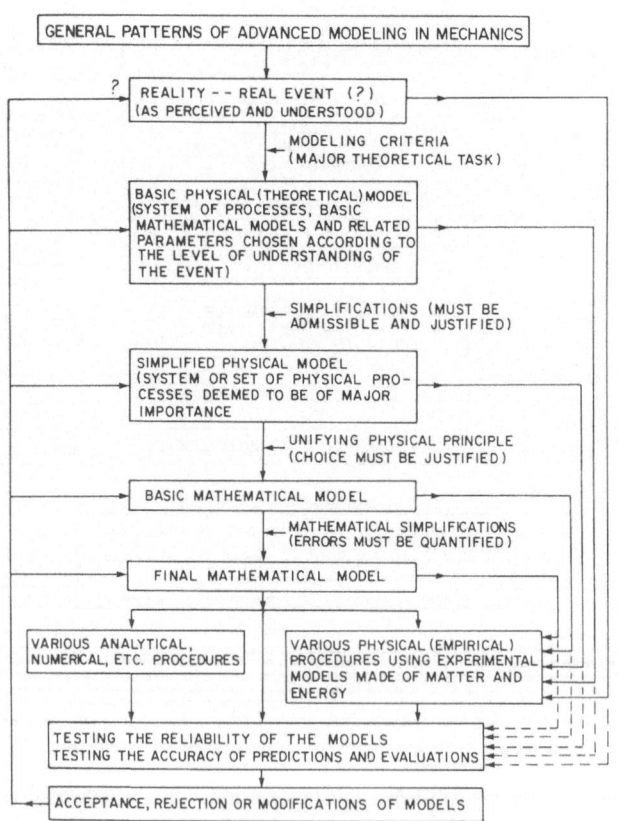

Fig. 1.2. Principle of modeling, and meaning of models in mechanics.

1. experimental data obtained by means of measurement systems;
2. theory of the experiment relating the quantity of interest to the observed quantity;
3. theory of the measurement system (modulation of energy flow) presented in terms of transfer functions and impedances;
4. models of responses of materials to various energy inputs (constitutive relations for materials);
5. models of the deformation-strain-stress states which are the object of research in analytical mechanics.

Experimental results are evaluated using experimental data within a framework provided by theories of the other four components of the system. Experimental results represent empirical evidence only when the theoretical framework of the experiment is correct (Figure 1.3), regardless of the precision of the measurement [22]. The basic problem of the modern, advanced experimental mechanics (AEM) is the reliability of the overall physical theory of experimentation, which encompasses the theories of the above-listed components; this distinguishes AEM from the traditional, phenomenological experimental mechanics.

Of particular significance in AEM is the reliability of the physical and mathematical models of deformation-strain-stress states being developed in analytical mechanics (AM). Within the framework of the physical-theoretical methodology, which forms the foundation of AEM, the reliability of predictions of the AM models pertains to the actual responses of real bodies and systems, made of real materials, and subjected to real energy inputs. The most accurate optical, electric, magnetic, thermal or mechanical measurements are useless or even misleading unless the AM models are reliable. Analogously, AM remains in the realm of speculation and develops relationships pertaining to the speculative, man-made universe, unless the empirical evidence produced by AEM is taken as a foundation of the concepts being developed, and as a measure for testing the predictions of AM models.

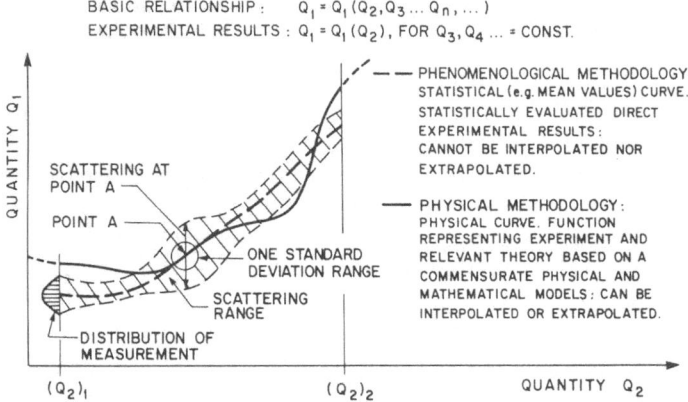

Fig. 1.3. Illustration of predictive powers of experimental procedures based on incompatible methodologies of evaluation of experimental data; phenomenological and physical.

Thus, experimental and analytical mechanics represent a major inherently interwoven component of applied mechanics.

Consequently, AEM can be described as a system of approaches, theories, techniques and procedures which satisfies the requirements of modern scientific methodology, including consideration of the prevailing paradigms, and which is developed to determine — reliably and as accurately as is necessary — the actual responses of real materials, bodies and systems to various inputs of energy (loads) in controlled and uncontrolled conditions. Thus, the major tasks of AEM are to provide reliable and accurate data which can be used for two interrelated purposes:

— to construct and to test the constitutive relations for materials, and to construct and to test the physical and mathematical models which can be taken as a rational basis of analytical solutions and procedures of AM;
— to test the predictive power of numerical solutions, to assess the reliability of procedures in engineering design, to supply data for risk analysis and to assess the reliability and accuracy of various experimental procedures in applied mechanics, including standardized testing.

Examples presented below illustrate the fundamental difference between the traditional and advanced experimental mechanics and the fundamental differences in the reliability and accuracy of experimental results.

1.2.9.1 Constitutive relations for engineering materials. The notion of inherent properties of hypothetical bodies can be as rigorous as the pertinent speculative model related to a hypothetical universe is rigorous. However, when the real universe is concerned, the notion of inherent properties represented by specific constants is a strong philosophical notion related to a particular philosophical system. To avoid any commitment to any philosophical belief, it is safer to use the terms "response" and "parameters of response"; these terms can be related rigorously to chosen physical and mathematical models and to a particular flow of a selected form of energy. Figure 1.4 illustrates the influence of the chosen theoretical framework on the meaning of evaluated material parameters. It should be noted that, according to accumulated knowledge, the overall processes of deformation and the origin of coupled responses depend on several mechanisms which are related to the structure of material. Existence and potential importance of the reversible alteration of temperature of real materials caused by elastic (reversible) deformation, called thermoelastic effect, is illustrated by Figure 1.5 [21]. Also, it was shown that some particular predictions of the phenomenological theory of plasticity pertaining to common engineering problems are at variance with the actual response [16]. It is well documented that all the responses of real bodies depend, often nonlinearly, on the spectral frequency of the detecting energy form [6, 19, 22, 37]. Nevertheless, various experimental methods and theories of some measurement instruments in experimental mechanics are based on the belief that such a frequency dependence either does not exist or can be represented by a simple linear, algebraic relation.

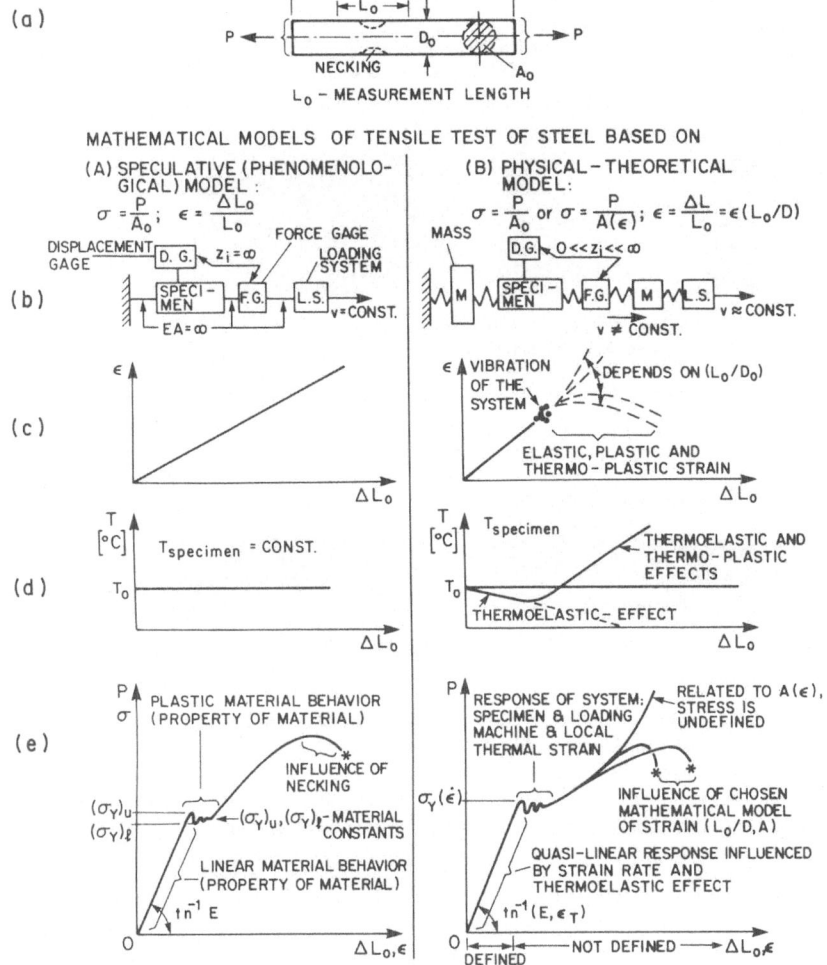

Fig. 1.4. Incompatible phenomenological and physical mathematical models of a simple tensile test of steel, and their incompatible predictions.

(a) Involved physical quantities. Subscripts "0" denote initial conditions; L_0 is the basis of the strain gage.

(b) Mathematical models of the testing system.

(c) The assumed and the actual relations between the elongation ΔL_0 and the average and local strains.

(d) The assumed and the actual relations between the specimen temperature and the specimen elongation.

(e) The phenomenological and the physical interpretations of the results of the tensile tests which depend on the choice of the mathematical models presented in (b), (c) and (d).

1.2.9.2 Speculative and physical models of stress states. The existence of three-dimensional stress states at the boundaries of plates, in particular, in regions of notches, cracks, and local loads, has been known for over half a century (see references in [46]). Such local stress states have been the subject of studies of

MATERIAL : POLYESTER RESIN, PALATAL P6 (BASF)
$$\alpha_T > 0$$
SPECIMEN : 11 mm x 25 mm x(240 + r = 50) mm
FREQUENCY : 5 Hz; MAX. DEFLECTION : 3.47 mm

(a)　　　PHOTOELASTIC STROBOSCOPIC RECORDINGS:
200 FLASHES PER PICTURE ; $\Delta t < 1$ ms ; $\lambda = 546$ nm

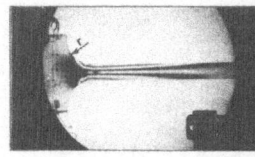 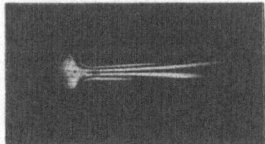

(b)　　　STRAIN & THERMOELASTIC RECORDINGS
STRAIN - UPPER BEAM (120 ohm GAGE)
TEMPERATURE - LOWER BEAM (0.025 mm CHROMEL-
CONSTANTAN THERMOCOUPLE)

SWEEP - 100 ms/ cm
STRAIN AMPLITUDE - 1350 x 10^{-6}/cm
THERMOCOUPLE SIGNAL - 2μV/cm (60μV/1°C)

(b1) STRAIN GAGE & THERMOCOUPLE 　(b2) STRAIN GAGE & THERMOCOUPLE
ON THE SAME SIDE OF BEAM 　　　　ON OPPOSITE SIDE OF BEAM

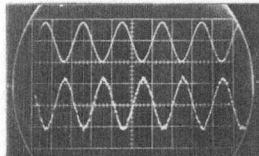 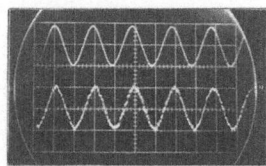

Fig. 1.5. An example of coupled responses of real materials: birefringence, strain, and temperature alterations in a vibrating beam, within elastic range.
(a) Photoelastic stroboscopic recordings.
(b) Strain and temperature recordings at selected points, Figs. (b1) and (b2) show that tensile stress lowers temperature of material when the thermal expansion coefficient is positive.

theorists and analysts for a long time [13]. Despite the accumulated knowledge, it is still common to study experimentally the stress analysis problems in plates using methods such as the transmission photoelasticity or the caustic methods which are based on the phenomenological concepts of a generalized plane stress state and the rectilinear light propagation in a stressed body. As a result, the evaluated results can be wrong by up to about 30%, in addition to being on the unsafe side [31, 36]. According to the empirical evidence, certain classes of fracture mechanics procedures are either logically or physically inadmissible [31, 35, 36, 37].

1.2.9.3 Speculative and physical models of energy flow through real bodies. The fact that the spectral frequency of the information-collecting energy flow is of major importance has been known in advanced research for a long time [6, 19, 22]. However, there still exists the prevailing belief, based on the speculative approach in experimental research and incompatible with the accumulated evidence [15, 37], that the energy propagation within stressed bodies is rectilinear. Various techniques of experimental mechanics are based on this assumption, which is wrong and leads to incorrect evaluation of accurate results.

1.3 Synthesis

A rational expansion of the theoretical foundation of the interrelated disciplines of experimental and analytical mechanics, based on verified empirical evidence, is a condition for further progress which would satisfy the intellectual requirements and the emerging and anticipated technological needs. At the present time, particularly painful in experimental mechanics is the lack of reliable analytical solutions of the linear and finite elasticity pertaining to local effects, and the lack of reliable theories on the basis of which reliable constitutive relations for materials can be constructed. The presently available empirical evidence of a descriptive character is incomplete and often misleading.

Regarding the choice of the methodology to be applied by theorists, analytists and experimentalists working in the field of engineering mechanics, the known technological achievements and costly failures prove that in engineering nothing is more practical than the physical-theoretical (or physical-scientific) methodology developed in the natural sciences. Such an approach made it possible to reach the ocean floor and the planets of our solar system, and facilitates enormously functioning of modern societies.

The speculative models, developed out of necessity, should be noncontradictory with the established and unrefuted basic scientific concepts and theories.

It appears that the major underlying issue of advanced engineering research is the need for engineering researchers with the character and mind of a modern Renaissance scientist. This imposes specific requirements upon the educational system.

The question is whether or not it is in the interest of modern technology, which, by definition, should be identical with the interest of society as a whole, to continue preserving two incompatible concepts of engineering education, engineering research, and engineering design:

- — a phenomenological one, based on acceptance of, or even belief in, engineering formulas which often contain physically inadmissible assumptions of a significant influence, or
- — a scientific-theoretical one, based on modern scientific methodology and utilizing models, theories, and procedures developed in science and mathematics.

The second framework is chosen in this work.

1.4 References

[1] Aben, H., Krasnowski, B., and Pindera, J. T., "On Nonrectilinear Light Propagation in Integral Photoelasticity of Bodies of Revolution" (in Russian), *Transactions of the Estonian Academy of Sciences* **31**(1), 1982, pp. 65—73.

[2] Aben, H., Krasnowski, B. R., and Pindera, J. T., "Nonrectilinear Light Propagation in Integrated Photoelasticity of Axisymmetric Bodies", *Transactions of the CSME* **8**(4), 1984, pp. 195—200.

[3] Brillouin, Leon, *Scientific Uncertainty and Information*, Academic Press, New York, 1964.

[4] Dean, R. C., "Truth in Publication", Trans. of the ASME, *Journal of Fluid Engineering* **99**(2), 1977, p. 270.

[5] Ehrenreich, H., "Electronic Theory for Materials Science", *Science* **235**, 1987, pp.1029–1035.

[6] Fitting, D. W. and Adler, L., *Ultrasonic Spectral Analysis for Nondestructive Evaluation*, Plenum Press, New York, 1981.

[7] Gurtin, Morton E., *Topics in Finite Elasticity*, Society for Industrial and Applied Mathematics, Philadelphia, 1981.

[8] Hirschfeld, "Instrumentation in the Next Decade", *Science* **230**, 1985, pp. 287–291.

[9] Kac, Mark, "Some Mathematical Models in Science", *Science* **166**, 1969, pp. 469–474.

[10] Krajewski, Wladyslaw, *Correspondence Principle and Growth of Science*, D. Reidel Publishing Company, Dordrecht, Holland, 1977.

[11] Kranya, U. E., Layzan, Ya. B., Uputus, Z. T., and Tutan, M. Ya., "Mechanoluminescence at Tensile Test of Plastics" (in Russian), *Mekhanika Polimerov*, No. 2, 1977, pp. 316-320.

[12] Kuhn, Thomas S., *The Structure of Scientific Revolution*, University of Chicago Press, Chicago, 1962, 1970.

[13] Ladevèze, Pierre (Ed.), *Local Effects in the Analysis of Structures*, Elsevier, New York, 1985.

[14] Martin, M. J. C., *Managing Technological Innovation and Entrepreneurship*, Reston Publishing Co., Reston (A. Prentice-Hall Co.), 1984.

[15] Pindera, Jerzy T., "Technique of Photoelastic Studies of Plane Stress States" (in Polish), Engineering Transactions, (Rozprawy Inzynierskie), *Polish Acad. Sciences* **3**, 1955, pp. 109–176.

[16] Pindera, Jerzy, T., "On Application of Brittle Coatings for Determination of Regions of Plastic Deformations" (in Polish), Engineering Transactions (Rozprawy Inzynierskie), *Polish Acad. Sciences* **5**, 1957, pp. 33–47.

[17] Pindera, J. T. and Straka, P., "Response of the Integrated Polariscope", *J. of Stress Analysis* **8**, 1973, pp. 65–76.

[18] Pindera, J. T. and Sze, Y., "Response of Elastic Plates in Flat Contact", in: J. T. Pindera *et al.* (Eds.), *Experimental Mechanics in Research and Development*, Study No. 9, Solid Mechanics Division, University of Waterloo, Ontario, Canada, 1973, pp. 617–635.

[19] Pindera, J. T. and Straka, P., "On Physical Measures of Rheological Responses of Some Materials in Wide Ranges of Temperature and Spectral Frequency", *Rheologica Acta* **13**, 1974, pp. 338–351.

[20] Pindera, J. T. and Mazurkiewicz, S. B., "Photoelastic Isodynes: A New Type of Stress-Modulated Light Intensity Distribution", *Mechanics Research Communications* **4**, 1977, pp. 247–252.

[21] Pindera, J. T., Straka, P., and Tschinke, M. F., "Actual Thermoelastic Response of Some Engineering Materials and Its Applicability in Investigations of Dynamic Response of Structures", *VDI-Berichte* **313**, 1978, pp. 579–584.

[22] Pindera, Jerzy T., "Foundations of Experimental Mechanics: Principles of Modelling, Observation and Experimentation', in: J. T. Pindera (Ed.), *New Physical Trends in Experimental Mechanics*, Springer-Verlag, Wien, 1981, pp. 188–236.

[23] Pindera, J. T., "Analytical Foundations of the Isodyne Photoelasticity", *Mechanics Research Communications* **8**, 1981, pp. 391–397.

[24] Pindera, J. T. and Mazurkiewicz, S. B., "Studies of Contact Problems Using Photoelastic Isodynes", *Experimental Mechanics* **21**, 1981, pp. 448–455.

[25] Pindera, J. T., Mazurkiewicz, S. B., and Krasnowski, B. R., "Determination of All Components of Plane Stress Field Using Simple Techniques of Differentiation of Photoelastic Isodynes", in: *Proc. 1981 Spring Meeting* (Dearbon), Society for Experimental Stress Analysis, Brookfield Center, 1981, pp. 35–40.

[26] Pindera, J. T., Issa, S. S., and Krasnowski, B. R., "Isodyne Coating in Strain Analysis", in: *Proc. 1981 Spring Meeting* (Dearborn), Society for Experimental Stress Analysis, Brookfield Center, 1981, pp. 111–117.

[27] Pindera, J. T., Krasnowski, B. R., and Pindera, M.-J., "An Analysis of Semi-Plane Stress States in Fracture Mechanics and Composite Structures Using Isodyne Photoelasticity", in: *Proc. of the 1982 Joint (JSME/SESA) Conference on Exp. Mechanics*, Part 1, May 1982, Oahu-Maui, Hawaii, SESA 1982, pp. 417—421.

[28] Pindera, J. T., Krasnowski, B. R., and Pindera, M.-J., "Determination of Interface Stresses on Composite Structures", in: *Proc. of the 1982 Joint (JSME/SESA) Conference on Exp. Mechanics*, Part 1, May 1982, Oahu-Maui, Hawaii, SESA, 1982, pp. 18—22.

[29] Pindera, J. T., Hecker, F. W., and Krasnowski, B. R., "Gradient Photoelasticity", *Mechanics Research Communications* **9**(3), 1982, pp. 197—204.

[30] Pindera, J. T., "New Development in Photoelastic Studies: Isodyne and Gradient Photoelasticity", *Optical Engineering* **21**(4), 1982, pp. 197—204.

[31] Pindera, J. T. and Krasnowski, B. R., "Determination of Stress Intensity Factors in Thin and Thick Plates Using Isodyne Photoelasticity", in: Leonard A. Simpson (Ed.), *Fracture Problems and Solutions in the Energy Industry*, Pergamon Press, 1982, pp. 147—156.

[32] Pindera, J. T. and Krasnowski, B. R., "Theory of Elastic and Photoelastic Isodynes", SMD-Paper No. 184, IEM-Paper No. 1. Solid Mechanics Division, University of Waterloo, October 1983, pp. 1—127.

[33] Pindera, J. T. (Ed.), "Modeling Problems in Crack Tip Mechanics" (Proc. of the Tenth Canadian Fracture Conference, August 24—26, 1983, University of Waterloo), Martinus Nijhoff, The Hague, The Netherlands, 1984.

[34] Pindera, J. T., "Isodyne Photoelasticity and Gradient Photoelasticity: Physical and Mathematical Models, Efficacy, Applications". *Mechanika Teoretyczna i Stosowana (Journal of Theoretical and Applied Mechanics)*, **22** (1/2), 1984, pp. 53–68.

[35] Pindera, J. T., Krasnowski, B. R., and Pindera, M.-J., "Theory of Elastic and Photoelastic Isodynes. Samples of Applications in Composite Structures", *Experimental Mechanics* **25**(3), 1985, pp. 272—281.

[36] Pindera, J. T. and Pindera, M.-J., "On the Methodologies of Stress Analysis of Composite Structures: Parts 1 and 2", *Theoretical and Applied Fracture Mechanics Journal* **6**(3), 1986, pp. 139—151 and 153—170.

[37] Pindera, J. T. and Hecker, F. W., "Basic Theories and Experimental Techniques of the Strain-Gradient Method", *Experimental Mechanics* **27**(3), 1987, pp. 314—327.

[38] Pindera, Jerzy T., "Advanced Experimental Mechanics in Modern Engineering Science and Technology", *Transactions of the CSME*, Vol. 11, No. 3, 1987, pp. 125—138.

[39] Pindera, J. T., "Local Effects — A Major Problem of Contemporary Stress/Strength Analysis of Homogeneous and Composite Structures" (Plenary Lecture), in: G. C. Sih, S. V. Hoa, and J. T. Pindera (Eds.), *Analytical and Testing Methodologies for Design with Advanced Materials* (Proceedings of the International Conference on Analytical and Testing Methodologies for Design with Advanced Materials, ATMAM '87, Concordia University, Montreal, August 26—28, 1987), North Holland, Amsterdam, 1988, pp. 29—55.

[40] Popper, Karl R., *The Logic of Scientific Discovery*, Harper and Row, New York, 1959, 1968.

[41] Robinson, A. L., "Physicists Try to Find Order in Chaos", *Science* **218**, 1982, pp. 554—556.

[42] Sedov, L. I., "On Prospective Trends and Problems in Mechanics of Continuous Media" (in Russian), *Prikladnaya Matematika i Mekhanika* **40**, 1976, pp. 963—980. English translation, Pergamon Press, 1977.

[43] Sih, G. C., "The State of Affairs Near the Crack Tip", in: Jerzy T. Pindera (Ed.), *Modelling Problems in Crack Tip Mechanics*, Martinus Nijhoff Publishers, Dordrecht, 1984, pp. 65—90.

[44] Sih, G. C., "Mechanics and Physics of Energy Density Theory", *Theoretical and Applied Fracture Mechanics* **4**(3), 1985, pp. 157—173.

[45] Starr, Ch. and Whippie, Ch., "Risk of Risk Decision", *Science* **8**, 1980, pp. 1114—1119.

[46] Thum, A., Petersen, C., and Svenson, O., *Verformung, Spannung und Kerbwirkung* (Deformation, Stress and Notch Action), VDI-Verlag, Dusseldorf, 1960.

2

Basic theoretical issues of stress analysis.
Accepted models

2.1 Introduction

Rational optimization of the performance of engineering structures requires that all pertinent physical quantities and parameters be defined unequivocally, and be determinable with satisfactory reliability and accuracy. For a design engineer, the measures of reliability and accuracy are the actual responses of the actual physical bodies, structures, and systems to the actual flow of energy, as discussed in Chapter 1.

As shown in Chapter 1, it is not possible to measure any physical quantity "as it is," because such a notion is meaningless. All observations and experiments resulting in measurements are performed within a defined — rationally or intuitively — theoretical frame of reference given by the chosen physical and mathematical models of the whole involved system. The level of pertinent knowledge of the experimenter or observer is a component of those models.

It is impossible to measure directly any physical quantity itself. All that can be done is to measure the values of the chosen parameters of a particular energy flow selected to detect the values and alteration of the physical quantity being the subject of measurement. The energy flow through the components of the total measurement system depends on [4, 18, 19]:

— component transfer functions relating the output signal to the input signal;
— output impedances which are measures of the ability of the component to supply a particular form of energy when maintaining linear characteristics;
— input impedances which are measures of the ability of the component to accept a particular form of energy;
— overall transfer function of the system which relates the measured signal to the physical quantity being measured.

The reliability of the overall transfer function with respect to the actual physical reality is the actual measure of the reliability of measurement. The accuracy of measurements pertains, or is related, to the resolution, repeatability, and sensitivity of the measurement system in prescribed conditions of experiment.

Two typical errors occur very often. The first type of error is rooted in confusing the precision of the instrument with the reliability of results of measurements. The second type of error is a result of a theoretically incorrect development of the overall transfer function of the measurement system. Such errors occur when the responses of some components of the system are physically and mathematically incompatible. This occurs often when the so-called hybrid techniques, or hyphenated techniques, or hyphenated instruments, are used. Quite often, such errors are made when simplified analytical solutions are combined with some experimentally obtained data pertaining to an actual physical situation which is incompatible with the chosen analytical solution.

Summarizing, the statement that "all experimental results are obtained within a chosen, or tacitly accepted, theoretical framework" is self-evident. When the theory of experiment is at variance with reality, the obtained experimental results are usually erroneous regardless of the accuracy of measurements.

The errors could be errors in magnitude, or in sign, or both. The most typical theoretical errors are related to:

— neglecting the local, three-dimensional stress states in plates, that is, neglecting or overlooking the local effects;
— neglecting the actual patterns of propagation of mechanical energy through elastic bodies in the form of strain-stress-thermal waves of various frequencies;
— neglecting the actual patterns of propagation of radiant energy through elastic and viscoelastic bodies;
— neglecting the condition that the foundations of theories taken as the basis of experimental methods must be physically admissible.

In other words, in experimental studies of stress states, and particularly in studies of local effects, three categories of errors can occur, rooted in the following issues:

— physical, logical and mathematical admissibility of chosen theoretical framework;
— consideration of the actual stress-strain-temperature-time states, particularly of the stress-strain states in real flat plates;
— recognition of the actual patterns of interaction between information-gathering and information-carrying energy forms, and stressed and non-stressed bodies — in particular, the influence of spectral frequencies of the applied energy forms.

One may accept as an axiom that the reliability of analytical and experimental procedures in applied mechanics, defined as the degree of correlation with the actual responses of real materials and systems, depends on the theoretical background of the accepted physical and mathematical models. The theoretical and empirical evidence presented in the following sections illustrates this point.

One can observe various fundamental incompatibilities or even contradictions between theories of various procedures of applied mechanics used concurrently in engineering. These incompatibilities are rooted in the basic dichotomy between two methodologies concurrently used in engineering research discussed in Chapter 1.

Theoretically correct evaluation of experimental data related to the stress/ strain/deformation states is one of the major problems of the experimental stress analysis. The theoretical correctness of the evaluation depends on the chosen mathematical model of the stress/strain states in the object under consideration. Depending on the accepted model, the experimenter — using the same experimental data — may evaluate experimental results which are qualitatively and quantitatively different, as follows from analysis of Figures 2.1, 2.2, and 2.3.

The objective of stress analysis — analytical, numerical, or experimental — is to supply reliable major component of a rational strength and deformation optimization of engineering materials and structures. Thus all the analytical solutions and procedures of theories of elasticity, viscoelasticity, or plasticity applied in engineering design must be tested for their reliability and applicability. A theory and the related analytical solution which are not testable do not satisfy a major condition for a scientific theory and only represent a speculative construct which may, or may not, reflect some features of reality.

Only some basic issues pertaining to the assessment and selection of mathematical models of stress-strain states are discussed in this Chapter. A more comprehensive presentation of the theory of modeling, observation, experimentation, measurements, and evaluation of experimental data is given in pertinent bibliography [18, 19].

2.2 Underlying physical and mathematical models of stress states in plates

The actual stress states in real plates of finite thickness are always three-dimensional with the exception of regions where the stress gradients are equal to zero. In particular, stresses in plates are pronounceably three-dimensional in regions of geometric or material discontinuities, rapidly changing boundary loads (contact problems), or local heat sources. Such local three-dimensional stress states, called commonly "local effects", are discussed below.

The underlying physical model of two-dimensional stress states in linear elastic plates and of related strain states consist of the following components:

- Data on boundary conditions given by the model of geometric and load boundary conditions, related to the concept of an infinitesimally thin plate;
- Constitutive material relations given by the model of a hypothetical, linear elastic body called Hooke's body;
- Data on particular relations between the stress components given by the model of two-dimensional stress states caused by infinitesimally small deformations, which is called linear elasticity.

2.2.1 Model of a plate

To avoid confusion the term "plate" denotes in this work a thin body having two parallel face surfaces and a narrow boundary surface normal to the face surfaces. This body is loaded by tractions active on the boundary surface and uniformly distributed across the plate thickness so that the resultants of local tractions are co-planar with the middle plane of the plate. No tractions act on the face surfaces of the plate. The stress state in such a plate is assumed to be two-dimensional (plane). Such plates, not subjected to bending moments and to shear forces causing curvature of the plate middle plane, are called "disks" in several languages.

Actually, the stress state in a plate of a finite thickness is plane only when the stress state is homogeneous. Noticeable stress gradients in the plate's plane are coupled with noticeably three-dimensional stress states regardless whether or not the plate is a "thin" or a "thick" plate. To allow the existence of a nonhomogeneous plane stress state in a plate, it is useful to introduce the concept of a hypothetical infinitesimally thin plate.

It is assumed that the middle plane of the plate is always plane and that the deformations are infinitesimally small; no buckling is allowed under compressive or tensile stresses.

Such plates, the "infinitesimally thin", or "thin", or "thick" plates are speculative, hypothetical concepts which do not have their counterparts in reality, but which are very useful as theoretical constructs because they are able to simulate quite reliably some responses of real plates within certain particular ranges, to be determined empirically.

2.2.2 Hooke's body

This is a mathematical model of a hypothetical body commonly used to simulate (describe and predict) the behavior (mechanical, electric, optic, etc.) of real materials. The concept of Hooke's body is based on the concepts of causality, athermal homogeneous material continuum, and linear superposition. The only responses of Hooke's body to mechanical and external thermal loads are deformations. Hooke's body represents a particular case of a general linear body. It is defined by

$$\sigma_{ij} = c_{ijk\ell}\varepsilon_{k\ell}, \tag{2.1a}$$

or

$$\varepsilon_{ij} = s_{ijk\ell}\sigma_{k\ell}, \tag{2.1b}$$

where $c_{ijk\ell}$ is the elastic stiffness tensor having the dimension [force/(length)^2] or [energy/(length)^3], and $s_{ijk\ell}$ is the elastic compliance tensor having the dimension [(length)^2/force]. Parameters of the components of $c_{ijk\ell}$ and $s_{ijk\ell}$ are constant and represent properties of the body. Symbols σ_{ij} and ε_{ij} denote components of the stress tensors and strain tensor. These components are defined as limiting values of the ratios of the normal and tangential components of surface traction over the surface element, or of the gradients of displacements of points of the material continuum with respect to corresponding coordinates of the chosen coordinate system, respectively. By definition, the field of displacement is continuous, and the notion of stress is unequivocal (unique) when the deformations are infinitesimal. In particular, the notion of Hooke's body implies a single mechanism of deformation of a homogeneous material continuum.

For an elastically homogeneous and isotropic body, which admits a work potential, the number of independent elastic constants of material representing the elastic stiffness and compliance tensors reduces from 36 to 2. For such materials, represented by Lame constants λ and μ, and engineering constants E, v and G, the relations (2.1a, b) can be presented in the form

$$\sigma_{ij} = \lambda\delta_{ij}\varepsilon_{kk} + 2\mu\varepsilon_{ij} \tag{2.2a}$$

and

$$\varepsilon_{ij} = -\frac{\lambda}{2\mu(3\lambda + 2\mu)}\,\delta_{ij}\sigma_{kk} + \frac{1}{2\mu}\,\sigma_{ij}, \tag{2.2b}$$

or

$$\sigma_{ij} = \frac{2vG}{1 - 2v}\,\delta_{ij}\varepsilon_{kk} + 2G\varepsilon_{ij}$$

$$= \frac{E}{(1 + v)(1 - 2v)}\,\delta_{ij}\varepsilon_{kk} + \frac{E}{1 + v}\,\varepsilon_{ij}, \tag{2.3a}$$

and

$$\varepsilon_{ij} = -\frac{v}{1 - 2v}\,\delta_{ij}\varepsilon_{kk} + \frac{1}{2G}\,\sigma_{ij} = \frac{1 + v}{E}\,\sigma_{ij} - \frac{v}{E}\,\delta_{ij}\sigma_{kk}, \tag{2.3b}$$

where δ_{ij} denotes Kronecker's delta, repeated subscript implies summation and ε_{kk} and σ_{kk} denote the invariants

$$\varepsilon_{kk} = \varepsilon_{xx} + \varepsilon_{yy} + \varepsilon_{zz},$$

$$\sigma_{kk} = \sigma_{xx} + \sigma_{yy} + \sigma_{zz}.$$

The material Lame constants λ and μ, the elastic modulus E called Young modulus, and the elastic modulus v called Poisson ratio or number, are related in the following manner:

$$\lambda = \frac{2v}{1 - 2v}\,G = \frac{v}{(1 - 2v)(1 + v)}\,E, \tag{2.4a}$$

$$\mu = G = \frac{1}{2(1 + v)}\,E. \tag{2.4b}$$

Of particular importance in experimental work is the quantity relating directly the volumetric deformation ε_{kk} to the sum of normal stress components σ_{kk}, called bulk modulus K,

$$\sigma_{kk} = K\varepsilon_{kk} \tag{2.4c}$$

where

$$K = \lambda + \frac{2}{3}\mu = \frac{2}{3}\frac{1+\nu}{1-2\nu} \quad G = \frac{1}{3}\frac{1}{1-2\nu}E \qquad (2.4d)$$

For most engineering materials the value of Poisson's ratio is positive; the thermodynamic considerations lead to the conclusion that Poisson's ratio is smaller than 0.5, but greater than -1:

$$-1 < \nu < 0.5. \qquad (2.5)$$

In fact, it was recently observed that some new materials can exhibit a negative value of Poisson's ratio when observed in a standard manner [10].

It is often overlooked that the relations (2.1a, b) in a general case do not comply with the principles of conservation of energy, because an adiabatic (isentropic), volumetric deformation related to volumetric stress produces temperature difference which can be approximated by the relation [7],

$$dT = -T_0(c_\varepsilon)^{-1}\alpha_T\,d\sigma_{ii}. \qquad (2.6)$$

The existence of this thermal effect, called thermoelastic effect, becomes obvious when the Hooke's body relations are replaced by a thermal equation of state and the entropy of the system is considered. The magnitude of the thermoelastic effect depends on the thermal expansion coefficient α_T and the specific heat at constant volume c_ε. It should be noted that more comprehensive theories of the thermoelastic coupling must consider the contribution of the distortional deformation caused by shear stresses, and the local effects at the grain boundaries [17].

The descriptive capability of Hooke's model is limited to cases for which the assumptions leading to expressions (2.1a, b) are acceptable within an acceptable margin of error, and when omission of the actual time dependence, of the actual nonlinear temperature dependence, and of the actual thermoelastic and electro-elastic couplings are of negligible consequence. In particular, one must note that some new materials, some of them with a negative Poisson's ratio may be beyond the conceptual framework of Hooke's model.

2.2.3 Plane linear theory of elasticity. Two-dimensional model

In the mathematical theory of elasticity — which is a branch of applied mathematics — it is not necessary to relate plane stress state to any particular physical body [2]. However, in practical engineering applications of stress analysis, it is customary to base the developed relations on the concept of a generalized plane stress state assumed to exist in a thin plate which is loaded at the boundary by forces (tractions) co-planar with the middle plane of the plate and uniformly distributed across the plate thickness [31]. When the Cartesian coordinate system has its z-coordinate normal to the plate surface, the basic equilibrium conditions, considering only surface forces, can be presented in the form

$$\frac{\partial}{\partial x}\,\sigma_{xx} + \frac{\partial}{\partial y}\,\sigma_{yx} = 0 \quad \text{and} \quad \frac{\partial}{\partial y}\,\partial_{xy} + \frac{\partial}{\partial y}\,\partial_{yy} = 0, \tag{2.7}$$

and the related conditions can be given in the form

$$\sigma_{zz} = \sigma_{xz} = \sigma_{yz} = 0, \quad \text{and} \quad \frac{\partial}{\partial z}\,\sigma_{ij} = 0, \quad \text{where } i, j = x, y, z. \tag{2.8}$$

The compatibility conditions in terms of stresses are usually presented in the form

$$\nabla^2(\sigma_{xx} + \sigma_{yy}) = 0. \tag{2.9}$$

The equilibrium and compatibility conditions are used as foundations of numerous particular solutions of plane infinitesimal elasticity. Some particular techniques of plane infinitesimal elasticity have been developed to predict and analyze stresses along boundaries of holes in thin plates subjected to tractions and to bending [5, 6, 26]. Many groups of analytical methods of stress analysis are based on the concept of stress functions, among which Airy stress function is of particular importance [2, 31].

The plàne stress states in real plates are sometimes called "disk problems".

2.2.4 Three-dimensional linear theory of elasticity. Three-dimensional model

The equilibrium conditions for elements of a three-dimensional body, considering only surface forces, can be completely presented in the form

$$\sigma_{ji,\,j} = 0, \quad \text{where } j, i = x, y, z, \tag{2.10a}$$

or

$$\frac{\partial}{\partial x}\,\sigma_{xx} + \frac{\partial}{\partial y}\,\sigma_{yx} + \frac{\partial}{\partial z}\,\sigma_{zx} = 0, \tag{2.10b}$$

$$\frac{\partial}{\partial x}\,\sigma_{xy} + \frac{\partial}{\partial y}\,\sigma_{yy} + \frac{\partial}{\partial z}\,\sigma_{zy} = 0, \tag{2.10c}$$

$$\frac{\partial}{\partial x}\,\sigma_{xz} + \frac{\partial}{\partial y}\,\sigma_{yz} + \frac{\partial}{\partial z}\,\sigma_{zz} = 0. \tag{2.10d}$$

No other conditions are imposed within the framework of linear (infinitesimal) theory of elasticity which requires that the local deformations are infinitesimally small.

When the body forces vanish as assumed above, the compatibility conditions in terms of stress can be presented as

$$\nabla^2 \sigma_{ij} + \frac{1}{1+\nu}\,\theta_{,ij} = 0, \tag{2.11a}$$

where θ denotes the stress invariant, and

$$\theta = \sigma_{kk} = \sigma_{xx} + \sigma_{yy} + \sigma_{zz}.$$

Often it is convenient to present the compatibility conditions (2.11a), valid for zero body forces, in the explicit form

$$\left.\begin{array}{ll} \nabla^2 \sigma_{xx} + \dfrac{1}{1+v}\,\dfrac{\partial^2 \theta}{\partial x^2} = 0; & \nabla^2 \sigma_{yz} + \dfrac{1}{1+v}\,\dfrac{\partial^2 \theta}{\partial y\,\partial z} = 0, \\[2mm] \nabla^2 \sigma_{yy} + \dfrac{1}{1+v}\,\dfrac{\partial^2 \theta}{\partial y^2} = 0; & \nabla^2 \sigma_{zx} + \dfrac{1}{1+v}\,\dfrac{\partial^2 \theta}{\partial z\,\partial x} = 0 \\[2mm] \nabla^2 \sigma_{zz} + \dfrac{1}{1+v}\,\dfrac{\partial^2 \theta}{\partial z^2} = 0; & \nabla^2 \sigma_{xy} + \dfrac{1}{1+v}\,\dfrac{\partial^2 \theta}{\partial x\,\partial y} = 0. \end{array}\right\} \qquad (2.12\text{b–g})$$

When a three-dimensional body is given in the shape of a thick plate loaded by tractions acting on the boundary surface and having the face surface free of tractions,

$$(\sigma_{zz})_s = (\sigma_{zx})_s = (\sigma_{zy})_s = 0,$$

the relations (2.12) yield (compare [2]),

$$\left(1 + \frac{1}{1+v}\right) \nabla^2 (\sigma_{xx} + \sigma_{yy} + \sigma_{zz}) = 0, \qquad (2.13\text{a})$$

and

$$\nabla^2 (\sigma_{yz} + \sigma_{zx} + \sigma_{xy}) + \frac{1}{1+v}\left(\frac{\partial^2}{\partial y\,\partial z} + \frac{\partial^2}{\partial z\,\partial x} + \frac{\partial^2}{\partial x\,\partial y}\right)\theta = 0. \qquad (2.13\text{b})$$

The relations (2.13a) and (2.13b) cannot be reduced to the relation (2.9), even when the face surface stresses are equal to zero,

$$(\sigma_{zz})_s = (\sigma_{zx})_s = (\sigma_{zy})_s = 0,$$

and when the plate thickness in the direction z approaches zero, that is when the normal stress component in the direction of plate thickness, σ_{zz}, approaches zero.

Evidently, the concept of a generalized plane stress state in a thin plate is not compatible with the general concepts of stresses in an elastic body. It appears that it is possible to circumvent this incompatibility by defining the plane stress state with reference to a hypothetical infinitesimally thin plate. It is true that the concept of an infinitesimally small, but not approaching zero and not finite, quantity, "an indivisible," rooted in the 18th century mathematics was never very clearly defined [1]. However, this concept is used successfully in thermodynamics denoting a small, virtual change [7]; in the case under consideration, it allows the elimination of the vagueness of the basic features of the concept of the generalized plane stress state such as "thin", "average", "small". It should be noted that the term "generalized plane stress state" does not denote a plane stress state that is generalized for the case of plates having finite thickness — it denotes an average through the plate thickness plane stress state. Within this concept the technique of averaging is specified, but the conditions of averaging are not specifically defined and — thus — the procedure is not quantifiable. In particular, in regions of rapid stress alterations and stress concentrations, the normal stresses in the direction of the plate thickness are neglected, and the influence of the measurable rotations of the plate surfaces on the equilibrium condition is neglected so the plane equilibrium conditions are applied.

Logically, the notion of the generalized plane stress state in plates is a misnomer because it does not introduce a more general notion of the stress state. On the contrary, it replaces the notion of a weakly three-dimensional stress state by a simplified notion of an average-across-the-thickness stress state. Regarding the local effects, this notion leads to an underestimation of the values of stresses in critical regions. However, as long as the actual stress state in a plate is weakly three-dimensional so that the resulting error in stress evaluation is technologically acceptable, the notion of the generalized (actually averaged) plane stress state is useful, provided that it is not applied close to a loaded or nonrectilinear boundary.

The admissibility of the notion of a generalized plane stress state in engineering design should be proven in each particular case because the practical consequences of the belief that the stress averaging across the thickness of a plate is admissible, without constraint, could be serious in practical engineering tasks. For example, according to the common notion of a generalized plane stress state not constrained by quantifiable conditions, the generalized stresses in a plate exhibiting a residual stress state symmetrical with respect to the plate middle plane are equal to zero. However, such residual stress states are often of significant technological importance. The simplest example is given by safety glass plates in which such stresses are essential to the strength of the glass plate and to the safe fracture patterns. Similar stress states occur in the regions of local effects. Thus the practical consequences of the belief that the stress averaging across-the-plate-thickness is admissible in engineering design could be very serious.

There is a strong tendency in practical engineering to assume that the relations of linear elasticity are directly applicable to real materials and real structures. It should be stressed that it is worthwhile in each particular practical application to recall some basic features of the applied relations to assess, or estimate, or — if possible — to quantify the ranges of their applicability.

2.3 Local effects

Contrary to some established and accepted procedures in the field of analytical and experimental stress analysis and fracture mechanics, the stress states in plates in regions of notches, crack tips, local loads, etc., are not two-dimensional. As mentioned earlier, they are clearly three-dimensional [33], and the differences between the actual maximal values of stress components and the corresponding maximal values calculated using analytical relations of plane elasticity can be up to 30 per cent, or more, on the unsafe side [20]. Thus the common procedures for the determination of stress components using the values of the stress intensity factors K_I and K_{II} calculated according to analytical relations of plane fracture mechanics may lead to even greater errors, usually on the unsafe side.

It has already been recognized in advanced research and design, primarily under the pressure of society and its economic and administrative institutions which have introduced the notion of technological hazard, that the plane elasticity applies rigorously only to infinitesimally thin plates and beams, or to the stress states in real plates, which are characterized by weak stress/strain gradients in the plate's plane. The fact that the actual stress/strain states in regions of high strain gradients are incompatible with the assumptions of plane elasticity was already recognized in aeronautical engineering long ago [27, 33, 34] and is recognized in advanced analytical research in mechanics [5, 6, 32, 36]. Lack of reliable data on the actual stress values, usually at the most critical points of engineering structures, has been one of the main causes of either unexpected failures, or of unnecessary, and sometimes dangerous, overdimensioning of structural components.

To maintain a convenient coherence between various analytical solutions for the plate problem, it is customary to denote the local but pronounced deviations from solutions of plane elasticity by the term "local effect" or by similar terms.

The term "local effect", [9] (edge effect, end effect, layer effect, boundary layer effect, Saint-Venant effect or Saint-Venant principle, free-edge effect, etc.), denotes various local, but pronounced, deviations from the assumed two-dimensional, but actually weakly three-dimensional, stress states in stretched plates, flat beams, etc., made of homogeneous or composite materials. Such deviations — characterized by the not negligible three-dimensional stress states in plates, varying with the thickness coordinate — occur most often at the boundaries of flat beams and of stretched or bent plates, particularly in the regions of notches, cracks, material discontinuities, local mechanical or thermal loads, etc. [21—23, 27—29, 32—36]. However, they may occur in any region of a plate where the gradients of stress/strain components are high [8]. The pertinent terminology is still in the stage of a rapid development. For example, the term "The Saint-Venant Principle" denotes a theorem which defines necessary and sufficient conditions ensuring the localization of the displacements and the stresses, formulated in terms of the rapidly decaying properties [9].

Local effects can be considered a general three-dimensional case of the so-called stress concentration effect, whose significance was recognized already at the turn of the century. The first, classical, theory of plane stress concentrations around, and in regions of, notches and holes in thin plates, and of three-dimensional stress states in regions of external and internal notches in circular cylindrical rods was presented 50 years ago [14]. An extensive two-dimensional analytical treatment of stress concentration and stress distribution in regions of notches and holes in thin plates is given in [26]. It should be noted that the problem of the actual three-dimensional stress states in plates, well known to experimentalists and to design engineers already in the late '30s [33], attracted subsequently the attention of analysts [32] who have expounded and generalized the ideas already presented in [14]. A review of some developing trends in the analysis of local effects is given in [9].

Summarizing, local effects in plates and beams can result in a significant increase of the actual values of maximal boundary stresses [14, 27, 33] with respect to predictions of the linear elastostatic solutions based on the concept of the generalized plane stress state. Actually, this increase is maximal in the middle plane of the plate and can reach up to about 30 percent above the corresponding surface stress [20]. In addition, the stress component in the direction of plate thickness can reach 10 percent of the maximal stress component at the boundary, so the actual value of the equivalent stress, determined using Huber—Mises—Hencky hypothesis, may be about 40 percent higher than the values given by a plane elastostatic solution. Such deviations cannot be acceptable for a design engineer who designs advanced structures.

It is also known that the notion of the existence of a plane stress state at the crack tip in a stretched plate containing stress singularity, which is a basic notion of classical linear fracture mechanics, leads to logical and mathematical contradictions [3]. The same can be said regarding stress states at boundaries of plates in bending. The involved physical processes are complicated and require a more theoretical, comprehensive treatment.

Let us illustrate this issue. The term "thin plate" is commonly used to denote a plate, whose thickness is one or more orders of magnitude smaller than any other characteristic dimension such as the radius of curvature of a notch, tip of a crack, width or length, etc. The stress state in such a hypothetically thin plate is illustrated by Figure 2.1a. Such thin plate problems occur in reality exceptionally seldom because a common mode of failure of such very thin plates is failure by instability if buckling of the plate is not prevented. The stress state in real plates in which values of the radii of curvature at the notch bottom, or at the crack tip, are of the same order of magnitude as the plate thickness or smaller is illustrated by Figure 2.1b.

The local effects in plates can be stationary, related to the geometry of boundary and fixed boundary loads, and nonstationary. The nonstationary local effects can propagate slowly under the influence of varying energy sources such as local input of thermal energy, or they can propagate with the sound velocity. The nonstationary local effects can cause fractures inside the plate, far away from the boundary [8].

Physical models simulating the stationary and slowly moving nonstationary local effects are, in fact, quite similar and are represented by essentially the same system of static or quasi-static relations, while the nonstationary local effects must be modeled differently, as dynamic phenomena. For the sake of convenience, three categories of problems can be distinguished:
  — stationary local effects, including the quasi-stationary effects;
  — nonstationary, transient local effects related to disturbances moving with the sound velocity;
  — hypervelocity local effects characterized by strong thermal effects.

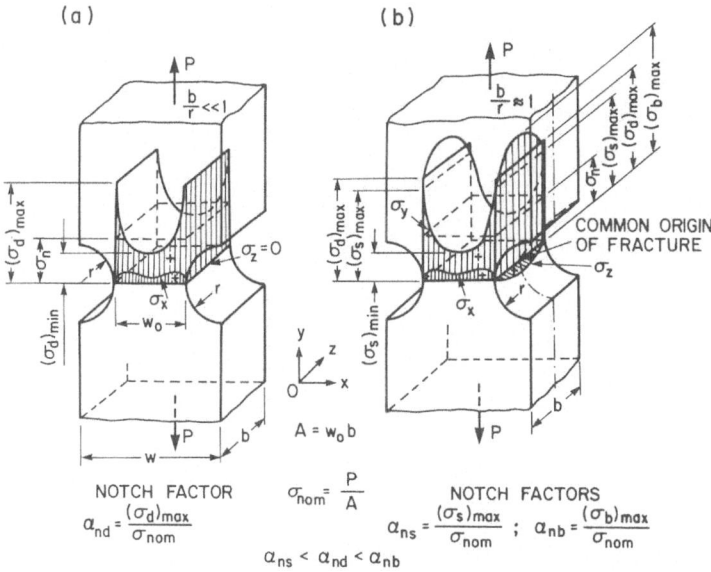

NOTCH FACTOR

$$\alpha_{nd} = \frac{(\sigma_d)_{max}}{\sigma_{nom}}$$

$$\sigma_{nom} = \frac{P}{A}$$

NOTCH FACTORS

$$\alpha_{ns} = \frac{(\sigma_s)_{max}}{\sigma_{nom}} \quad ; \quad \alpha_{nb} = \frac{(\sigma_b)_{max}}{\sigma_{nom}}$$

$$\alpha_{ns} < \alpha_{nd} < \alpha_{nb}$$

Fig. 2.1. Two incompatible models of stress states in plates in regions of cracks and notches. Both models are concurrently used in the analytical and experimental procedures of stress analysis, which are applied in fracture mechanics and mechanics of composite materials and structures. (a) Two-dimensional model of stresses related to the concept of stresses in an infinitesimally thin plate or to the concept of averaged stresses in a thin plate (generalized plane stress state). (b) Three-dimensional model of stresses related to the actual stress states in real homogeneous plates of finite thickness.
Models (a) and (b) lead to different notions of the stress concentration factor.

A sample of a stationary local effect related to a sharp notch in a beam under four-point load is presented in Figure 2.2. The radius of curvature of the notch bottom is 0.2 mm — this simulates the size of a region of the crack tip in steel, beneath which the concept of locally homogeneous stress is no longer applicable because of a finite size of grains of metallic alloys. The alteration of the normal stress component at the crack tip with the alteration of the thickness coordinate is pronounced, and the related tensile stress component along the crack bottom is not negligible [20]. This example is discussed in Part 2, Chapter 11.

Of particular practical interest are the local effects at the plate boundary in the vicinity of cracks or sharp notches, which are characterized by the local alteration of the magnitude and of the sign of the boundary stress, Figure 2.3.

The nonstationary local effects have attracted the attention of design engineers and R&D engineers and researchers for several decades because of their practical importance, particularly in the mining and in the defense industry. Perusing the huge amount of experimental data published in the open literature, one may notice that the theoretical foundation used for the evaluation and generalization of quite precise experimental data is rather weak. This makes it difficult to assess rationally to what extent the performed model experiments

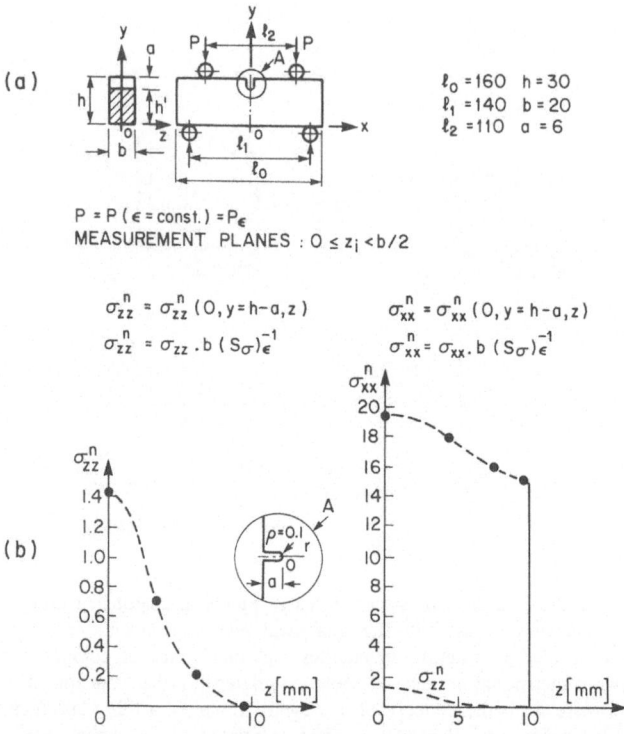

P = P(ε = const.) = P_ε
MEASUREMENT PLANES : 0 ≤ z_i < b/2

$$\sigma_{zz}^{n} = \sigma_{zz}^{n}(0, y = h-a, z)$$

$$\sigma_{zz}^{n} = \sigma_{zz} \cdot b (S_\sigma)_\epsilon^{-1}$$

$$\sigma_{xx}^{n} = \sigma_{xx}^{n}(0, y = h-a, z)$$

$$\sigma_{xx}^{n} = \sigma_{xx} \cdot b (S_\sigma)_\epsilon^{-1}$$

Fig. 2.2. Three-dimensional stress state at the tip of a crack in a beam subjected to pure bending, according to results obtained using the amplitude modulation isodyne technique. (a) Geometry of the specimen. (b) Variation of the normal stress components at the crack tip with the thickness coordinate z: normal stress component in direction of the beam axis x (right, top curve), and normal stress component in direction of the beam thickness z, along the crack tip (left, and bottom curve at the right).

represent the actual dynamic events occurring in real bodies. In particular, the common physical models of propagation of elastic disturbance in real bodies caused by various loading functions (impact, explosion, thermal shock) are very simplified. It is usually neglected that any real shock load produces a system of harmonic waves of various frequencies, propagating through a body with various velocities, which, in turn, are influenced by stationary and transient stress states, and are attenuated differently. Thus the group velocity of elastic waves usually differs from the phase velocities and is not constant in a general case.

An early example of an experimental study of transient local effects is given in Figure 2.4, taken from paper [8] published 30 years ago. This is an example of a dynamic fracture caused by the interference of two separate elastic wave fronts. Another kind of dynamic fracture related to nonstationary local effects is caused by the focusing of a single wave front which results in high transient stress concentrations. Such focusing can be produced by a focusing boundary which suitably reflects the impinging elastic wave fronts or — which is often overlooked — by the local inhomogeneities which bend the paths of propaga-

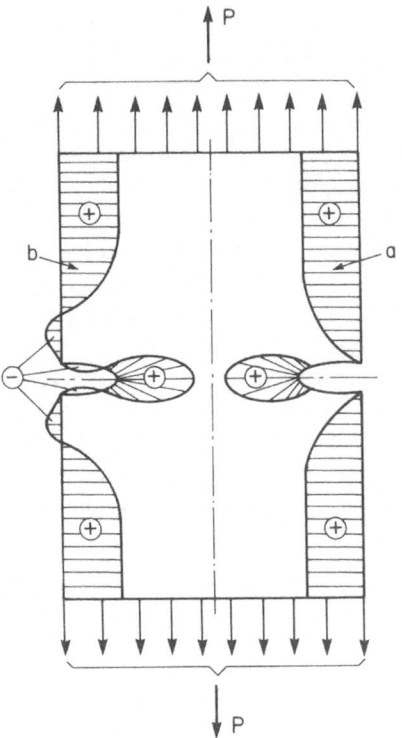

Fig. 2.3. Local effects of the boundary discontinuities in a plate under a tensile load. a. common approximation; b. actual patterns of boundary stresses.

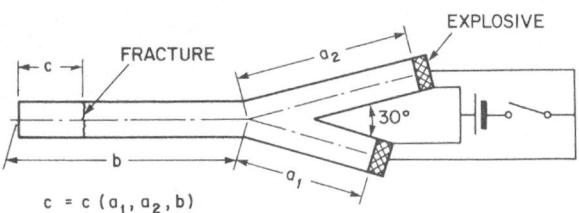

Fig. 2.4. An example of a nonstationary local effect. Dynamic fracture caused by interference of two elastic wave fronts.

tion of energy of elastic waves. It is easy to note that the thermo-elastic effect (elastic strain-coupled temperature alteration) may increase the magnitude of the local stress concentration to a larger or smaller extent.

At the present time it is common to deal with stationary local effects in flat plates by constructing analytical relations which consist of an interior component and the layer components. The common general condition is that only physically admissible solutions are accepted, namely, those with finite energy or finite power. The interior component of the solution represents the behaviour of the plate as a whole, while the layer component of the solution only represents the local effects which decay rapidly, usually exponentially. However, one must note that such a scheme is not compatible with the notion of the nonlocal stress fields.

The elastostatic approach to local effects outlined above is not applicable to nonstationary local effects caused by the interference or focusing of elastic waves, because the underlying physical models of dynamic phenomena must be much more comprehensive, as discussed above.

A particular class of local effects is represented by effects caused by material discontinuities which occur at the lamination surfaces, fiber-matrix interfaces, fiber tips, cracks, defects, delaminations, etc. A sample of such a local effect is given in Figure 2.5 and Figure 2.6 [11].

Local effects in composite structures and composite materials are of crucial practical importance with respect to the reliability of the optimization procedures used in structural design [23, 35]. Of particular significance are the edge effects in composite structures and the related shear and normal stresses in lamination planes, because these stresses lead to the development of local and

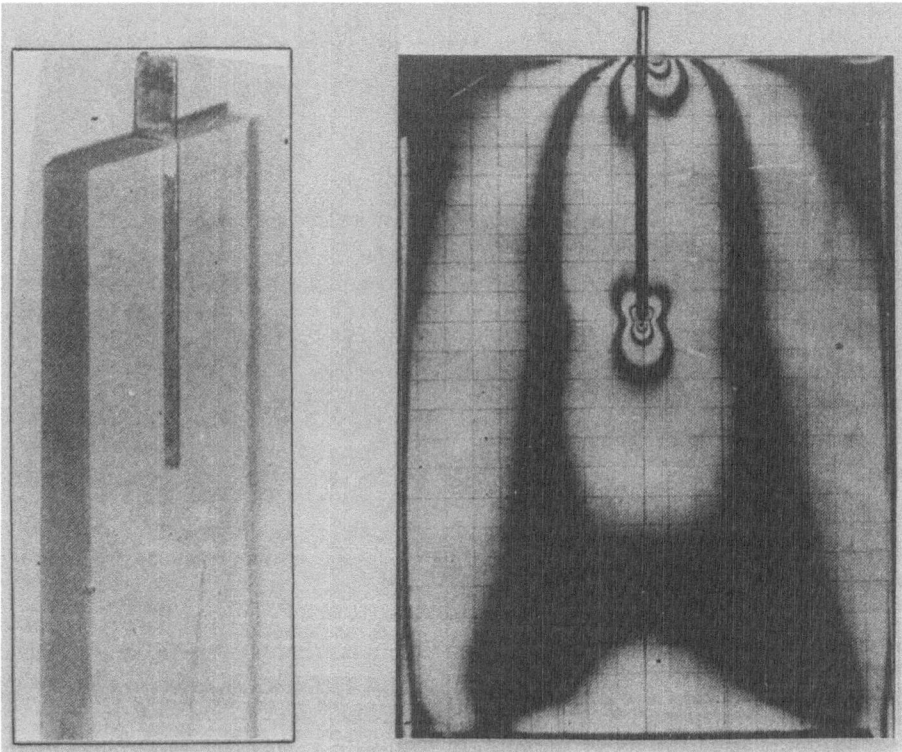

Fig. 2.5. Steel fibre in a polyester resin matrix in the form of a plate. Left: specimen after test, showing a local delamination. Right: specimen during the test showing effect of the local delamination. The density distribution of the transmission isochromatics indicates two separate regions of local effects in matrix, when the fibre is loaded by a tensile force. The visible local delamination was caused by the local three-dimensional effect at the plate boundary.

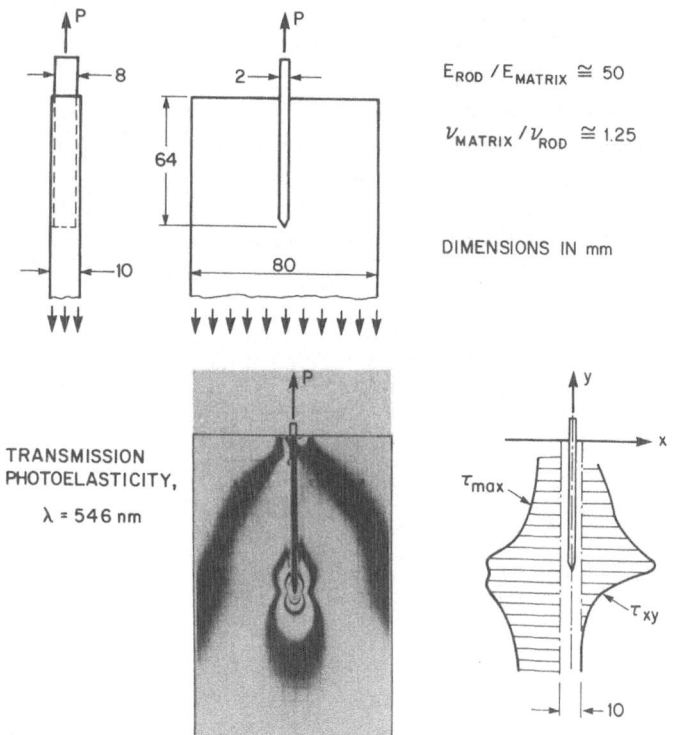

$E_{ROD} / E_{MATRIX} \cong 50$

$\nu_{MATRIX} / \nu_{ROD} \cong 1.25$

DIMENSIONS IN mm

TRANSMISSION
PHOTOELASTICITY,

$\lambda = 546$ nm

Fig. 2.6. Averaged through the thickness photoelastic analysis of the load transfer between a relatively rigid fibre of a finite length and the surrounding matrix of a finite thickness. The critical stress concentration at the fibre tip causes the tensile fracture in the matrix which propagates catastrophically under the angles of 80—85° with respect to the fibre.

global delamination defects. This problem is compounded by the fact that for practical reasons the analytical relations developed to describe and predict the behaviour of composite materials and structures are usually simplified. To ensure that the developed relations are sufficiently reliable, it is necessary to test them experimentally. It may be repeated that this imposes serious demands on the reliability of the applied experimental methods and procedures, in particular on the reliability of the theoretical background of experimental procedures. An up-to-date review of analytical and numerical research in the field of local effects is presented in [9].

It appears necessary to mention that research pertaining to local effects is not homogeneous theoretically. One may observe that quite a significant number of papers present experimental results which were obtained to prove the existence of stress singularities despite the fact that the boundlessly increasing stress values are physically inadmissible and that the assumptions of the infinitesimal (linear) deformations and of the strain values and strain gradients increasing boundlessly exclude each other. The thermal effects coupled with nonlinear deformations are still effectively excluded from considerations, with some noticeable exceptions, despite the fact that an isothermal analysis of large, irreversible deformations often leads to a misunderstanding of the mechanism of the local inelastic deformations.

Local yielding and the related geometric instability of a body represent a particular category of local effects. It has been shown that to understand the mechanism of local yielding, such as the necking of a cylindrical steel specimen under tensile load, or the randomly distributed regions of plastic deformation produced in a flat steel specimen by a tensile load, Figure 2.7 [15], it is necessary to incorporate the related thermal phenomena into the basic physical

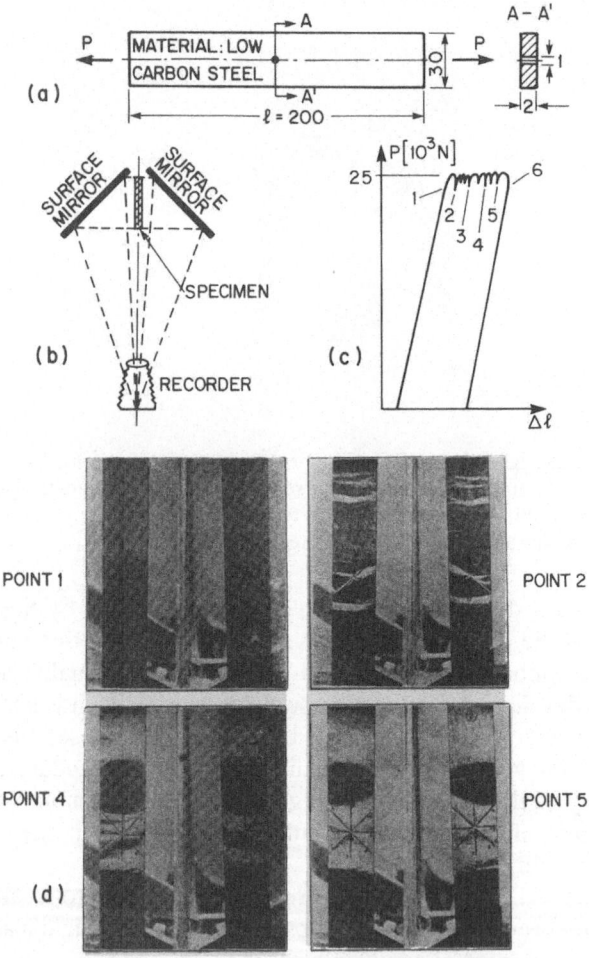

Fig. 2.7. Flat specimen made of low-carbon steel, under a tensile load. Development of typical numerous nonhomogeneous plastically deformed regions, and the corresponding tensile diagram. (a) Geometry of specimen. (b) Scheme of recording system. (c) Load-total elongation diagram. Numbers identify points at which recordings were made. (d) Recordings of elastic (dark) and plastic (light) regions at various values of total plastic deformation. Recordings show that process of plastic deformation is usually stochastic.

model, and to accept a stochastic model of inelastic deformations rather than a deterministic one.

The local effects related to hypervelocity impacts, which are of utmost importance in some advanced technical applications, are beyond the scope of this work.

2.4 Practical conclusions

Obviously, with regard to plates of finite thickness the condition (2.8),

$$\frac{\partial \sigma_{xy}}{\partial z}, \frac{\partial \sigma_{xz}}{\partial z} = 0, \; \frac{\partial \sigma_{yz}}{\partial z} = 0, \; \frac{\partial \sigma_{xx}}{\partial z} = 0; \; \frac{\partial \sigma_{yy}}{\partial z} = 0, \; \frac{\partial \sigma_{zz}}{\partial z} = 0,$$

is only satisfied when the stress field is homogeneous.

Such a restriction imposed on real plates of finite thickness yields several conditions regarding the load distribution and geometry of the plate boundary. For instance:

— the gradients of the intensity of boundary loads must be negligible;
— the minimal values of the radii of curvature of plate boundary, ρ_{min} are related to the plate thickness b by the condition:

$$b/\rho_{min} \ll 1. \tag{2.14}$$

Relations (2.7) and (2.8) characterize the common mathematical model of a hypothetical plate which is usually denoted by the term "thin plate"; in this text the term "thin plate" is used as defined above.

In a general case, the stress states in real plates and beams made of real materials and loaded by real forces are basically three-dimensional. Sometimes they are weakly three-dimensional, sometimes they are strongly three-dimensional. The problem is how "weak" is "weak", how "strong" is "strong", how "small" is "small", how "large" is "large", or how "wrong" is "wrong". To answer such questions it is necessary to quantify such terms as "weak" or "strong", and to establish numerical values of acceptable deviations based on the concept of risk analysis.

The concept of a homogeneous material continuum, which is the major component of Hooke's body, neglects existence of discrete structure of real materials with all the consequences resulting thereof [10, 12–15, 24, 25, 27, 30]. From the physical point of view, the continuum mechanics notion of stress at a point collapses when the length of the region of the body under observation is less than 20 grain sizes [25]. The assumption of a single mechanism of deformation is physically a very drastic assumption, particularly when the inelastic deformations of real bodies are considered [12, 15, 16].

The thermodynamic processes coupled with the deformation states may, under certain circumstances, become a major concern to the designer [7, 17, 19]. The thermoelastic effect may become of major importance at the temperature of outer space, below the temperature of 100 K, where the value of the specific heat of materials decreases rapidly.

It must be noted that all real engineering materials creep under load and that this creep may be pronounced under certain conditions. In particular, creep of all polymeric materials is not negligible and is often quite complicated [16]. Thus the notion of a linear elastic body can be applied to real materials, such as steel or aluminum alloys, under certain conditions only; analogously, the notion of a linear viscoelastic body must be applied prudently for high polymers such as epoxy or polyester resins.

Inelastic material deformations in real materials are still incompletely simulated in various analytical solutions of the theory of plasticity. It occurs that predictions of some analytical procedures deviate quantitatively and also qualitatively from the real responses of steel or aluminum alloys [15].

2.5 References

[1]   Aleksandrov, A. D., Kolgomorov, A. N., and Lavrent'ev, M. A., *Mathematics, Its Content, Methods and Meaning* (translated from the Russian edition: "Matematika, ye Soderzhanye, Metody i Znachenye", Izdatielstvo Akademii Nauk SSSR, Moskva, 1956), The MIT Press, Cambridge, MA, 1981.

[2]   Borg, S. F., *Matrix-Tensor Methods in Continuum Mechanics*, D. Van Nostrand Company, Princeton, 1963.

[3]   Bui, H. D., "Théorie linéare de la rupture", *Revue Française de Mécanique* **3** (1983), pp. 3—7.

[4]   Doeblin, Ernest O., *Measurement Systems: Application and Design*, McGraw-Hill Book Company, New York, 1983.

[5]   Ellyin, F., Lind, N. C., and Sherbourne, A. N., "Elastic Stress Field in a Plate with a Skew Hole", *J. of Eng. Mechanics Division, ASCE* **92** (EM1), 1966, pp. 1—10.

[6]   Ellyin, F. and Sherbourne, A. N., "Effect of Skew Penetration on Stress Concentration", *J. of Eng. Mechanics Division, ASCE* **94** (EM6), 1968, pp. 1317—1336.

[7]   Kestin, Joseph, *A Course in Thermodynamics*, Vols. I and II, McGraw-Hill Book Company, New York, 1979.

[8]   Khanukayev, A. N., "A Study of the Effect of Interference of Two Waves Successively Reflected from the Free End of a Rod" (in Russian), in: S. P. Shikhobalov (Ed.), *Polarization-Optical Method of Stress Analysis*, University of Leningrad, Leningrad, 1960, pp. 253—256.

[9]   Ladevèze, Pierre (Ed.), *Local Effects in the Analysis of Structures*, Elsevier, New York, 1985.

[10] Lakes, Roderick, "Foam Structures with a Negative Poisson's Ratio", *Science* **235**, 1987, pp. 1038—1040.

[11] Lewicki, B. and Pindera, J. T., "Photoelastic Models of Reinforced Structures" (in Polish), *Archiwum Inzynierii Ladowej*, Vol. II (4), 1956, pp. 381—418.

[12] Mohamed, F. A. and Soliman, M. S., "On the Creep Behaviour of Uranium Dioxide", *Materials Science and Engineering* **53**, 1982, pp. 184—190.

[13] Müller, R. K., "Der Einfluss der Messlänge auf die Ergebnisse bei Dehnmessungen an Beton", *Beton* **14**(5), 1964, pp. 205—208.

[14] Neuber, H., *Kerbspannungslehre* (Notch Stresses), Verlag von Julius Springer, Berlin, 1937.

[15] Pindera, J. T., *On Application of Brittle Coatings for Determination of Regions of Plastic Deformations* (in Polish), Engineering Transactions (Rozprawy Inzynierskie), Polish Acad. of Sciences, Vol. V (1), 1957, pp. 33—47.

[16] Pindera, Jerzy T., *Rheological Properties of Some Polyester Resins*, Part I, II, III (in Polish), Engineering Transactions (Rozprawy Inzynierskie) Polish Acad. Sciences, Vol. VII (3 and 4), 1959, pp. 361—411, 481—520, 521—540.

[17] Pindera, J. T., Straka, P., and Tschinke, M. F., "Actual Thermoelastic Response of Some Engineering Materials and Its Applicability in Investigations of Dynamic Response of Structures", *VDI-Berichte*, Nr. 313, 1978, pp. 579—584.

[18] Pindera, J. T., Contemporary Trends in Experimental Mechanics: Foundations, Methods, Applications", in: J. T. Pindera *et al.* (Eds), *Experimental Mechanics in Research and Development*, Solid Mechanics Division, University of Waterloo, Study No. 9, Waterloo, 1973, pp. 143—168.

[19] Pindera, Jerzy T., "Foundations of Experimental Mechanics: Principles of Modelling, Observation and Experimentation", in: J. T. Pindera (Ed.), *New Physical Trends in Experimental Mechanics*, Springer-Verlag, Wien, 1981, pp. 188—236.

[20] Pindera, J. T. and Krasnowski, B. R., "Determination of Stress Intensity Factors in Thin and Thick Plates Using Isodyne Photoelasticity", in: Simpson, Leonard A. (Ed.), *Fracture Problems &Solutions in the Energy Industry*, Pergamon Press, 1982, pp. 147—156.

[21] Pindera, J. T. (Ed.), *Modelling Problems in Crack Tip Mechanics* (Proc. of the Tenth Canadian Fracture Conference, August 24—26, 1983, University of Waterloo), Martinus Nijhoff, The Hague, The Netherlands, 1984.

[22] Pindera, J. T., Krasnowski, B. R., and Pindera, M.-J., "Theory of Elastic and Photoelastic Isodynes. Samples of Applications in Composite Structures", *Experimental Mechanics* **25**(3), 1985, pp. 272—281.

[23] Pipes, R. B. and Pagano, N. J., "Interlaminar Stresses in Composite Laminates under Uniform Axial Extension", *J. Composite Materials* **4**, 1970, pp. 538—548.

[24] Provan, J. W., "The Micromechanics of Fatigue Crack Initiation", in: Pindera, Jerzy T. (Ed.), *Modelling Problems in Crack Tip Mechanics*, Martinus Nijhoff Publishers, Dordrecht, 1984, pp. 131—154.

[25] Rohrbach, Ch., "Dehnungsmessstreifen mit metallischen Träger als schnell messbereites, feuchtigkeitsunempfindliches Messelement für Dehnnungsmessungen auf Beton", *Der Bauingenieur* **33**, 1958, pp. 265—268.

[26] Savin, G. N., *Kontsentratsya Napriazhenii Okolo Otverstii* (Stress Concentration around Holes), Gosud. Izd. Tehkniko-Teoreticheskoy Literatury, Moskva, 1951.

[27] Siebel, E., *Handbuch der Werkstoffprüfung*, Vols. 1 and 2, Springer-Verlag, Berlin, 1958 and 1955.

[28] Sih, G. C., Williams, M. L., and Swedlow, J. L., *Three-Dimensional Stress Distribution Near a Sharp Crack in a Plate of Finite Thickness*, Air Force Materials Laboratory, Wright Patterson Air Force Base, AFML-TR, 1966, pp. 66—242.

[29] Sih, G. C., "A Review of the Three-Dimensional Stress Problem for a Cracked Plate", *International Journal of Fracture Mechanics* **7**, 1971, pp. 39—61.

[30] Sih, G. C., "The State of Affairs Near the Crack Tip", in: Pindera, Jerzy T. (Ed.), *Modelling Problems in Crack Tip Mechanics*, Martinus Nijhoff Publishers, Dordrecht, 1984, pp. 65—90.

[31] Sokolnikoff, I. S., *Mathematical Theory of Elasticity*, McGraw-Hill, 1956.

[32] Sternberg, E. and Sadovsky, M. A., "Three-Dimensional Solution for the Stress Concentration around a Circular Hole in a Plate of Arbitrary Thickness", *Journal of Applied Mechanics, ASME* **16**(1), 1949, pp. 27—38.

[33] Thum, A. and Svenson, O., "Die Verformungs- und Beanspruchungsverhältnisse von glatten und gekerbten Stäben, Scheiben and Platten in Abhängigkeit von deren Dicke und Belastungsart", *Forschung Ing. Wes.* **13**, 1942, pp. 1—11.

[34] Thum, A., Petersen, C., and Svenson, O., *Verformung, Spannung und Kerbwirkung* (Deformation, Stress and Notch Action), VDI-Verlag, Dusseldorf, 1960.

[35] Wang, S. S. and Choi, I., "Boundary-Layer Effects in Composite Laminates: Part II — Free-Edge Stress Solutions and Basic Characteristics", *J. Applied Mechanics* **49**, 1982, pp. 549—560.

[36] Youngdahl, C. K. and Sternberg, E., "Three-Dimensional Stress Concentration Around a Cylindrical Hole in a Semi-Infinite Elastic Body", *Journal of Applied Mechanics, ASME* **33**(4), 1966, pp. 855—865.

Theory of analytical isodynes

3.1 Introduction

In this Chapter the basic theory of analytical elastic isodynes is presented. The theory of the plane elastic analytical isodynes is presented in a more comprehensive form because of the practical significance of this class of isodynes. Only the essential components of differential isodynes, pertaining to the symmetrical three-dimensional stress states in plates are presented. This theory is still incompletely developed, and the errors resulting from the simplifying assumptions are still not quantifiable.

The theory of plane elastic analytical isodynes is developed on the basis of the notion of a two-dimensional stress state which exists in an infinitesimally thin plate subjected to the infinitesimally small deformation (a linear problem). The plate is loaded only at the boundary surface by tractions coplanar with the plate and having the dimension of force intensity (force/length). The plate is able to deform in its plane only, by definition any geometric instability is excluded, so no buckling or out-of plane bending is permitted. The plate is made of a linear elastic material which deforms athermally; this material represents the simplest notion of a hypothetical homogeneous and isotropic material continuum. In such a material, a deformation state may produce mechanical inhomogeneity but not mechanical anisotropy.

The purpose of the theory of analytical isodynes is to provide a theoretical basis for various particular theories and techniques of experimental isodynes, in particular of the optical isodynes presented in Chapter 5. Development of the concept of analytical isodynes is given in References [1—5]. The presented relations pertain to linear elastic or linear viscoelastic states of deformation, therefore the word "elastic" in the term "analytical elastic isodynes" is often omitted. The same pertains to the word "analytical".

3.2 Concept of plane analytical isodynes

The concept of analytical isodynes is related to the plane stress fields in linear elastic plates. It is generalized to the actual three-dimensional local stress fields in plates which occur in regions where the gradients of the stress components are not negligible. This occurs in regions of stress concentrations around notches, cracks, voids, delaminations, or contact loads, or, shortly, in regions of local effects. The concept of isodynes is applicable, in some cases, to general three-dimensional stress states.

3.2.1 Mathematical model. Analytical relations

The stress state in a mathematical model of a hypothetical plate presented in Section 2.2.1 is conveniently characterised by the Airy stress function, $\Phi(x, y)$ which — in the absence of body forces — yields the known expressions for the stress components with respect to an arbitrarily chosen Cartesian coordinate system (x, y, z),

$$\frac{\partial \Phi(x, y)}{\partial x} = \Phi_x(x, y); \quad \frac{\partial \Phi(x, y)}{\partial y} = \Phi_y(x, y), \tag{3.1a}$$

$$\sigma_{xx} = \frac{\partial^2 \Phi}{\partial y^2} = \frac{\partial}{\partial y} \Phi_y; \quad \sigma_{yy} = \frac{\partial^2 \Phi}{\partial x^2} = \frac{\partial}{\partial x} \Phi_x, \tag{3.1b}$$

$$\sigma_{xy} = \sigma_{yx} = -\frac{\partial^2 \Phi}{\partial x \, \partial y} = -\frac{\partial}{\partial y} \Phi_x = -\frac{\partial}{\partial x} \Phi_y, \tag{3.1c}$$

and, by definition,

$$\sigma_{zz} = \sigma_{xz} = \sigma_{yz} = \sigma_{zx} = \sigma_{zy} = 0. \tag{3.1d}$$

Obviously, to comply with the basic condition that all relations which describe physical events must be dimensionally homogeneous, Airy stress function $\Phi(x, y)$ must have the dimension of force [force].

It is convenient to consider the $\Phi(x, y)$-function as a surface; consequently the first partial derivatives of Airy functions, $\Phi_x(x, y)$ and $\Phi_y(x, y)$, are related surfaces whose shapes depend on the direction of differentiation.

There exist particular relations between the functions $\Phi_i(x, y)$, stress components $\sigma_{ij}(x, y)$, and the intensities of normal forces p_x and p_y acting on the cross-sections collinear with the corresponding directions of differentiation, x and y:

$$\int \sigma_{yy}(x, y) \, dx = \int \frac{\partial^2 \Phi(x, y)}{\partial x^2} \, dx$$

$$= \Phi_x(x, y) + f_x(y) = p_y(x, y), \tag{3.2a}$$

and

$$\int \sigma_{xx}(x, y) \, dy = \int \frac{\partial^2 \Phi(x, y)}{\partial y^2} \, dy$$

$$= \Phi_y(x, y) + f_y(x) = p_x(x, y). \tag{3.2b}$$

The functions p_y and p_x are continuous in the whole region defined by the plate boundary, and can be presented in a general form as functions $p_x(x, y)$ and $p_y(x, y)$. These functions are obviously related to the directions of differentiation of the function $\Phi(x, y)$: these directions are called "characteristic directions".

The plane containing and related to the x-characteristic direction is called the (x, y)-characteristic plane. Analogously, the plane containing and related to the y-characteristic direction is called the (y, x)-characteristic plane. In the case of plane stress existing in an infinitesimally thin plate, the plane of the plate and the x- and y-characteristic planes are coplanar. Using the above geometric interpretation, the intensities of normal forces acting on specified cross-sections given by the functions $p_y(x, y)$ and $p_x(x, y)$ represent surfaces spanned above the plate, which are related to the chosen characteristic directions; they will be called x- and y-surfaces of plane elastic isodynes, respectively.

The functions $p_x(x, y)$ and $p_y(x, y)$ can be used to introduce a new family of characteristic lines of plane stress field called elastic isodynes which are related to particular values of total normal forces acting on corresponding loci of points. Along these lines the total normal force acting on corresponding cross-sections collinear with the characteristic direction is constant.

Thus, the equations of two families of elastic isodynes related to the x- and y-directions and presented in terms of normal force intensities can be formulated as:

$$p_y(x, y) = \text{const.} \tag{3.3a}$$

$$p_x(x, y) = \text{const.} \tag{3.3b}$$

where the function $p_y(x, y)$ is related to x-characteristic direction and $p_x(x, y)$ is related to y-characteristic direction. The dimension of quantities p_x and p_y is [force/length].

Geometrically, isodynes represent the loci of points on the surfaces $p_x(x, y)$ and $p_y(x, y)$ where the values of p_x and p_y are constant.

It is convenient to present equations of the plane elastic isodyne surfaces in a dimensionless form, as $p_{ny}(x, y)$ and $p_{nx}(x, y)$:

$$p_y(x, y) = S_s p_{ny}(x, y), \tag{3.4a}$$

and

$$p_x(x, y) = S_s p_{nx}(x, y), \tag{3.4b}$$

where S_s is a suitably chosen coefficient related to the material of the plate, having the dimension [force/length].

Thus, elastic isodynes can be presented in parametric form

$$p_{ny}(x, y) = m_{sx}, \tag{3.4c}$$

and

$$p_{nx}(x, y) = m_{sy}, \tag{3.4d}$$

where m_s denotes the order of isodyne, and the subscripts x and y in m_{sx} and m_{sy} denote the characteristic directions. Obviously,

$$-\infty < m_{sx}, m_{sy} < \infty.$$

It follows from relations (3.3) and (3.4) that relations (3.2) represent cross-sections through the corresponding isodyne fields, in the corresponding characteristic directions, at constant values of the parameters $y = y_0$ or $x = x_0$, respectively:

$$p_y(x, y_0) = S_s m_{sxx}(x, y_0), \tag{3.5a}$$

$$p_x(x_0, y) = S_s m_{syy}(x_0, y), \tag{3.5b}$$

where $m_{sxx}(x, y_0)$ which is called isodyne function in the x-characteristic direction for a chosen $y = y_0$, is the cross-section through the x-isodyne surface; and analogously, $m_{syy}(x_0, y)$, which is called isodyne function in the y-characteristic direction for a chosen $x = x_0$, is the cross-section through the y-isodyne surface.

The functions $p_y(x, y)$ and $p_x(x, y)$ can be utilized to determine the values of normal stress components using relations (3.1) and (3.2):

$$\sigma_{yy}(x, y) = \frac{\partial}{\partial x} \Phi_x(x, y) = \frac{\partial}{\partial x} p_y(x, y) = S_s \frac{\partial}{\partial x} m_{sx}(x, y), \tag{3.6a}$$

and

$$\sigma_{xx}(x, y) = \frac{\partial}{\partial y} \Phi_y(x, y) = \frac{\partial}{\partial y} p_x(x, y) = S_s \frac{\partial}{\partial y} m_{sy}(x, y), \tag{3.6b}$$

since

$$\frac{d}{dx}[f_x(y)] = \frac{d}{dy}[f_y(x)] = 0. \tag{3.6c}$$

The functions $f_x(y)$ and $f_y(x)$ can be easily determined from the boundary conditions as is shown later.

It follows from (3.6) that the normal stress components σ_{xx} and σ_{yy} are proportional to the slopes of the isodyne functions, which are the cross-sections through the isodyne surfaces related to the x- and y-characteristic directions, as defined earlier.

Obviously, the characteristic directions can be chosen according to convenience; — each corresponding family of isodyne surfaces and isodyne functions yields values of the related normal stress components.

Introducing the order of isodyne, m_{sx} or m_{xy}, as a parameter, the isodyne fields, related to a Cartesian coordinate system taken as characteristic directions, can be presented in the following form:

$$\int \frac{\partial^2 \Phi(x, y)}{\partial x^2} \, dx = p_y(x, y, m_{sx}) = S_s m_{sxx}(x, y) = S_s P_{ny}(x, y, m_{sx}). \tag{3.7a}$$

$$\int \frac{\partial^2 \Phi(x, y)}{\partial y^2} \, dy = p_x(x, y, m_{sy}) = S_s m_{syy}(x, y) = S_s P_{nx}(x, y, m_{sy}) \tag{3.7b}$$

Of course, the mathematical model of elastic isodynes represented by relations (3.7) yields values of normal stress components according to (3.6) only when the basic assumptions of Airy stress functions are satisfied.

The functions $\Phi_x(x, y)$ and $\Phi_y(x, y)$ can yield so called "generalized plane elastic isodynes" presented in the form:

$$\Phi_x(x, y, m_{sx}) = p_y(x, y, m_{sx}) - f_x(y), \tag{3.8a}$$

$$\Phi_y(x, y, m_{sy}) = p_x(x, y, m_{sy}) - f_y(x). \tag{3.8b}$$

As already mentioned, the functions $f_x(y)$ and $f_y(x)$ can be determined from the given boundary conditions. In most practical cases it is possible to choose the characteristic directions in such a manner that the functions $f_x(y)$ and $f_y(x)$ are constant or vanish, and the isodynes are given by the first derivatives of Airy stress function. Thus, when these functions vanish,

$$f_x(y) = f_y(x) = 0.$$

Then

$$p_y(x, y, m_{sx}) = \Phi_x(x, y, m_{sx}) = \frac{\partial}{\partial x}\, \Phi(x, y, m_{sx}), \tag{3.9a}$$

and

$$p_x(s, y, m_{sy}) = \Phi_y(x, y, m_{sy}) = \frac{\partial}{\partial y}\, \Phi(x, y, m_{sy}). \tag{3.9b}$$

It follows from the Airy stress function, represented by relations (3.1), that it is possible to obtain information on the shear stress components by taking cross-sections through isodyne surfaces in directions normal to the characteristic directions. Thus in accordance with relations (3.2a, b), (3.7a, b) and (3.1c), the shear stress components σ_{xy} and σ_{yx}, and their intensities t_{xy} and t_{yx}, can be described by the following relations:

$$\sigma_{xy} = -\frac{\partial}{\partial y}\, \Phi_x = -S_s\, \frac{\partial}{\partial y}\, p_{ny}(x, y) + \frac{\partial}{\partial y}\, f_x(y)$$

$$= -S_s\, \frac{\partial}{\partial y}\, m_{sx}(x, y) + \frac{d}{dy}\, f_x(y) = \frac{\partial}{\partial y}\, t_{xy}(x, y), \tag{3.10a}$$

$$\sigma_{yx} = -\frac{\partial}{\partial x}\, \Phi_y = -S_s\, \frac{\partial}{\partial x}\, p_{nx}(x, y) + \frac{\partial}{\partial x}\, f_y(x)$$

$$= -S_s\, \frac{\partial}{\partial x}\, m_{sy}(x, y) + \frac{d}{dx}\, f_y(x) = \frac{\partial}{\partial x}\, t_{yx}(x, y). \tag{3.10b}$$

Consequently,

$$t_{xy} = t_{xy}(x, y, m_{sx}) = \int \sigma_{xy}\,\mathrm{d}y = -\int \frac{\partial \Phi_x}{\partial y}\,\mathrm{d}y$$

$$= -S_s m_{sx}(x, y) + f_x(y) + C_1, \qquad (3.11a)$$

$$t_{yx} = t_{yx}(x, y, m_{sy}) = \int \sigma_{xy}\,\mathrm{d}x = -\int \frac{\partial \Phi_y}{\partial x}\,\mathrm{d}x$$

$$= -S_s m_{sy}(x, y) + f_y(x) + C_2. \qquad (3.11b)$$

Obviously, when the stress state in a plate of the thickness b_0 is plane, then the total shear forces T_{xy} and T_{yx} are

$$T_{xy} = b_0 t_{xy}, \text{ and } T_{yx} = b_0 t_{yx}. \qquad (3.12a, b)$$

At any chosen point of a plane stress field $\sigma_{xy} = \sigma_{yx}$, thus,

$$S_s \left[\frac{\partial}{\partial y} m_{sx}(x, y) - \frac{\partial}{\partial x} m_{sy}(x, y) \right] = \frac{\mathrm{d}}{\mathrm{d}y} f_x(y) - \frac{\mathrm{d}}{\mathrm{d}x} f_y(x). \qquad (3.13)$$

Relations (3.13) can be used to check the accuracy of determination of the functions $f_x(y)$ and $f_y(x)$.

Determination of the slopes of the x- and y-isodynes in the directions y, and x, respectively, is identical with differentiation of the isodyne functions obtained as cross-sections of the x- and y-isodyne surfaces with the planes normal to the characteristic directions x and y.

Thus the surfaces of plane elastic x- and y-isodynes yield the following isodyne functions:

the x-isodyne surfaces yield:

m_{sxx}: xx-isodyne functions whose slopes are proportional to σ_{yy};
m_{sxy}: xy-isodyne functions whose slopes are proportional to σ_{yx} and to the slopes of the integration function $f_x(y)$;

the y-isodyne surfaces yield:

m_{syy}: yy-isodyne functions whose slopes are directly proportional to σ_{xx};
m_{syx}: yx-isodyne functions whose slopes are proportional to σ_{xy} and to the slopes of the integration functions $f_y(x)$.

As mentioned earlier, it is usually easy to choose the cross-sections (x_0, y) and (x, y_0) in such a manner that the slopes of the integration functions $f_x(y)$ and $f_y(x)$ vanish.

According to relations (3.6a, b) and (3.10a, b), two families of isodynes — related to two different isodyne surfaces which are characterized by two, preferably orthogonal, characteristic directions, for instance, the surfaces

$p_{nx}(x, y)$ and $p_{ny}(x, y)$ or shortly (x, y)-isodyne surfaces — yield four pieces of information on the values of the normal and shear stress components,

$$\sigma_{xx}, \sigma_{yy}, \sigma_{xy} \text{ and } \sigma_{yx}, \text{ where } \sigma_{xy} = \sigma_{yx}.$$

In addition, the fields of (x, y)-isodynes are related by the condition

$$\sigma_{xx} + \sigma_{yy} = \sigma_1 + \sigma_2. \tag{3.14}$$

Thus, the equation of isopachics or of the geometric loci of points at which the sums of normal stresses are constants may be presented in the form:

$$\frac{\partial}{\partial x} \Phi_x + \frac{\partial}{\partial y} \Phi_y = S_s \left(\frac{\partial}{\partial x} p_{ny} + \frac{\partial}{\partial y} p_{nx} \right) = S_s m_i = \text{const.} \tag{3.15}$$

where m_i denotes the order of isopachics and S_s denotes the elastic coefficient of isopachics.

When the boundary conditions are such that the functions $f_x(y)$, $f_y(x)$, and the constants C_1 and C_2 vanish, then the elastic isodynes are identical with the first derivatives of Airy stress function.

It is convenient to use the term "characteristic lines of plane stress fields" to denote a set of functions characterizing the stress fields, such as the isostatics, isoclinics, isochromatics, isopachics, etc. Within such a framework the elastic isodynes represent a new family of the characteristic lines of plane stress fields.

3.2.2 Summary

(a) Four pieces of information given by two conjugated isodyne surfaces, (x, y)- and (y, x)-isodyne surfaces, describe three independent stress components, $\sigma_{xx}, \sigma_{yy}, \sigma_{xy} = \sigma_{yx}$; one piece of information is redundant and can be used to increase the accuracy of experimental evaluation of stress components. Various known techniques of differentiation of isodyne fields can be applied or adapted, to determine the quantities of interest.

(b) When the boundary conditions are known, two isodyne fields related to two different characteristic directions are, in general, sufficient to determine numerically the stress function. In this approach the Airy stress function represents a surface having the particular property that the slopes of this surface in characteristic directions are proportional to the total normal forces acting on corresponding cross-sections through the object, and their curvatures are related to the stress components.

(c) Basic concepts of elastic isodynes and related definitions and notation are summarized in Figure 3.1. The isodynes plotted in the plane (x, y) are vertical projections of the isodynes lying on the isodyne surfaces, $p_{nx}(x, y)$ or $p_{ny}(x, y)$. Using the Mohr-circle transformation, it is possible to relate the stress components related to an (x, y)-coordinate system to the stress components related to a convenient (x', y')-coordinate system to further increase the accuracy of the stress analysis.

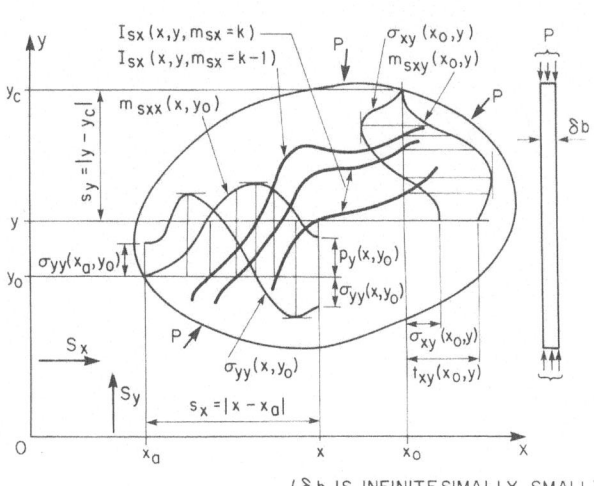

FIELD OF x – ISODYNES : $I_{sx} = I_{sx}(x, y, m_{sx})$

Fig. 3.1. Concept of plane analytic isodynes. Basic parameters.

3.3 Determination of integration functions. Boundary conditions

3.3.1 General approach

The geometric and loading boundary conditions are used to determine the integration functions and integration constants, $f_x(y)$, $f_y(x)$, C_1, C_2, Figure 3.2. The presentation is based on Reference [3].

It is convenient to use the following forms of boundary equation:

$$s(x, y) = 0 \tag{3.16a}$$
$$s(x, y) = x - s_x(y) = 0 \tag{3.16b}$$
$$s(x, y) = y - s_y(x) = 0. \tag{3.16c}$$

It is also convenient to relate the boundary equation $s(x, y)$ to characteristic directions and, consequently, to present it in the form of two boundary segments, the entrance segment and the exit segment.

With respect to the x-characteristic direction, S_x, the s-boundary can be represented by (Figure 3.2):

- entrance segment, $A_x - A_y - B_x : x = s_{x1}(y)$ (3.16d)
- exit segment, $B_x - B_y - A_x : x = s_{x2}(y)$ (3.16e)

With respect to S_y the s-boundary can be represented by:

- entrance segment, $B_y - A_x - A_y : y = s_{y1}(x)$ (3.16f)
- exit segment, $A_y - B_x - B_y : y = s_{y2}(x)$ (3.16g)

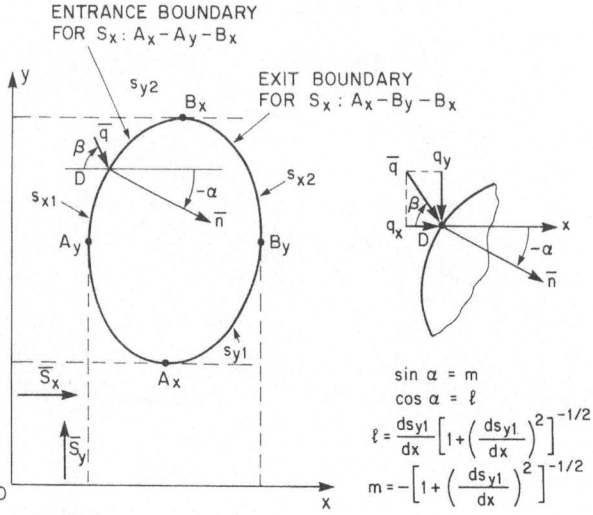

Fig. 3.2. Equilibrium conditions at the entrance boundary. Definitions.

To relate equations (3.2a, b) to the known boundary conditions, it is convenient to present them in the following manner:

$$p_y(x, y) = \int_{s_{x1}}^{x} \sigma_{yy}(x, y) \, dx = \int_{s_{x1}}^{x} \frac{\partial^2}{\partial x^2} \Phi(x, y) \, dx$$

$$= \Phi_x(x, y) - \Phi_x[s_{x1}(y), y] \qquad (3.17a)$$

and

$$p_x(x, y) = \int_{s_{11}}^{y} \sigma_{xx}(x, y) \, dy = \int_{s_{11}}^{y} \frac{\partial^2}{\partial y^2} \Phi(x, y) \, dy$$

$$= \Phi_y(x, y) - \Phi_y[x, s_{y1}(x)]. \qquad (3.17b)$$

Using relations (3.17a, b) one can determine the functions $f_x(y)$ and $f_y(x)$, defined by (3.2a, b):

$$f_x(y) = -\Phi_x[s_{x1}(y), y] \qquad (3.18a)$$

$$f_y(x) = -\Phi_y[x, s_{y1}(x)]. \qquad (3.18b)$$

Relations (3.18a, b) present a general form of the integration functions. To determine the particular forms of these functions, it is necessary to use specific boundary conditions.

The same result can be obtained by presenting relations (3.2a, b) in terms of boundary coordinates.

Equilibrium conditions at a boundary point D, Figure 3.2, can be presented in the form:

$$\ell\sigma_{xx} + m\sigma_{yx} = q_x \tag{3.19a}$$

$$m\sigma_{yy} + \ell\sigma_{xy} = q_y \tag{3.19b}$$

where q_x and q_y are components of the traction $\bar{q}$, distributed along the boundary s, and m and ℓ are directional parameters: $m = \sin\alpha$, and $\ell = \cos\alpha$, where α denotes the angle between the x-characteristic direction and the normal to the boundary, $\bar{n}$, at the point D.

Relation (3.19a, b) can be rewritten in terms of the unknown functions $\Phi_x(x, y)$ and $\Phi_y(x, y)$:

$$\ell\frac{\partial}{\partial y}\Phi_y - m\frac{\partial}{\partial x}\Phi_y = q_x \tag{3.20a}$$

$$m\frac{\partial}{\partial x}\Phi_x - \ell\frac{\partial}{\partial y}\Phi_x = q_y. \tag{3.20b}$$

This is a system of two uncoupled partial differential equations of the first order, which describes conditions at the boundary s.

Two cases are considered. In the first case the boundary conditions are a known function of the boundary coordinates; in the second case, the boundary conditions are not known.

3.3.2 Case of known boundary conditions

It has been shown that the functions $f_x(y)$ and $f_y(x)$ which are related to functions Φ_x and Φ_y determined at the entrance boundary, (3.18a, b), are constant along corresponding characteristic cross-sections, (x, y_0) and (x_0, y), in the region bounded by the line $s(x, y)$. Thus, it is sufficient to determine these functions at the entrance or exit boundary.

The geometry of boundary at an arbitrary point S, Figure 3.3, can be given by:

$$\cos\alpha = \ell = \frac{dy}{ds} \tag{3.21a}$$

$$\sin\alpha = m = -\frac{dx}{ds}. \tag{3.21b}$$

Using (3.21a, b), the boundary equations (3.20a, b) can be presented in the form:

$$\frac{\partial\Phi_y}{\partial y}\frac{dy}{ds} + \frac{\partial\Phi_y}{\partial x}\frac{dx}{ds} = \bar{q}\cdot\bar{i} \tag{3.22a}$$

$$-\frac{\partial\Phi_x}{\partial x}\frac{dx}{ds} - \frac{\partial\Phi_x}{\partial y}\frac{dy}{ds} = \bar{q}\cdot\bar{j} \tag{3.22b}$$

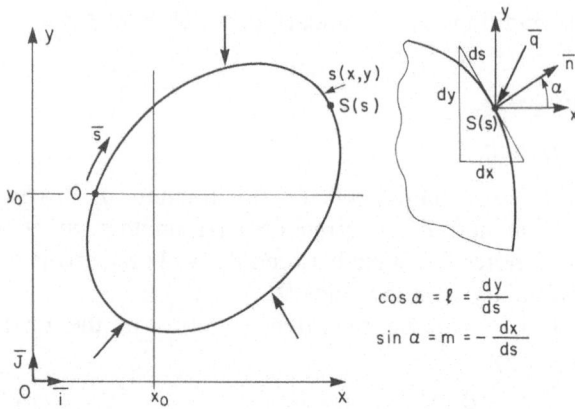

Fig. 3.3. Boundary loading conditions used in determination of the integration functions.

or

$$\frac{d}{ds}(\Phi_y) = \bar{q} \cdot \bar{i},$$ (3.23a)

and

$$\frac{d}{ds}(\Phi_x) = -\bar{q} \cdot \bar{j}.$$ (3.23b)

Integration of (3.23a, b) yields

$$\Phi_y(s) = \int \bar{q} \cdot \bar{i} \, ds + C_1,$$ (3.24a)

$$\Phi_x(s) = -\int \bar{q} \cdot \bar{j} \, ds + C_2,$$ (3.24b)

where C_1 and C_2 are arbitrary constants.

Consequently, the integrals (3.24a, b) can be determined for an arbitrary boundary segment s, starting from an arbitrary point S_0 ($s = 0$):

$$\Phi_y(s) = \int \bar{q} \cdot \bar{i} \, ds = P_x(s) = P_x[x, s_{y1}(x)];$$ (3.25a)

$$\Phi_x(s) = -\int_0^s \bar{q} \cdot \bar{j} \, ds = -P_y(s) = -P_y[s_{x1}(y), y];$$ (3.25b)

where P_x and P_y denote the x-component and y-component of the resultant force intensity acting on the boundary segment $0 - s(x, y)$, that is on the boundary $S_0 - S_1 - S$, Figure 3.4a.

Relations (3.25a, b) are used to determine the unknown integration functions

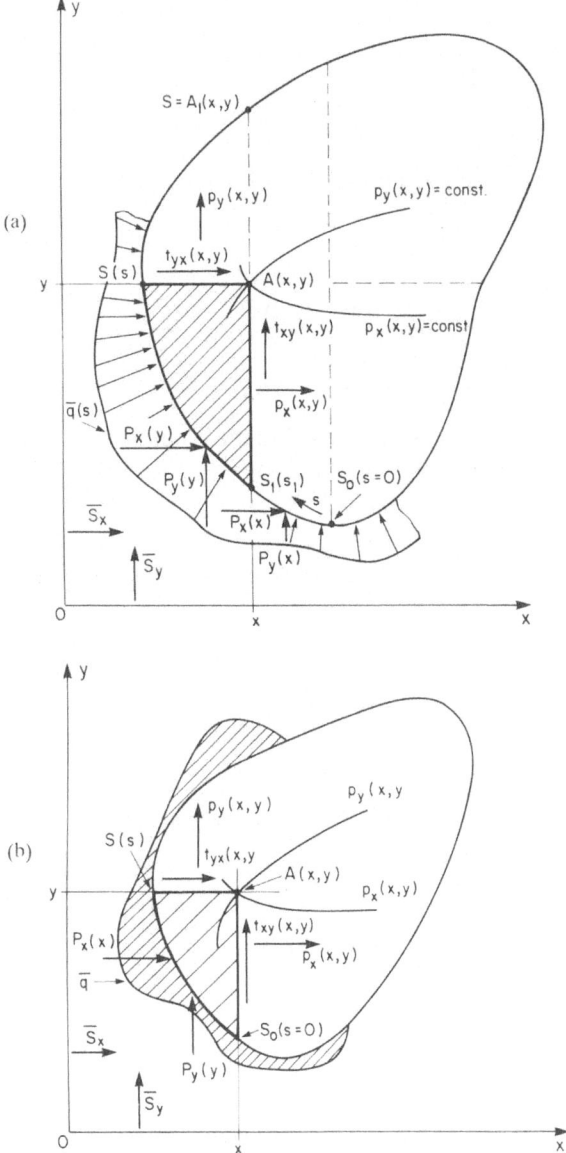

Fig. 3.4. Equilibrium relations for an arbitrary segment of a plate.
(a) For the segment $S_1 - S - A$; (b) For the segment $S_0 - S - A$.

$f_x(y)$ and $f_y(x)$, which are given by relations (3.18a, b). For the entrance section of the boundary line the relations (3.25a, b) and (3.18a, b) lead to

$$f_x(y) = P_y(y), \tag{3.26a}$$

$$f_y(x) = P_x(x). \tag{3.26b}$$

Thus the relations for normal force intensities (3.2a, b) can be formulated as follows:

$$\int_{S_{x1}}^{x} \sigma_{yy}(x, y)\, dx = \int_{S_{x1}}^{x} \frac{\partial^2}{\partial x^2}\, \Phi(x, y)\, dx$$

$$= \Phi_x(x, y) + P_y(y) = p_y(x, y), \tag{3.27a}$$

and

$$\int_{S_{y1}}^{y} \sigma_{xx}(x, y)\, dy = \int_{S_{y1}}^{y} \frac{\partial^2}{\partial y^2}\, \Phi(x, y)\, dy$$

$$= \Phi_y(x, y) - P_x(x) = p_x(x, y). \tag{3.27b}$$

Relations (3.27a, b) apply to all points within the boundary $s(x, y)$.

Analogously, the relations for shear force intensities, (3.11a, b) can be formulated using (3.2a, b) and (3.27a, b), as:

$$\int \sigma_{xy}(x, y)\, dy = -\int \frac{\partial}{\partial y}\, \Phi_x(x, y)\, dy = -\Phi_x(x, y) + g_y(x)$$

$$= -\int \frac{\partial}{\partial y}\, [P_y(x, y) - f_x(y)]\, dy$$

$$= -p_y(x, y) + f_x(y) + g_y(x)$$

$$= -[p_y(x, y) - P_y(y)] + g_y(x) = t_{xy}(x, y), \tag{3.28a}$$

and

$$\int \sigma_{yx}(x, y)\, dx = -\int \frac{\partial}{\partial x}\, \Phi_y(x, y)\, dx = -\Phi_y(x, y) + g_x(y)$$

$$= -\int \frac{\partial}{\partial x}\, [P_x(x, y) - f_y(x)]\, dx$$

$$= -p_x(x, y) + f_y(x) + g_x(y)$$

$$= -[p_x(x, y) + P_x(x)] + g_x(y) = t_{yx}(x, y). \tag{3.28b}$$

The integration functions $g_x(y)$ and $g_y(x)$ can be calculated by integrating (3.28a, b),

$$\int_{S_{y1}}^{y} \sigma_{xy}(x, y)\, dy = -\int_{S_{y1}}^{y} \frac{\partial}{\partial y}\, \Phi_x(x, y)\, dy$$

$$= -\Phi_x(x, y) + \Phi_x[x, s_{y1}(x)], \tag{3.28c}$$

and

$$\int_{S_{x1}}^{x} \sigma_{yx}(x, y)\, dx = -\int_{S_{x1}}^{x} \frac{\partial}{\partial x} \Phi_y(x, y)\, dx$$

$$= -\Phi_y(x, y) + \Phi_y[s_{x1}(y), y].$$ (3.28d)

From (3.25a, b) one obtains:

$$\Phi_x[x, s_{y1}(x)] = -P_y[x, s_{y1}(x)] = -P_y(x)$$ (3.28e)

and

$$\Phi_Y[s_{x1}(y), y] = P_x[s_{x1}(y), y] = P_x(y).$$ (3.28f)

Using relations (3.28a, b) and relations (3.28c, d, e, f) one evaluates the functions $g_x(y)$ and $g_y(x)$:

$$g_x(y) = P_x(y),$$ (3.28g)

$$g_y(x) = -P_y(x).$$ (3.28h)

Thus it follows from (3.28a, b) that

$$t_{xy}(x, y) + p_y(x, y) - P_y(y) + P_y(x) = 0,$$ (3.29a)

$$t_{yx}(x, y) + p_x(x, y) + P_x(x) - P_x(y) = 0.$$ (3.29b)

Relations (3.29a, b) can be understood as equilibrium conditions for a segment of plate bounded by the characteristic lines and the boundary segment $S_1 - S - A$, Figure 3.4a.

Relations (3.29a, b) simplify to

$$t_{xy}(x, y) + p_y(x, y) - P_y(y) = 0$$ (3.29c)

$$t_{yx}(x, y) + p_x(x, y) - P_x(x) = 0$$ (3.29d)

when the point S_0 ($s = 0$) is chosen as in Figure 3.4b, that is when the point S_0 ($s = 0$) coincides with the point S_1 [$s = s_{y1}(x)$], Figure 3.4a.

Relations (3.29c, d), (3.27a, b) and (3.28a, b) yield:

$$t_{xy}(x, y) = -\Phi_x(x, y),$$ (3.30a)

$$p_x(x, y) = \Phi_y(x, y).$$ (3.30b)

When the points S_1 and S are on the same characteristic line, that is when the point A lies on the boundary (x_s, y_s) and is identical with the point $S(s)$, $S(s) \equiv A_1(x, y)$, Figure 3.4a, relations (3.29a, b) yield:

— for S_x-characteristic direction:

$$P_x(x) = -t_{yx}(x_2, y_2) = \Phi_y(x_2, y_2),$$ (3.31a)

$$P_y(y) = p_y(x_2, y_2) = \Phi_x(x_2, y_2);$$ (3.31b)

— for S_y-characteristic direction:

$$P_x(x) = p_x(x_2, y_2) = \Phi_y(x_2, y_2),$$ (3.31c)

$$P_y(y) = -t_{xy}(x_2, y_2) = \Phi_y(x_2, y_2)$$ (3.31d)

where (x_2, y_2) denote the points at the exit boundary.

Relations (3.31a, b, c, d) can be used to determine completely the isodyne field and to check the accuracy of the analytical or experimental evaluation of the stress field.

3.3.3 Case of unknown boundary conditions

Very often the boundary forces, $\bar{q}(s)$ are not known explicitly. This case occurs when the boundary conditions are given in terms of displacements, or in regions of elastic or rigid contacts when not only the distribution of normal forces is not known, but also unknown shear stresses are produced by the elastic deformations. In such cases it is not possible to determine functions $f_x(y)$ and $f_y(x)$, (3.26a, b), and the functions t_{yx} and t_{xy} (3.28a, b). However, it is still possible to relate the shear stress components σ_{yx} and σ_{xy} to the elastic isodynes p_y and p_x.

To show this we refer to equations (3.17a, b), reproduced below for convenience.

$$p_y(x, y) = \int_{s_{11}}^{x} \frac{\partial^2}{\partial x^2} \Phi(x, y)\, dx = \Phi_x(x, y) - \Phi_x[s_{x1}(y), y],$$ (3.17a)

and

$$p_x(x, y) = \int_{s_{y1}(x)} \frac{\partial^2}{\partial y^2} \Phi(x, y)\, dy = \Phi_y(x, y) - \Phi_y[x, s_{y1}(x)].$$ (3.17b)

Differentiation of (3.17a, b) with respect to y and x yields:

$$\frac{\partial}{\partial y} p_y(x, y) = \frac{\partial^2}{\partial x\, \partial y} \Phi(x, y) - \frac{\partial^2}{\partial x\, \partial y} \Phi[s_{x1}(y), y] -$$

$$- \frac{\partial^2}{\partial x\, \partial s_{x1}} \Phi[s_{x1}(y), y] \frac{d}{dy} s_{x1}(y);$$ (3.32a)

and

$$\frac{\partial}{\partial x} p_x(x, y) = \frac{\partial^2}{\partial x\, \partial y} \Phi(x, y) - \frac{\partial^2}{\partial x\, \partial y} \Phi[x, s_{x1}(x)] -$$

$$- \frac{\partial^2}{\partial y\, \partial s_{y1}} \Phi[x, s_{y1}(x)] \frac{d}{dx} s_{y1}(x).$$ (3.32b)

At the boundary $x = s_{x1}(y)$ and $y = s_{y1}(x)$, thus

$$\frac{\partial}{\partial y} p_y[s_{x1}(y), y] = - \frac{\partial^2}{\partial x_1\, \partial s_{x1}} \Phi[s_{x1}(y), y] \frac{d}{dy} s_{x1}(y),$$ (3.33a)

and

$$\frac{\partial}{\partial x} p_x[x, s_{y1}(x)] = -\frac{\partial^2}{\partial y \, \partial s_{y1}} \Phi[x, s_{y1}(x)] \frac{d}{dx} s_{y1}(x). \tag{3.33b}$$

Consequently, equations (3.32a, b) can be rewritten, using the relations (3.33a, b) and replacing the second mixed derivatives of the stress function Φ by the corresponding shear stress components σ_{xy} or σ_{yx}, as follows:

$$\frac{\partial}{\partial y} p_y(x, y) - \frac{\partial}{\partial y} p_y[s_{x1}(y), y] = \sigma_{xy}(x, y) - \sigma_{xy}[s_{x1}(y), y], \tag{3.34a}$$

and

$$\frac{\partial}{\partial x} p_x(x, y) - \frac{\partial}{\partial x} p_x[x, s_{y1}(x)] = \sigma_{yx}(x, y) - \sigma_{yx}[x, s_{y1}(x)]. \tag{3.34b}$$

Relations (3.34a, b) represent a set of two equations with the three unknown quantities $\sigma_{xy}(x, y)$, $\sigma_{xy}[s_{x1}(y), y]$, $\sigma_{yx}[x, s_{y1}(x)]$, Figure 3.5.

In general, such a set of equations cannot be solved unequivocally on the basis of two known orthogonal isodyne fields, without additional information on one of the unknown shear stress components. It follows that — in order to determine the internal and boundary shear stresses — it is sufficient to choose one and only one characteristic section in such a manner that at one of the boundary points of the characteristic section the shear stress component is known. The simplest way of satisfying this requirement is to choose the characteristic section normal to the boundary at the point where the shear stress is equal to zero.

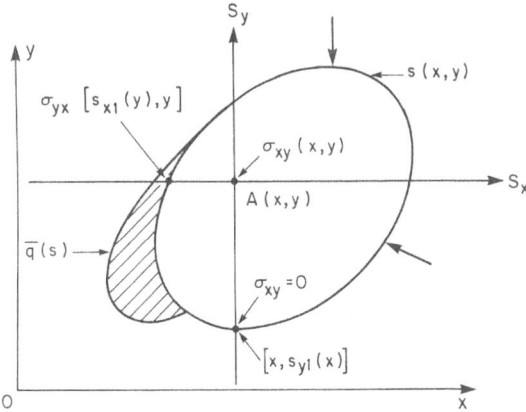

Fig. 3.5. Determination of shear stress components at the boundary and inside of the region of plate, starting from a chosen point at the boundary. Pertinent geometric parameters.

3.4 Properties of plane analytical isodynes

Isodyne fields described by relations (3.7a, b) and (3.8a, b) represent continuous surfaces which are conjugated by the angles between their characteristic directions. The simplest cases of conjugated directions are orthogonal. For such cases the relationships between the surfaces of plane elastic isodynes are very particular, and result in simple expressions for the components of stress tensor and related quantities (force intensities, total forces, strain energy, etc.). That is the reason why such a case has been chosen as a framework of the presented relationships.

Consequently, such orthogonally (through the orthogonal characteristic directions) conjugated isodyne surfaces are treated here as coupled units. The basic conditions satisfied by such conjugated PEI-surfaces according to relations (3.15) and (3.1c) are:

$$\frac{\partial}{\partial x} p_{ny} + \frac{\partial}{\partial y} p_{nx} = \text{const.} = m_i \tag{3.35}$$

$$\frac{\partial}{\partial y} \Phi_x - \frac{\partial}{\partial x} \Phi_y = 0. \tag{3.36}$$

Only two particular properties of the conjugated isodyne surfaces are discussed below: isodyne sources, and isodyne orders.

Properties of isodyne surfaces associated with the conjugate cross-sections through the isodyne fields in directions parallel and normal to characteristic directions are related closer to the theory of some experimental techniques than to the theory of plane elastic isodynes, and therefore are discussed in the pertinent section of Chapter 5.

3.4.1 Sources and ridges

The isodyne surfaces exhibit, in a general case, local or global maxima or minima. It follows from equations (3.6a, b) and (3.10a, b) that at such extreme points of the isodyne surfaces, the related normal and/or shear stress components are equal to zero. The term "source", (Z), is used to denote such points. Consequently, the corresponding stress components have opposite signs at the opposite sides of the source with respect to the characteristic directions.

3.4.1.1 True sources. The term "true source" (TZ), Figure 3.6, denotes here all the extreme values of the plane elastic analytical isodyne surfaces (or functions), that means it denotes the location of points in the (x, y)-plane at which the derivatives of an isodyne surface are equal to zero for all directions:

$$\frac{\partial}{\partial x'} p_x(x, y) = \frac{\partial}{\partial y'} p_x(x, y) = 0 \tag{3.37}$$

where x', y' is an arbitrary direction.

At the true source $(x, y)_z$, both the normal and shear stress components

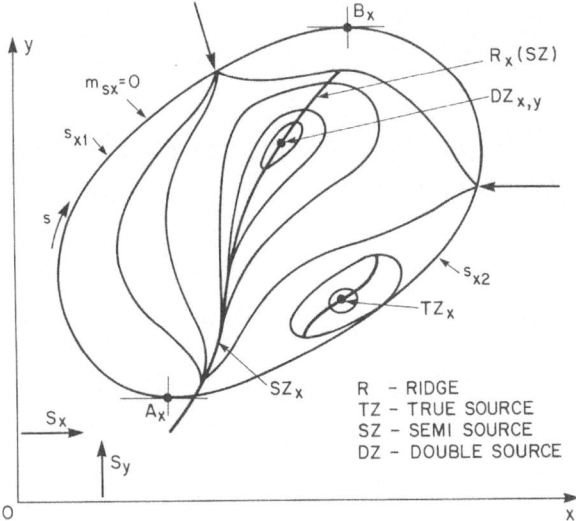

Fig. 3.6. Properties of plane elastic isodyne surfaces: sources, ridges, zero order isodyne at the entrance boundary.

related to the characteristic direction are equal to zero. For example, at the true source on the y-isodyne surface related to the y-characteristic direction, denoted as the point $(x_z, y_z)_y$,

$$\sigma_{xx}(x_z, y_z) = 0. \tag{3.38a}$$

Analogously, for the x-isodyne surface,

$$\sigma_{yy}(x_z, y_z) = 0. \tag{3.38b}$$

3.4.1.2 Double sources. The term "double source" (DZ) denotes the points in the (x, y) plane at which both conjugate plane elastic isodyne surfaces, the x-isodyne surface and the y-isodyne surface, exhibit extremal values. In other words, the (x, y)-coordinates of the true sources on both conjugate isodyne surfaces coincide:

$$(x_z, y_z)_x = (x_z, y_z)_y. \tag{3.39}$$

At double sources both normal stress components related to characteristic directions are equal to zero:

$$\sigma_{xx}(x_z, y_z) = \sigma_{yy}(x_z, y_z) = 0. \tag{3.40}$$

3.4.1.3 Semi sources. The term "semi source" (SZ), Figure 3.6, denotes the points in the (x, y) plane at which only the derivatives of the plane elastic analytical isodyne surfaces in the related characteristic directions are equal to zero:

$$\frac{\partial}{\partial y} p_x(x_z, y_z) = 0, \tag{3.41a}$$

or

$$\frac{\partial}{\partial x} P_y(x_z, y_z) = 0. \tag{3.41b}$$

Semi sources indicate points at which the corresponding normal stress components are equal to zero.

$$\sigma_{yy}(x_z, y_z) = 0 \tag{3.42a}$$

or

$$\sigma_{xx}(x_z, y_z) = 0. \tag{3.42b}$$

3.4.1.4 Ridges. The term "ridge" (R), Figure 3.6, denotes geometric loci of semi sources, and, by definition, all the sources of higher order, double and true sources. Thus, at the ridge,

$$\sigma_{xx}(x_z, y_z) = 0 \tag{3.43a}$$

and

$$\sigma_{yy}(x_z, y_z) = 0. \tag{3.43b}$$

Ridges characterize, very elegantly, the major patterns of plane stress states.

All sources and ridges do not depend on the level of load. There exist an analogy between the sources and ridges and the singular points and lines of plane stress fields, pertaining to isostatics.

3.4.1.5 Sign of sources and ridges. As had been stated above, sources and ridges are geometric loci of points where the isodyne surfaces exhibit extremes along characteristic directions. It is convenient to distinguish between positive sources, where the cross-section through the isodyne surface has a local maximum and negative sources where the cross-section through the isodyne surface has a local minimum. Consequently, positive sources separate regions where the normal or the shear stress components are positive from the regions where the stress components are negative, along the characteristic direction. Analogous relations exist in the vicinity of negative sources.

In general, the normal or the shear stress components can alter their signs only at the corresponding sources.

The sign of sources can be determined using equilibrium conditions.

3.4.2 Order of isodynes

From the geometric point of view, isodynes are cross-sections of the isodyne surfaces cut by the planes parallel to the characteristic plane called the reference plane, in this case the (x, y) plane.

The isodynes in the plane (x, y) are vertical projections of isodynes on the isodyne surfaces. The density of those cross-sections is basically arbitrary;

however, it is convenient to choose the spacing of the cross-sections in such a way that the value of the total normal force intensity, $p_y(x, y)$, related to each isodyne changes by a constant factor from one isodyne to the next isodyne. The isodyne surfaces can be conveniently described using the normalized form of the expression for isodynes.

$$p_{ny}(x, y) = m_{sx}(x, y), \tag{3.4c}$$

$$p_{nx}(x, y) = m_{sy}(x, y), \tag{3.4d}$$

where p_{nx} and p_{ny} represent the values of the total normal force intensities p_x and p_y divided by a normalizing factor S_s, (3.4a, b).

In this presentation the isodyne orders m_{sx} or m_{sy} can assume any real number.

$$-\infty < m_{sx}, m_{sy} < +\infty. \tag{3.44}$$

However, it is convenient to distinguish integer values of the isodyne order and to call them integer isodyne orders, M, and to choose accordingly the value of the normalizing factor S_s.

Thus

$$m_{sx} = (M + \Delta M)_{sx}, \tag{3.45a}$$

and

$$m_{sy} = (M + \Delta M)_{sy}, \tag{3.45b}$$

where

$$M = 0, \pm 1, \pm 2, \pm 3, \ldots, \tag{3.45c}$$

and

$$0 < |\Delta M| < 1. \tag{3.45d}$$

The lines in the (x, y)-plane defined as

$$M_{sx} = \text{const.} \tag{3.46}$$

represent a normalized form of the isodyne field related to the x-characteristic direction, and also represent the constant elevation lines in the normalized isodyne surfaces, with respect to (x, y) plane.

The order of isodynes m_{sx} of m_{sy} can be readily determined using sources, ridges, and the conditions at the entrance boundary, Figure 3.6.

Obviously, isodynes can assume positive and/or negative signs, depending on the boundary condition and the shape of the boundary of the plate.

It is obvious that the entrance boundary, Figure 3.2, represents the zero-order isodyne, excluding the points of application of concentrated loads. The order of the isodyne at the exit boundary, Figure 3.2, follows from the total boundary conditions (3.25a, b; 3.31a, d):

$$(m_{sx})_{s2} = \frac{1}{S_s} P_y(s_{y2}, y) = P_{ny}(s_{y2}, y)$$

$$= \frac{1}{S_s} \int_{s1}^{s2} \bar{q} \cdot \bar{j} \, ds = -\frac{1}{S_s} P_y(y), \tag{3.47a}$$

$$(m_{sy})_{s2} = \frac{1}{S_s} P_x(x, s_{x2}) = P_{nx}(x, s_{x2})$$

$$= \frac{1}{S_s} \int_{s1}^{s2} \bar{q} \cdot \bar{i} \, ds = -\frac{1}{S_s} P_x(x), \tag{3.47b}$$

where the values of $P_x(x)$ and $P_y(y)$ can be easily determined, the value of S_s is chosen conveniently, and $s1$ and $s2$ are entrance and exit points on the boundary.

Of particular interest are the orthogonally conjugated, zero-order isodynes

$$m_{sx} = m_{sx}(x, y) = 0, \tag{3.48a}$$
$$m_{sy} = m_{sy}(x, y) = 0 \tag{3.48b}$$

which can be identified using positive and negative sources, ridges and boundaries as references.

As has been said before, the entrance boundary lines are zero-order isodynes,

$$s_{x1}(x, y) = m_{sx} = 0 \tag{3.49a}$$
$$s_{y1}(x, y) = m_{sy} = 0. \tag{3.49b}$$

Obviously, the locations of zero-order isodynes and of sources and ridges do not depend on the scale factors with regard to the geometry and the load level.

3.4.3 Particular relations at the boundary

At the boundary, as at any point within the bounded region, plane elastic isodynes yield values of normal and shear stress components related to orthogonally conjugated characteristic directions. In a particular case of a traction-free boundary point at which the characteristic direction is collinear with the normal to the boundary $\bar{n}$, the derivatives of the plane elastic analytical isodyne surface with respect to the orthogonally conjugated characteristic directions yield

$$\frac{\partial p_y(x, y)}{\partial x} = \sigma_{yy}(s) = \sigma_s, \tag{3.50a}$$

$$\frac{\partial p_y(x, y)}{\partial y} = \sigma_{xy}(s) = 0; \tag{3.50b}$$

and

$$\frac{\partial p_x(x,\ y)}{\partial y} = \sigma_{xx}(s) = \sigma_s,$$ (3.50c)

$$\frac{\partial p_x(x,\ y)}{\partial x} = \sigma_{yx}(s) = 0.$$ (3.50d)

3.4.4 Plane stress tensor in terms of isodyne quantities

When the isodyne fields related to an arbitrary orthogonal coordinate system $(x,\ y)$ are known,

$$p_x = p_x(x,\ y),$$

$$p_y = p_y(x,\ y),$$

then the plane stress tensor σ_{ij} can be presented in the form

$$\{\sigma_{ij}\} = \begin{bmatrix} \dfrac{\partial}{\partial y} p_x(x,\ y); & \dfrac{\partial}{\partial x} [p_x(x,\ y) + f_y(x)] \\[2mm] \dfrac{\partial}{\partial y} [p_y(x,\ y) + f_x(y)]; & \dfrac{\partial}{\partial x} p_y(x,\ y) \end{bmatrix},$$ (3.51)

where functions $f_x(y)$ and $f_y(x)$ are the integration functions (3.2a, b).
   When the traction $\bar{q}$ at the boundary s is known, $\bar{q} = \bar{q}(s)$, then the integration functions can be determined according to relations (3.25a, b) and (3.26a, b)

$$f_x(y) = -\int_0^s \bar{q} \cdot \bar{j} \, ds,$$ (3.52a)

$$f_y(x) = \int_0^s \bar{q} \cdot \bar{i} \, ds.$$ (3.52b)

   When the boundary conditions in general are not known, with the exception of at least one boundary point where the shear stress component is known, then the relations (3.34a, b) can be used to determine the derivatives of the unknown integration functions in equations (3.51),

$$\frac{\partial}{\partial x} f_y(x) = \frac{d}{dx} f_y(x) = -\frac{\partial}{\partial x} p_x(x,\ s_{y1}) + \sigma_{yx}(x,\ s_{y1}),$$ (3.53a)

$$\frac{\partial}{\partial y} f_x(y) = -\frac{\partial}{dy} f_x(y) = -\frac{\partial}{\partial y} p_y(s_{x1},\ y) + \sigma_{xy}(x_{x1},\ y),$$ (3.53b)

where $(x,\ s_{y1})$ and $(s_{x1},\ y)$ denote coordinates of the points at the entrance boundary s_1.

Obviously, relations (3.53a, b) are also valid when one of the orthogonal characteristic directions, say the x-direction, is normal to the boundary at the point (s_{x1}, y). In such a case, the shear stress components σ_{yx} and σ_{xy} become the boundary shear stresses $\bar{q} \cdot \bar{s}$ at this point.

It has been shown in section 3.3.2 that it is sufficient to know the boundary shear stress $\bar{q} \cdot \bar{s}$ at only one point to determine the values of the stress components $\sigma_{yx}(x, s_{y1})$ and $\sigma_{xy}(s_{x1}, y)$ at all points along the boundary.

3.5 The concept of differential analytical isodynes

As discussed earlier, stress states in real plates of finite thickness are noticeably three-dimensional in regions of notches and cracks when the radius of curvature of the boundary is of the order of magnitude of the plate thickness or less. Also, in regions of contact loads, stress states in real plates deviate strongly from the two-dimensional model. It is easy to show that in such cases the concept of isodynes is also applicable and that some pertinent analytical relations are applicable if the characteristic surfaces are coplanar with, or parallel and normal to, the plate's middle plane. This follows from the fact that the actual values of the stress components coplanar with the plate's middle plane deviate, on the average, not more than 30% from the values at the surface, and the value of the stress component normal to the plate's middle plane is usually less than 10% of the maximal value of related stress component in the characteristic plane. Consequently, the angle between the normal to the characteristic plane and the third axis of the stress tensor oriented toward the surface of the plate is relatively small.

For such stress states one can define one more family of characteristic lines of stress fields in plates. They are called differential elastic isodynes. The differential analytical elastic isodynes, in analogy to the plane analytical elastic isodynes, are defined as geometric loci of points in the characteristic planes of real, finite thickness plates along which the differences of the total normal force intensities acting on the characteristic sections between two differential isodynes in directions within and normal to the characteristic plane are constant, and are proportional to an increase of the order of isodyne, Δm_s, along the characteristic section, Δs:

$$
\Delta p_{n1,3} = \Delta p_{n1} - \Delta p_{n3} = \int_s^{s+\Delta s} \sigma_{n1} \, ds - \int_s^{s+\Delta s} \sigma_{n3} \, ds
$$

$$
= \int_s^{s+\Delta s} (\sigma_{n1} - \sigma_{n3}) \, ds = \Delta m_{s1,3} = \text{const.} \tag{3.54}
$$

It is convenient to define three main characteristic planes related to stress states in real plates of finite thickness, when the (x, y) plane of a Cartesian coordinate system is coplanar with the middle plane of the real plate:

- the (x, y)- or (y, x)-characteristic planes containing the x- or y-characteristic directions, respectively;
- the (x, z)- or (z, x)-characteristic planes, containing the x- or y-characteristic directions, respectively;
- the (y, z)- or (z, y)-characteristic planes, containing the y- or z-characteristic directions, respectively.

When the plane (x, y) of the Cartesian coordinate system coplanar with the middle plane of the plate is parallel to the characteristic planes, and the axes x, y, and z are characteristic directions, the differential isodynes can be given by:

For x-characteristic direction within the (x, y) characteristic plane,

$$\Delta p_{y,z}(x, y, z = z_i) = \Delta p_y - \Delta p_z = \int_x^{x+\Delta x} (\sigma_{yy} - \sigma_{zz})\,dx$$

$$= S_s \Delta m_{sx(z)}(x, y, z = z_i) = \text{const.} \qquad (3.55a)$$

For y-characteristic direction within the (y, x) characteristic plane,

$$\Delta p_{x,z}(x, y, z = z_1) = \Delta p_x - \Delta p_z = \int_y^{y+\Delta y} (\sigma_{xx} - \sigma_{zz})\,dy$$

$$= S_s \Delta m_{sy(z)}(x, y, z = z_i) = \text{const.} \qquad (3.55b)$$

Obviously, relation (3.55a) represents a surface defined as a difference between elevations of two related surfaces of x-isodynes.

It is easy to show that, in general, relations (3.10) and (3.11) cannot be applied to analyze properties of the surfaces of differential isodynes and to draw conclusions about the magnitude of shear stresses and shear force intensities.

3.6 References

[1] Pindera, J. T., "Analytical Foundations of the Isodyne Photoelasticity", *Mechanics Research Communications* **8**(6), 1981, pp. 391—397.
[2] Pindera, J. T., "New Development in Photoelastic Studies: Isodyne and Gradient Photoelasticity", *Optical Engineering* **21**(4), 1982, pp. 672—678.
[3] Pindera, J. T. and Krasnowski, B. R., *Theory of Elastic and Photoelastic Isodynes*, SMD-Paper No. 184, IEM-Paper No. 1. Solid Mechanics Division, University of Waterloo, October 1983, pp. 1—127.
[4] Pindera, J. T., "Isodyne Photoelasticity and Gradient Photoelasticity: Physical and Mathematical Models, Efficacy, Applications", *Mechanika Teoretyczna i Stosowana* (Journal of Theoretical and Applied Mechanics) **22**(1/2), 1984, pp. 53—68.
[5] Pindera, J. T., Krasnowski, B. R., and Pindera, M.-J., "Theory of Elastic and Photoelastic Isodynes. Samples of Applications in Composite Structures", *Experimental Mechanics* **25**(3), September 1985, pp. 272—281.

4

Models of interaction between radiation and matter

4.1 Introduction

This Chapter presents and discusses models of particular interactions between radiation and matter, and models of some pertinent measurement procedures of interest in the theory and techniques of isodynes. These models are components of the overall model of energy flow which is produced to collect and carry information on optical isodynes presented in Chapter 5.

In order to present the basic relations for optical isodynes in a possibly simplest form, physical models of real bodies and of birefringence responses of real bodies are simplified to such an extent that they are able to simulate reliably only the least complicated responses of some real bodies which occur in particular ranges of values of major physical parameters. Such an approach leads to simple mathematical models of the involved phenomena which can be presented in the form of ordinary linear differential equations [20, 24]. According to experience, such a limited description is still able to simulate reliably the birefringence responses of some classes of solid and liquid bodies, such as crystals and glasses, and the linear viscoelastic polymers in limited ranges of deformation, stress, temperature, and spectral bands of the birefringence detecting radiation.

Obviously, the applicability of the derived relations outside of the above discussed constraints must be tested and proven in each case. It can be shown that unacceptable extrapolation of the simplified assumptions of the presented linear theory can be readily detected by means of optical isodynes: for instance, the limits of reliability of two-dimensional solutions of linear elasticity in regions of high strain gradients are clearly indicated by the alteration of the optical isodyne fields.

Theories and techniques of optical measurements are subjected to the same conditions and constrains as all observations, experiments, and measurements dealing with real bodies and real processes. Within such a framework the physical admissibility and the rigorousness of the derived analytical relations pertains to real processes occurring in real bodies, whereas the mathematical admissibility pertains to requirements of mathematical logic and rigorousness, and of compatibility between assumptions and conclusions.

4.2 Basic components of pertinent elementary physical and mathematical models of interaction between radiation and matter

The presented relations pertain to situations at chosen points, thus they describe local states with all the consequences resulting thereof. Extrapolations of these relations, based essentially on the principle of local state, to describe states of continuous systems requires various assumptions or limiting conditions. Such assumptions, or disregard of limiting conditions, may invalidate the correctness of the derived equations if the conditions of theoretical (physical) and/or mathematical admissibility are not observed.

4.2.1 Phenomenological models of stress/strain-induced birefringence

Thus far, no theoretical models of the stress/strain-induced birefringence in real materials such as ceramics or polymers are available. However, some speculative-phenomenological models are able to simulate the actual processes in satisfactorily wide ranges of alteration of the involved physical parameters.

At the present time the most general phenomenological model of the stress/strain-induced birefringence in solids is that presented by G. N. Ramachandran and S. Ramaseshan in 1961 [34], and this model was utilized in the pertinent bibliography [18—30]. This model describes light propagation in hypothetical deformed bodies, which do not exhibit either selective absorption (dichroism), nor optical activity. The body is transparent at the wavelength of radiation used and not absorbing; that is, the general expression for the refractive index,

$$n = n_0 + ik, \tag{4.1}$$

where k denotes absorption index related to conductivity, simplifies to

$$n = n_0. \tag{4.2}$$

The basic assumptions of this model are presented by Ramachandran and Ramaseshan in the following form:

1. "In a homogeneously deformed solid, all the laws of propagation of light derived for homogeneous anisotropic solids are valid. The effect of the deformation is only to alter the parameters contained in these laws of propagation, that is to alter the magnitudes and directions of the principal axes of the optical ellipsoid of a solid".

2. "When the strain is within elastic limits, the variation of the optical parameters of the solid due to the deformation can be expressed as a homogeneous linear function of the six stress components, σ_{xx}, σ_{yy}, σ_{zz}, σ_{yx}, σ_{zx}, σ_{xy}, or the six strain components, ε_{xx}, ε_{yy}, ε_{zz}, ε_{yz}, ε_{zx}, ε_{xy}". This hypothesis is a generalization of the empirical Brewster's law.

As the most convenient optical parameter altered by the stress or strain state the index tensor a_{ij} (tensor of wave normals), connecting the displacement vector D with the electric vector E in Maxwell's relations, is chosen. It may be recalled that according to relations of crystal optics [6, 34] two waves can propagate through non-absorbing and non-optically active crystals. Both waves are linearly polarised and they propagate with their vibrations along the principal axes of the elliptic section of the index ellipsoid normal to the wave normal. For anisotropic crystalline, optically non-active bodies the behavior of which can be satisfactorily described by the relation

$$\bar{E} = [a]\bar{D},\tag{4.3}$$

the index ellipsoid (ellipsoid of wave normals) can be presented in the form

$$a_{11}x^2 + a_{22}y^2 + a_{33}z^2 + 2a_{23}yz + 2a_{31}zx + 2a_{12}xy = 1,\tag{4.4}$$

or more conveniently with reference to an arbitrary coordinate system,

$$a_{xx}x^2 + a_{yy}y^2 + a_{zz}z^2 + 2a_{yz}yz + 2a_{zx}zx + 2a_{xy}xy = 1.\tag{4.5}$$

Denoting the values of a_{ij} in the undeformed crystal by a_{ij}^0 and in the deformed crystal by a_{ij}, the change $\Delta a_{ij} = (a_{ij} - a_{ij}^0)$ can be presented as a linear homogeneous function of the stress or strain components.

In terms of stress components, the above relation may be written as

$$\Delta a_{xx} = a_{xx} - a_{xx}^0 = q_{11}\sigma_{xx} + q_{12}\sigma_{yy} + q_{13}\sigma_{zz} +$$
$$+ q_{14}\sigma_{yz} + q_{15}\sigma_{zx} + q_{16}\sigma_{xy},\tag{4.6a}$$

$$\Delta a_{yy} = a_{yy} - a_{yy}^0 = q_{21}\sigma_{xx} + q_{22}\sigma_{yy} + q_{23}\sigma_{zz} +$$
$$+ q_{24}\sigma_{yz} + q_{25}\sigma_{zx} + q_{26}\sigma_{xy},\tag{4.6b}$$

$$\Delta a_{zz} = a_{zz} - a_{zz}^0 = q_{21}\sigma_{xx} + q_{32}\sigma_{yy} + q_{33}\sigma_{zz} +$$
$$+ q_{34}\sigma_{yz} + q_{35}\sigma_{zx} + q_{36}\sigma_{xy},\tag{4.6c}$$

$$\Delta a_{yz} = a_{yz} - a_{yz}^0 = q_{41}\sigma_{xx} + q_{42}\sigma_{yy} + q_{43}\sigma_{zz} +$$
$$+ q_{44}\sigma_{yz} + q_{45}\sigma_{zx} + q_{46}\sigma_{xy},\tag{4.6d}$$

$$\Delta a_{zx} = a_{zx} - a_{zx}^0 = q_{51}\sigma_{xx} + q_{52}\sigma_{yy} + q_{53}\sigma_{zz} +$$
$$+ q_{54}\sigma_{yz} + q_{55}\sigma_{zx} + q_{56}\sigma_{xy},\tag{4.6e}$$

$$\Delta a_{xy} = a_{xy} - a_{xy}^0 = q_{61}\sigma_{xx} + q_{62}\sigma_{yy} + q_{63}\sigma_{zz} +$$
$$+ q_{64}\sigma_{yz} + q_{65}\sigma_{zx} + q_{66}\sigma_{xy}.\tag{4.6f}$$

In terms of strain components, these relations are:

$$\Delta a_{xx} = a_{xx} - a_{xx}^0 = p_{11}\varepsilon_{xx} + p_{12}\varepsilon_{yy} + p_{13}\varepsilon_{zz} +$$
$$+ p_{14}\varepsilon_{yz} + p_{15}\varepsilon_{zx} + p_{16}\varepsilon_{xy},\ \text{etc.}\tag{4.7}$$

In crystal optics, the constants q_{ij} are called the piezo-optic constants and the constants p_{ij} are called the elasto-optic constants. No information is built-in in this model regarding the dependence of the constants q_{ij} and p_{ij} on the spectral frequency of energy used to detect the changes in values of the index tensor a_{ij}. On the other hand it is known that the values of components of the dielectric tensor ε_{ij} depend on the frequency of the electric field. Since the index tensor and the dielectric tensor are inversely related,

$$[a] = [\varepsilon]^{-1}, \tag{4.8}$$

the components of the index tensor and their derivatives do depend on the frequency (f) of the electric field, in all spectral bands

$$a_{ij} = a_{ij}(f). \tag{4.9}$$

It is convenient to relate the alterations of the index tensor to the principal axes of the index tensor (optical ellipsoid) of the undeformed crystal, (x_0, y_0, z_0).

The index ellipsoid in this case can be presented in the form

$$a_{11}^0 x_0^2 + a_{22}^0 y_0^2 + a_{33}^0 z_0^2 = 1. \tag{4.10}$$

An arbitrary stress state transforms it into:

$$a_{11} x_0^2 + a_{22} y_0^2 + a_{33} z_0^2 + 2 a_{23} y_0 z_0 + 2 a_{31} z_0 x_0 + 2 a_{12} x_0 y_0 = 1, \tag{4.11}$$

with respect to the original principal axes (x_0, y_0, z_0). Relation (4.11) can be presented in the same form as relation (4.10), by referring the alteration of the index tensor components to the new set of principal axes x', y', z',

$$a_{11}' x'^2 + a_{22}' y'^2 + a_{33}' z'^2 = 1. \tag{4.12}$$

The number of the non-zero piezo-optic or elasto-optic constants q_{ij} and p_{ij} depends on the order of symmetry of anisotropic body (crystal). In the most general case, the number of p_{ij} or q_{ij} constants is 45. Cubic crystals exhibit three independent photo-elastic constants and three independent elasto-optic constants; whereas isotropic bodies are described by only two independent piezo-optic or elasto-optic constants for the given frequency of the electric field.

In the particular case of hydrostatic pressure, relations (4.6) simplify to:

$$\sigma_{xx} = \sigma_{yy} = \sigma_{zz} = p, \tag{4.13a}$$

$$\sigma_{xy} = \sigma_{yz} = \sigma_{zx} = 0, \tag{4.13b}$$

$$\Delta a_{xx} = a_{xx} - a_{xx}^0 = (q_{11} + q_{12} + q_{13})p, \tag{4.14a}$$

$$\Delta a_{yy} = a_{yy} - a_{yy}^0 = (q_{21} + q_{22} + q_{23})p, \tag{4.14b}$$

$$\Delta a_{zz} = a_{xx} - a_{zz}^0 = (q_{31} + q_{32} + q_{33})p, \tag{4.14c}$$

$$\Delta a_{yz} = a_{yz} - a_{yz}^0 = (q_{41} + q_{42} + q_{43})p, \tag{4.14d}$$

$$\Delta a_{zx} = a_{zx} - a_{zx}^0 = (q_{51} + q_{52} + q_{53})p, \tag{4.14e}$$

$$\Delta a_{xy} = a_{xy} - a_{xy}^0 = (q_{61} + q_{62} + q_{63})p. \tag{4.14f}$$

It follows from (4.14a–f) that the hydrostatic pressure does not alter the directions of optic axes when the crystal (an anisotropic body) is symmetric. In a general case of optical anisotropy the hydrostatic pressure rotates the optic axes. In the above-mentioned cases the state of anisotropy, in particular the crystal symmetry, is not altered by the hydrostatic pressure.

Of particular interest is the behaviour of anisotropic bodies which may be classified as cubic crystals belonging to the XIth photoelastic class described by one optical, three elastic, and three photoelastic constants. (The term "photo-elastic constant" denotes either the piezo-optic or stress-optic constant q, or the elasto-optic or strain optic constant p).

The optical anisotropy described by

$$\Delta a_{ij} = a_{ij} - a_{ij}^0 \tag{4.15}$$

as in relations (4.14) can also be represented by the differences of related quantities, the dielectric tensor $[\varepsilon]$ or the refractive index matrix $[n]$.

The second order dielectric and index tensors are simply related, (4.8), that is,

$$a_{ij} = \frac{\varepsilon^{ij}}{|\varepsilon|}, \tag{4.16}$$

where ε^{ij} is the minor of ε_{ij} in the determinant $|\varepsilon|$. Thus for isotropic bodies when $[\varepsilon] \equiv \varepsilon_0$, the relation (4.16) reduces to

$$a_0 = \frac{1}{\varepsilon_0}.$$

The refractive index n, defined as the ratio of light velocity in vacuum, c, and in a body, v,

$$n = \frac{c}{v}, \tag{4.17}$$

can be related to the dielectric coefficient in an isotropic body by means of Maxwell's relation

$$v = c(\varepsilon\mu)^{-1/2}, \tag{4.18}$$

where μ denotes the magnetic permeability, and both ε and μ are normalized with respect to the corresponding parameters of free space, and thus are dimensionless. Thus, for nonconductive bodies,

$$n = (\varepsilon)^{-1/2}, \tag{4.19}$$

and

$$n_{ij} = (\varepsilon_{ij})^{1/2}. \tag{4.20}$$

For such photoelastic class of cubic crystals, the number of independent photoelastic constants not equal to zero is three, since the three crystallographic principal axes are equivalent,

$$q_{11} = q_{22} = q_{33}, \tag{4.21a}$$

$$q_{12} = q_{21} = q_{32} = q_{23} = q_{31} = q_{13}, \tag{4.21b}$$

$$q_{44} = q_{55} = q_{66}, \tag{4.21c}$$

and the number of optical constants for the undeformed crystal is equal to one,

$$a_{xx}^0 = a_{yy}^0 = a_{zz}^0 = \frac{1}{\varepsilon_0} = \frac{1}{n_0^2}. \tag{4.22}$$

In such a case, the relations (4.6) simplify to

$$\Delta a_{xx} = a_{xx} - a^0 = (q_{11} - q_{12})\sigma_{xx} + q_{12}(\sigma_{xx} + \sigma_{yy} + \sigma_{zz}), \tag{4.23a}$$

$$\Delta a_{yy} = a_{yy} - a^0 = (q_{11} - q_{12})\sigma_{yy} + q_{12}(\sigma_{xx} + \sigma_{yy} + \sigma_{zz}), \tag{4.23b}$$

$$\Delta a_{zz} = a_{zz} - a^0 = (q_{11} - q_{12})\sigma_{zz} + q_{12}(\sigma_{xx} + \sigma_{yy} + \sigma_{zz}), \tag{4.23c}$$

$$a_{yz} = q_{44}\sigma_{yz}, \tag{4.23d}$$

$$a_{zx} = q_{44}\sigma_{zx}, \tag{4.23e}$$

$$a_{xy} = q_{44}\sigma_{xy}. \tag{4.23f}$$

Relations (4.23) show that optically isotropic cubic crystals which are characterized by three independent photoelastic constants become, in general, biaxial crystals. It is clear that the birefringence (represented by components of the index tensor) in any direction is proportional to the stress level. It can also be shown that the magnitude of the angle between optical axes depends on the orientation of the stress tensor with respect to the crystallographic axes of the cubic crystal, and does not depend on the magnitude of the stress tensor.

It is easy to deduce from relations (4.6) that for isotropic bodies these relations must be invariant with respect to the axes of a chosen Cartesian coordinate system. Thus, for such bodies, the number of independent photoelastic constants q_{ij} reduces to two,

$$q_{11} = q_{22} = q_{33}, \tag{4.24a}$$

$$q_{12} = q_{21} = q_{32} = q_{23} = q_{31} = q_{13}, \tag{4.24b}$$

since, in this case,

$$2q_{55} = 2q_{66} = 2q_{44} = q_{11} - q_{12}. \tag{4.24c}$$

Since for isotropic bodies $\varepsilon_x = \varepsilon_y = \varepsilon_z = \varepsilon_0$, that is, for isotropic bodies all directions are equivalent, it is correct to choose any, most convenient, coordinate systems without limiting the generality of analysis. Obviously, it is convenient to align the coordinate system with the principal directions of stress tensor (1, 2, 3) to present the general relation (4.6), when applied to initially isotropic bodies, in the simplest possible form:

$$\Delta a_1 = a_1 - \frac{1}{\varepsilon_0} = q_{11}\sigma_1 + q_{12}(\sigma_2 + \sigma_3), \tag{4.25a}$$

$$\Delta a_2 = a_2 - \frac{1}{\varepsilon_0} = q_{11}\sigma_2 + q_{12}(\sigma_3 + \sigma_1), \tag{4.25b}$$

$$\Delta a_3 = a_3 - \frac{1}{\varepsilon_0} = q_{11}\sigma_3 + q_{12}(\sigma_1 + \sigma_2), \tag{4.25c}$$

where σ_1, σ_2, σ_3, denote principal stresses; a_1, a_2, a_3, denote principal components of the index tensor (ellipsoid of wave normals); and the deviatoric components of the index tensor, a_{yz}, a_{zx}, a_{xy}, are equal to zero by definition. Thus the principal axes of the index tensor, or of the ellipsoid of wave normals, are identical with those of the stress tensor when the body is initially isotropic, both mechanically and optically.

In the case presented by equations (4.25), principal indices of refraction are simply related to the values of principal components of index tensor,

$$n_1 = (a_1)^{-1/2};\ n_2 = (a_2)^{-1/2};\ n_3 = (a_3)^{-1/2}. \tag{4.26a, b, c}$$

The relations between the alteration of index tensor and the strain components can be presented in a form analogous to (4.25),

$$\Delta a_1 = a_1 - \frac{1}{\varepsilon_0} = p_{11}\varepsilon_1 + p_{12}(\varepsilon_2 + \varepsilon_3), \tag{4.27a}$$

$$\Delta a_2 = a_2 - \frac{1}{\varepsilon_0} = p_{11}\varepsilon_2 + p_{12}(\varepsilon_3 + \varepsilon_1), \tag{4.27b}$$

$$\Delta a_3 = a_3 - \frac{1}{\varepsilon_0} = p_{11}\varepsilon_3 + p_{12}(\varepsilon_1 + \varepsilon_2), \tag{4.27c}$$

where ε_1, ε_2, ε_3 denote principal components of the strain tensor.

In practical applications of relations (4.25) or (4.27), use is made of the differences between the principal components of the index tensor:

$$\Delta a_{1,2} = a_1 - a_2 = (q_{11} - q_{12})(\sigma_1 - \sigma_2) = q_\sigma(\sigma_1 - \sigma_2), \tag{4.28a}$$

$$\Delta a_{2,3} = a_2 - a_3 = (q_{11} - q_{12})(\sigma_2 - \sigma_3) = q_\sigma(\sigma_2 - \sigma_3), \tag{4.28b}$$

$$\Delta a_{3,1} = a_3 - a_1 = (q_{11} - q_{12})(\sigma_3 - \sigma_1) = q_\sigma(\sigma_3 - \sigma_1), \tag{4.28c}$$

and

$$\Delta a_{1,2} = a_1 - a_2 = (p_{11} - p_{12})(\varepsilon_1 - \varepsilon_2) = p_\varepsilon(\varepsilon_1 - \varepsilon_2), \tag{4.29a}$$

$$\Delta a_{2,3} = a_2 - a_3 = (p_{11} - p_{12})(\varepsilon_2 - \varepsilon_3) = p_\varepsilon(\varepsilon_2 - \varepsilon_3), \tag{4.29b}$$

$$\Delta a_{3,1} = a_3 - a_1 = (p_{11} - p_{12})(\varepsilon_3 - \varepsilon_1) = p_\varepsilon(\varepsilon_3 - \varepsilon_1). \tag{4.29c}$$

Obviously, the coefficients p_{ij} and q_{ij} depend on the frequency of electric vector in a manner depending on the atomic structure of a body.

The quantities a_{ij} are difficult to measure. However, it is very easy to measure the quantities $\Delta a_{i,j}$. Also, it is not difficult to measure the quantities Δa_{ii} by relating them to the corresponding values of principal indices of refraction, in accordance with relations (4.26).

Referring to principal indices of refraction, relations (4.28) can be presented and approximated in terms of the so-called relative retardation, defined as differences of principal indices of refraction,

$$\Delta n_{i,j} = n_i - n_j, \tag{4.30}$$

in the following manner: when

$$\frac{n_2 - n_1}{n_1} = \frac{\Delta n_{1,2}}{n_1} \ll 1: \tag{4.31}$$

then the equations (4.26) and (4.28a) yield

$$\frac{1}{n_1^2} - \frac{1}{n_2^2} = (n_2 - n_1)2n_1^{-3} = q_\sigma(\sigma_1 - \sigma_2), \tag{4.32}$$

or

$$\Delta n_{2,1} = n_2 - n_1 = C_\sigma(\sigma_1 - \sigma_2). \tag{4.33a}$$

Analogously,

$$\Delta n_{3,2} = n_3 - n_2 = C_\sigma(\sigma_1 - \sigma_2), \tag{4.33b}$$

$$\Delta n_{1,3} = n_1 - n_3 = C_\sigma(\sigma_3 - \sigma_1). \tag{4.33c}$$

where

$$C_\sigma = q_\sigma \frac{n_1^3}{2}. \tag{4.34}$$

Obviously,

$$C_\sigma = \frac{1}{2} q_\sigma(\lambda) n_1^3(\lambda) = C_\sigma(\lambda), \tag{4.35}$$

where λ denotes the wavelength of the birefringence detecting radiation.

The quantity C_σ is called here the photoelastic stress coefficient. Similarly:

$$\Delta n_{2,1} = n_2 - n_1 = C_\varepsilon(\varepsilon_1 - \varepsilon_2). \tag{4.36a}$$

$$\Delta n_{3,2} = n_3 - n_2 = C_\varepsilon(\varepsilon_2 - \varepsilon_3). \tag{4.36b}$$

$$\Delta n_{1,3} = n_1 - n_3 = C_\varepsilon(\varepsilon_3 - \varepsilon_1). \tag{4.36c}$$

where

$$C_\varepsilon = p_\varepsilon \frac{n_1^3}{2}, \tag{4.37}$$

and

$$C_\varepsilon = \frac{1}{2} p_\varepsilon(\lambda) n_1^3(\lambda) = C_\varepsilon(\lambda). \tag{4.38}$$

The quantity C_ε is called here the photoelastic strain coefficient.

The quantity $\Delta n_{i,j}$ is a measure of birefringence or of relative retardation expressed in terms of refractive indices.

Relations (4.34) and (4.37) show that the quantities C_σ and C_ε can be considered constant with respect to stresses and strains when

$$\frac{n_1 - n_0}{n_0} \ll 1. \tag{4.39}$$

Sometimes it is more convenient or necessary to use relations (4.25) or (4.27) as a theoretical basis of measurements. In such cases, the so-called absolute retardation measurements are performed, the results of which are expressed in terms of the differences of the original and induced indices of refraction,

$$\Delta n_{i,0} = n_i - n_0. \tag{4.40}$$

When the condition given by (4.39) is satisfied, relations (4.25) and (4.27) can be approximated by:

$$\Delta n_{1,0} = n_1 - n_0 = C_{1\sigma}\sigma_1 + C_{2\sigma}(\sigma_2 + \sigma_3), \tag{4.41a}$$
$$\Delta n_{2,0} = n_2 - n_0 = C_{1\sigma}\sigma_2 + C_{2\sigma}(\sigma_3 + \sigma_1), \tag{4.41b}$$
$$\Delta n_{3,0} = n_3 - n_0 = C_{1\sigma}\sigma_3 + C_{2\sigma}(\sigma_1 + \sigma_2), \tag{4.41c}$$

where

$$C_{1\sigma} = \frac{n_0^3}{2} q_{11}, \ C_{2\sigma} = \frac{n_0^3}{2} q_{12}, \tag{4.42a, b}$$

and $C_{1\sigma}$ and $C_{2\sigma}$ are called absolute photoelastic stress coefficients. Obviously,

$$C_{1\sigma} = \frac{1}{2} n_0^3(\lambda) q_{11}(\lambda) = C_{1\sigma}(\lambda), \tag{4.43a}$$

and

$$C_{2\sigma} = \frac{1}{2} n_0^3(\lambda) q_{12}(\lambda) = C_{2\sigma}(\lambda), \tag{4.43b}$$

Analogously,

$$\Delta n_{1,0} = n_1 - n_0 = C_{1\varepsilon}\varepsilon_1 + C_{2\varepsilon}(\varepsilon_2 + \varepsilon_3), \tag{4.44a}$$
$$\Delta n_{2,0} = n_2 - n_0 = C_{1\varepsilon}\varepsilon_2 + C_{2\varepsilon}(\varepsilon_3 + \varepsilon_1), \tag{4.44b}$$
$$\Delta n_{3,0} = n_3 - n_0 = C_{1\varepsilon}\varepsilon_3 + C_{2\varepsilon}(\varepsilon_1 + \varepsilon_2), \tag{4.44c}$$

where $C_{1\varepsilon}$ and $C_{2\varepsilon}$ are called absolute photoelastic strain coefficients. Obviously,

$$C_{1\varepsilon} = \frac{1}{2} n_0^3(\lambda) p_{11}(\lambda) = C_{1\varepsilon}(\lambda), \tag{4.45a}$$

$$C_{2\varepsilon} = \frac{1}{2} n_0^3(\lambda) p_{12}(\lambda) = C_{2\varepsilon}(\lambda). \tag{4.45b}$$

Relations (4.33), (4.36), (4.41) and (4.44) represent phenomenological constitutive relations for the stress, or strain, induced birefringence at a point (local state presentation).

4.2.1.1 Limitations of the derived relations. No information is built into the basic physical model on the spectral dependence of all the photoelastic coefficients; the very simplified physical models based on the concept of vibrating dipoles are only able to describe some selected parts of the empirical functional relations. It appears that a deeper insight into the mechanism of birefringence is needed to describe satisfactorily the known relationships and to predict the wavelength-dependence of the birefringence outside of the already investigated spectral region. The above phenomenological relations do not contain any information on absorption effects (e.g. dichroic effect), orientational effects at the molecular level, dispersion of optic axes, etc. The above information was neglected at the design of the mathematical model of phenomenon, and therefore the derived relations cannot yield any related data.

The above phenomenological relations pertain to specific physical states at a point, that is they pertain to local states. It may be recalled that the principle of local states within a continuous system ignores the eventual influence of the values of local gradients. Thus the above phenomenological relations supply no information on the patterns of light propagation inside a birefringent body, and therefore cannot be integrated with respect to the spatial coordinates without additional information, or without some explicitly specified assumptions whose admissibility must be tested. Specifically, the state of a continuous system is described by a number of continuous functions which are functions of position x, y, z, or representing time-dependent fields $f = f(x, y, z, t)$. Gradients of those functions at particular points characterizing local states influence patterns of the flow of energy used to detect and measure the local states of birefringence (see Section 4.2.4).

The constitutive birefringent stress-optic relations for the stress/strain-induced birefringence at a point presented in this Chapter pertain to materials which belong to the constitutive class of materials whose electro-magnetic behavior is satisfactorily described by the classical linear constitutive relations connecting the electric displacement vector $\bar{D}$, the electric vector $\bar{E}$, the magnetic induction vector $\bar{B}$, and the magnetic intensity vector $\bar{H}$,

$$D_i = \varepsilon_{ij} E_j, \tag{4.46a}$$

$$B_i = \mu H_i, \tag{4.46b}$$

where ε_{ij} are the components of the dielectric (permittivity) tensor and μ is magnetic permeability, constant equal to one.

Various materials whose behavior cannot be satisfactorily approximated by the constitutive class of hypothetical materials represented by relations (4.46) may exhibit strong birefringent stress-optic effects. Such materials, being in a different constitutive class, may be still used in stress birefringence observations and experiments provided that the theoretical basis of the constitutive relations (4.33), (4.36), (4.41) and (4.44) is suitably expanded. For instance, for materials such as germanium and silicon transparent in near infrared, the refractive index is a complex number as presented in (4.1), or for materials such as ceramics transparent in microwave range, the magnetic permeability may not be constant.

The most important limitation of the derived relations follows from the fact that the basic relations (4.6) and (4.7) pertain to a local state and consequently, only describe the local state of optical anisotropy. It must be clearly stated that it is theoretically incorrect to describe the response of the whole system (the global state) using only the relations which describe the local state. One of the reasons is that, by definition, equations describing a local state do not contain terms accounting for the influence of the gradients of the involved physical quantities. In this particular case equations (4.6) or (4.7) are insufficient to predict the path of the energy form chosen to detect and quantify the stress/strain-induced optical anisotropy.

4.2.2 Models of scattering of radiant energy

The patterns of scattering of light, and of electromagnetic radiation in general, are inherently related to the structure of matter and therefore carry valuable information on the structure and state of scattering bodies. Numerous physical and phenomenological models of scattering have been constructed as bases of various scattering theories. The pertinent bibliography is extensive [4, 9, 12]. For the purpose of this work only the Rayleigh model of scattering is presented and discussed. This model can be accepted as satisfactory with regard to the present level of sophistication of stress analysis procedures and procedures used for testing constitutive material relations, in particular for viscoelastic materials. It should be pointed out that scattering by anisotropic bodies is beyond the predictive capacity of the Rayleigh model of scattering.

Scattering is one of the basic phenomena which is produced when radiation interacts with matter. The first comprehensive explanation was presented by Lord Rayleigh, and scattering by spheres which are small compared to the wavelength ($a/\lambda < 0.05$) has been called Rayleigh scattering. The practical application of this phenomenon in photoelasticity was proposed by Weller in 1941 [35].

If polarized radiation reaches an isotropic, homogeneous sphere of arbitrary size, a dipole is created which oscillates synchronously and in the same direction as the vibrating electromagnetic field [12]. In other words the electrons are displaced with respect to the nuclei, and there is partial orientation of any permanent dipoles that may be present. The oscillating dipole radiates electromagnetic energy and this secondary emission constitutes the scattering. The scattered radiation is polarized in the same sense as the dipole.

The scattered intensity is assumed to be proportional to the intensity of the incident radiation and to be dependent on the following major parameters: the volume of the spherical particle V; the distance to the point of observation r; the wavelength of radiation λ; and the refractive indices of the particle and medium n_1, n_2,

$$\frac{I}{I_0} = f(V, r, \lambda, n_1, n_2). \tag{4.47}$$

It follows from dimensional analysis of the accepted physical model formulated above that equation (4.47) should be presented in the form

$$\frac{I}{I_0} = I'(n_1, n_2)\, \frac{V^2}{r^2\lambda^4}. \tag{4.48}$$

The wavelength of the scattered radiation is identical with the wavelength of the incident radiation.

For the far field zone of observation, the scattered wave becomes a transverse wave as a result of the rapid decay of the longitudinal component. The scattered wave is in general elliptically polarized. If a coordinate system as in Figure 4.1 is assumed, it is convenient to present the intensity of the scattered radiation propagating in the r direction by its components polarized in the θ

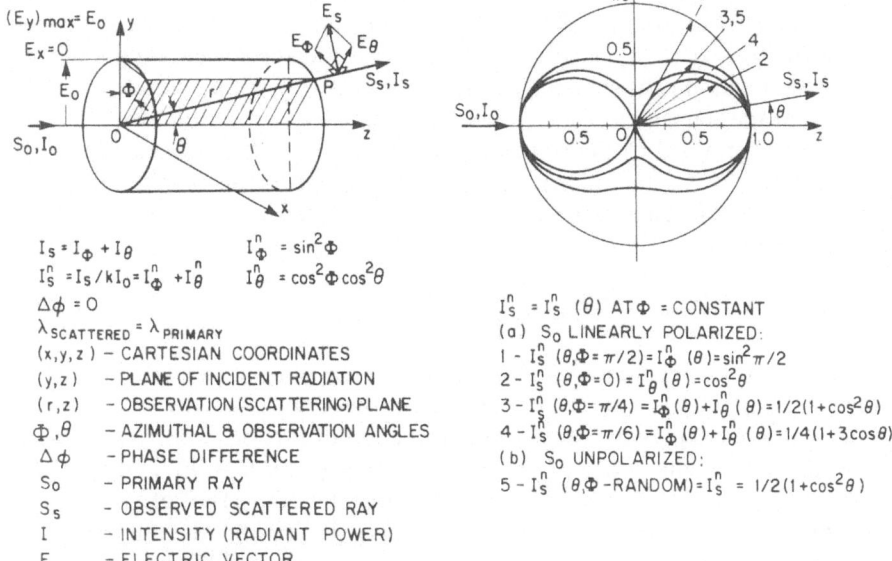

$$I_s = I_\Phi + I_\theta \qquad I_\Phi^n = \sin^2\Phi$$
$$I_s^n = I_s / kI_0 = I_\Phi^n + I_\theta^n \qquad I_\theta^n = \cos^2\Phi\cos^2\theta$$
$$\Delta\phi = 0$$

$\lambda_{SCATTERED} = \lambda_{PRIMARY}$

(x,y,z) — CARTESIAN COORDINATES
(y,z) — PLANE OF INCIDENT RADIATION
(r,z) — OBSERVATION (SCATTERING) PLANE
Φ,θ — AZIMUTHAL & OBSERVATION ANGLES
$\Delta\phi$ — PHASE DIFFERENCE
S_0 — PRIMARY RAY
S_s — OBSERVED SCATTERED RAY
I — INTENSITY (RADIANT POWER)
E — ELECTRIC VECTOR

$I_s^n = I_s^n(\theta)$ AT Φ = CONSTANT
(a) S_0 LINEARLY POLARIZED:
1 - $I_s^n (\theta,\Phi=\pi/2) = I_\Phi^n (\theta) = \sin^2\pi/2$
2 - $I_s^n (\theta,\Phi=0) = I_\theta^n (\theta) = \cos^2\theta$
3 - $I_s^n (\theta,\Phi=\pi/4) = I_\Phi^n (\theta) + I_\theta^n (\theta) = 1/2(1+\cos^2\theta)$
4 - $I_s^n (\theta,\Phi=\pi/6) = I_\Phi^n (\theta) + I_\theta^n (\theta) = 1/4(1+3\cos\theta)$
(b) S_0 UNPOLARIZED:
5 - $I_s^n (\theta,\Phi -RANDOM) = I_s^n = 1/2(1+\cos^2\theta)$

Fig. 4.1. Rayleigh model of light scattering at a point. Mathematical model, definitions, distributions of intensity of scattered light.

and ϕ azimuths perpendicular and parallel to the scattering plane determined by the r and z directions. In general, a phase difference between these components results in elliptical polarization of the scattered radiation.

In the Rayleigh model of scattering only the effect of the electric vector is considered. For incident linearly polarized light, the scattered light consists of two linearly polarized components and the phase difference is equal to zero.

For some media the scattering becomes a more complicated function of the wavelength than is represented by Rayleigh's mathematical model and shows some kind of resonance maximum. This fact might be a significant factor in designing photoelastic scattered-light measurements. When the incident radiant energy is high, the scattered light may contain radiation of a longer wavelength than the wavelength of the primary beam.

For a particular homogeneous and initially isotropic body located in vacuum or air, and illuminated by monochromatic linearly polarized radiation, the scattered light may be represented by two linearly polarized components I_Φ and I_θ:

$$I_\Phi = I_0 K \sin^2 \Phi, \qquad (4.49)$$

$$I_\theta = I_0 K \cos^2 \Phi \cos^2 \theta, \qquad (4.50)$$

$$\Delta\phi = 0, \qquad (4.51)$$

where K denotes a coefficient which is constant for a given material and given wavelength, and where I_0, the radiant power per unit area or intensity of the primary beam at the point of scattering, has been modulated by the stress components along the optical path between the point of the entry into the body and the point of scattering. Symbol $\Delta\phi$ denotes the phase difference between components I_θ and I_Φ of the scattered beam. This radiant power and the related state of modulation can be conveniently represented by the amplitudes and the azimuths of the related electric vectors, describing two wave fronts polarized in the planes determined by the principal directions and exhibiting a certain phase difference (relative retardation). The azimuth of the plane of vibration with respect to the scattering plane of the resultant scattered Rayleigh radiation is determined by

$$\tan\Phi_R = \frac{\sqrt{I_\Phi}}{\sqrt{I_\theta}} = \frac{\tan\Phi}{\cos\theta}. \tag{4.52}$$

It follows from relations (4.49)–(4.52) that the state of polarization of the primary beam (which can be modulated by the principal stress components acting in the plane perpendicular to the direction of propagation) can be easily analyzed when the directions of observation are given by $\theta = 90°$ and $\Phi = 90°$ or $0°$ (see Figure 4.1). For such values of parameters the plane of vibration of the scattered beams is normal to the primary beam and the intensities of scattered beams assume extreme values.

Relations (4.49) and (4.50) can be rewritten as follows:

$$I_\Phi = k'KE^2\sin^2\Phi = K'E^2\sin^2\Phi, \tag{4.53}$$

$$I_\theta = k'KE^2\cos^2\Phi\cos^2\theta = K'E^2\cos^2\Phi\cos^2\theta, \tag{4.54}$$

or

$$I_\theta = K'E^2(1 - \sin^2\Phi)\cos^2\theta = (KI_0 - I_\Phi)\cos^2\theta, \tag{4.55}$$

where E denotes the amplitude of the electric vector at the point under consideration, Φ the angle between this vector and the scattering plane, θ the angle of observation within the scattering plane, and k' a related factor of proportionality.

It is shown in Section 4.3 that the amplitude and the azimuth of the vector E of the primary beam, or the amplitudes and relative phase difference of the component vectors E_1 and E_2 representing the primary beam as the source of scattered radiation, are modulated by the stress components. Consequently, the scattered light intensities are modulated by the stress components acting along the path of the primary beam. Thus the scattered light intensities and their gradients carry local information on the stress state at particular points, and the global information on the stress state along the path of the primary beam.

It may be noted that the scattering of elastic waves (ultrasonic radiation) also carries valuable information on the scattering materials or the scattering objects.

4.2.3 Models of propagation of radiant energy across a boundary

The state of polarization and the phase of the light ray entering a body at other than normal incidence may be altered depending on the angle of incidence and the values of the indices of refraction. The pertinent relations can be derived using the electromagnetic light theory and considering the conditions at the boundary between two media. Without any loss of information of interest, the modulation of the radiant power when the linearly polarized wave front is resolved into a reflected front and a refracted front on the boundary of two optical media can be conveniently presented by the known Fresnel equations [6, 10].

For external reflection presented in Figure 4.2, the relations between the amplitudes and phases of the linearly polarized incident electric vector $\overline{E}$, and the reflected and refracted vectors $\overline{R}$ and $\overline{T}$ presented by their components parallel and normal to the plane of incidence can be formulated as follows:

$$\frac{R_p}{E_p} = \frac{\tan(\delta - \delta')}{\tan(\delta + \delta')} = \frac{n_{12} \cos \delta - \cos \delta'}{n_{12} \cos \delta + \cos \delta'}, \tag{4.56a}$$

$$\Delta\phi(R_p) = 0 \quad \text{for} \quad \delta < \delta_p = \arctan n_{12}, \tag{4.56b}$$

$$\Delta\phi(R_p) = \pi \quad \text{for} \quad \delta > \delta_p,$$

$$\frac{R_n}{E_n} = -\frac{\sin(\delta - \delta')}{\sin(\delta + \delta')} = \frac{\cos \delta - n_{12} \cos \delta'}{\cos \delta + n_{12} \cos \delta'}, \tag{4.56c}$$

$$\Delta\phi(R_n) = \pi \quad \text{for} \quad 0 < \delta < \pi. \tag{4.56d}$$

$$\frac{T_p}{E_p} = \frac{2 \sin \delta' \cos \delta}{\sin(\delta + \delta') \cos (\delta - \delta')} = \frac{2 \cos \delta}{n_{12} \cos \delta + \cos \delta'}, \tag{4.56e}$$

$$\Delta\phi(T_p) = 0 \quad \text{for} \quad 0 < \delta < \pi, \tag{4.56f}$$

$$\frac{T_n}{E_n} = \frac{2 \sin \delta' \cos \delta}{\sin(\delta + \delta')} = \frac{2 \cos \delta}{\cos \delta + n_{12} \cos \delta'}, \tag{4.56g}$$

$$\Delta\phi(T_n) = 0 \quad \text{for} \quad 0 < \delta < \pi, \tag{4.56h}$$

where δ_p denotes the polarizing angle or the Brewster angle, and the quantity n_{12} is defined as

$$n_{12} = \frac{n_2}{n_1} = \frac{\sqrt{\varepsilon_2\mu_2}}{\sqrt{\varepsilon_1\mu_1}} = \frac{\sin \delta}{\sin \delta'}. \tag{4.56i}$$

It follows that for external reflection presented in Figure 4.2, the reflected and the refracted beams are linearly polarized. However, the azimuth of the plane of vibration of the incident beam (angle between the plane of vibration and the normal to the plane of incidence) is not equal to the azimuths of the planes of vibration of the reflected and refracted beams. The azimuths of the reflected and refracted linearly polarized beams and the intensity of these beams depend on the angle of incidence and on the relative indices of refraction of both media.

Relations (4.56) describe the propagation of the scattered beam from an optically less dense medium to a denser one. When the light beam crosses the boundary between the denser and the less dense medium, i.e., when internal reflection is considered, relations (4.56) are only valid when the angle δ is replaced by δ' and vice versa. The phase changes of the parallel components

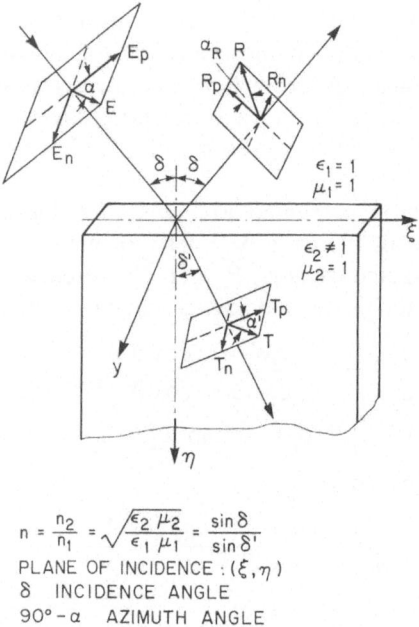

$$n = \frac{n_2}{n_1} = \sqrt{\frac{\epsilon_2 \, \mu_2}{\epsilon_1 \, \mu_1}} = \frac{\sin \delta}{\sin \delta'}$$

PLANE OF INCIDENCE : (ξ, η)
δ INCIDENCE ANGLE
$90° - \alpha$ AZIMUTH ANGLE

Fig. 4.2. Reflection and refraction of a monochromatic linearly polarized light beam presented in terms of Fresnel's model.

and the normal components of the internally refracted beams are continuous functions of the angle of incidence; consequently, the internally reflected beam is elliptically polarized. This fact imposes a serious theoretical constraint on all measurement techniques using reflected light, such as the technique of oblique incidence.

The values of the critical angles for all optical components used in the oblique incidence techniques are important parameters since the signal/noise ratio for the arrangements presented in Figure 4.2 depends on the critical angles, or the angles of total reflection, δ_t, determined by

$$\sin \delta_t = n_{21} = \frac{n_1}{n_2}. \tag{4.57}$$

The phenomenon of the radiant power transfer as presented in Figure 4.2 and relations (4.56) is somehow simplified. In fact, there is always penetration of energy from an optically denser to an optically rarer medium at, and above, the critical angle of the total reflection [36]. The pentrating electromagnetic wave, called evanescent wave, is attenuated rapidly, and its amplitude is negligible at the distance larger than one wavelength from the boundary between two media, Figure 4.3. Nevertheless, this distance is sufficiently large to be utilized in various, valuable measurement techniques, as demonstrated already over 20 years ago [3]. The evanescent wave can be used as a source of information by applying suitable, noncontacting coupling. The application of evanescent waves in modern, information collecting and processing system is extensive.

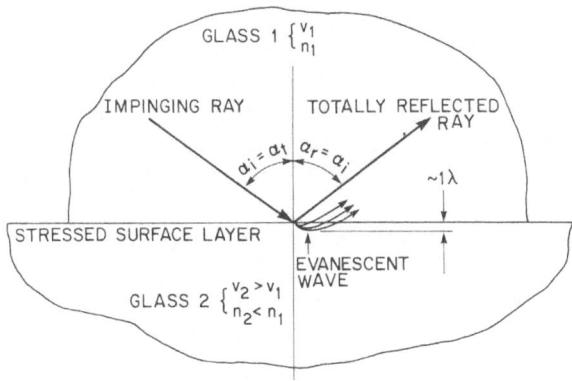

Fig. 4.3. Evanescent wave at total reflection.

Summarizing, it is very impractical to disregard the theory of light propagation across a boundary between two media. In the best case the noise level in the system increases unnecessarily; in the worst case not only the accuracy of the result, but even its reliability, can be seriously impaired, as follows from relations (4.56).

4.2.4 Models of light propagation through inhomogeneous and anisotropic bodies. Strain gradient optical effect

Various modern optical measurement techniques in applied physics and high technology either recognize or utilize the fact that the path of light beam propagating through an inhomogeneous body is not rectilinear, and that the light beam propagating through an inhomogeneous and anisotropic body is resolved into two noncollinear conjugated beams propagating along separate nonrectilinear paths. Various methods of optical tomography developed during the last decade utilize the fact that the gradients of the index of refraction in the system under study deflect the light beam. The feasibility and usefulness of the optical tomography reconstruction from phase measurements has been demonstrated using the beam-deflection techniques, interferometry, and holography.

In contrast, the theories and techniques of various analytical-experimental methods of applied mechanics such as the transmission caustics methods or oblique incidence methods are based on the assumption that the path of light beam propagating through an inhomogeneous and optically anisotropic body is rectilinear, and that the paths of both conjugated, ordinary and extraordinary, light beams are always collinear. This matter is of a significant theoretical and practical importance and deserves particular attention.

It has been known for a very long time that light propagation in nonhomogeneous media is not rectilinear: the light path is curved in the direction of higher density of the medium. The best-known example of this phenomenon is the mirage which occurs in deserts, on highway pavements and in nonhomogeneous aqueous solutions. Similar local effects observed in glass and in air are called shadow effects or Schlieren effects.

Using the wave theory of light, the mirage or shadow phenomenon is explained in the following manner. Alternation of the density results in alteration of the optical density, i.e., of the light velocity. Consequently, such local alterations cause distortion of the wavefront which, in turn, causes alteration of the direction of light propagation.

It has also been known to specialists for about 40 years that light beams propagating through solid bodies with load-induced inhomogeneity and anisotropy are resolved into two differently curved beams and eventually become spatially separated.

In the field of applied mechanics basic study of the major consequences of this phenomenon was presented about 40 years ago by Bokshtein [5]. While investigating the resolving power of photoelastic systems, Bokshtein observed that photoelastic fringes disappear in the regions of concentrated loads. The explanation of this fact, given in Ref. [5], is that in regions of high stress gradients, both conjugated rays are curved differently and become spatially separated. Consequently, these rays do not interfere simply as is predicted by the elementary physical model of photoelastic measurements, since this elementary model is based on the assumption of rectilinear light propagation. Unfortunately, this paper was virtually unknown outside of the author's country until the middle of the 1950s.

In 1955 Pindera [17] presented evidence that the light beams which propagate through a body exhibiting residual optical effect at the boundary are curved, even at normal incidence, Figure 4.4. The author describes the influence

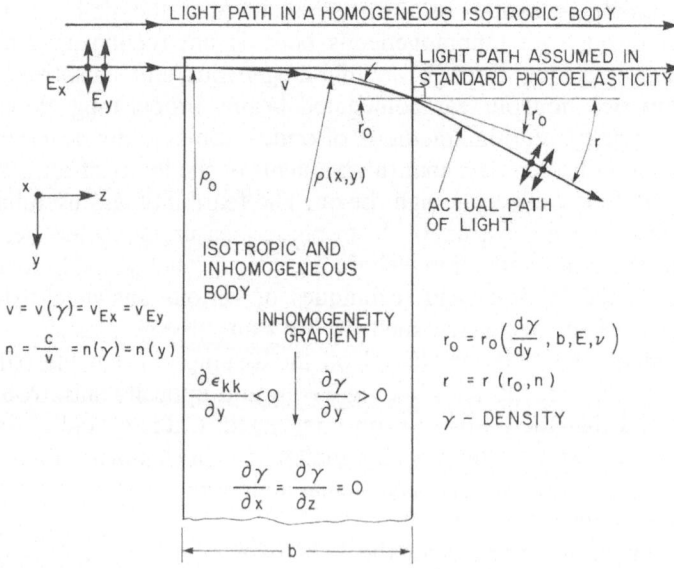

Fig. 4.4. Curvature of a light beam caused by the inhomogeneity gradient in an inhomogeneous isotropic body.

of the curvature of light beams produced by residual optical anisotropy, analogous to frozen stresses, on the recordings of isochromatics as functions of the parameters of the optical system. A sample of the empirical evidence given in Ref. [17] is presented in Figure 4.5. The character of the light-intensity distributions presented in Figure 4.5 for the natural and polarized collimated light beams can be readily explained in terms of the dependence of the curvature of the light path on the gradient of optical density. This interpretation explains the fact that the width of the black bands visible in collimated light (c) depends on the aperture.

When analyzing the empirical evidence on isochromatics field in contact regions [17, 25], it is easy to see that within the contact region the results of the effect of light curvature are different when diffused and collimated light is used. One can observe that in such cases some interference fringes appear in the regions which are obscured in a parallel light beam. Obviously, these interference fringes cannot be isochromatics as defined by the theory of photoelasticity under the assumption that the light propagation is rectilinear. It should

SPECIMEN : BEAM 12 mm x 12 mm x 88 mm
MATERIAL : CATALIN 800 WITH RESIDUAL BIREFRINGENCE
LOAD : NONE

✳ NATURAL LIGHT, WHITE   (d) DIFFUSED LIGHT
(ǁ) CIRCULARY POLARIZED   (c) COLLIMATED LIGHT
    LIGHT   λ = 546 nm   SMALL APERTURE

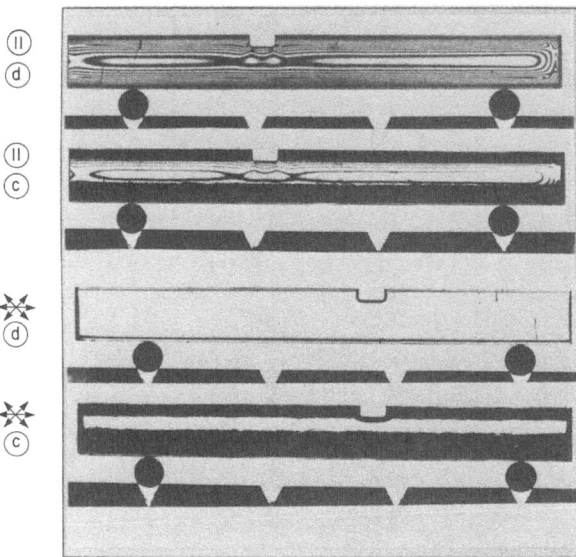

Fig. 4.5. Optical effects caused by curvature of the light path within the body with residual birefringence, recorded using a transmission polariscope. Bending of collimated light beams causes loss of information in darkened regions.
Bending of diffused light beams modifies photoelastic isochromatics.

be mentioned that the 'lens effect' caused by distortion of the parallelism of the plate faces in regions of contact can be eliminated by means of a matching immersion liquid. The remaining 'index lens effect' is caused by the curvature of light beams and the spatial separation of the coherent rays only. Similar effects occur in regions of sharp notches, crack tips, etc.

It may be noted that observations of the noticeable curvature of light beams propagating through photoelastic specimens made independently by the authors of Refs. [5] and [17], the evidence of which is given in Refs. [24, 26, 27, 29, 30], was the result of only partially successful attempts to increase the resolving power of the transmission polariscope conducted by Bokshtein, and during 1948—1953 by Pindera. The results presented in Ref. [17] were obtained using a highly collimated light beam and small aperture of a photographic camera; it has been observed that the 'skin effect' increasing with time limits the resolving power of the transmission polariscope. Consequently, it is concluded in Ref. [17] that the curvature of light beams, which is particularly noticeable at all surfaces of photoelastic specimens, places inherent limits on the resolution of optical systems used in photoelasticity. The empirically collected evidence regarding light propagation in inhomogeneous-isotropic bodies and in inhomo-geneous-anisotropic bodies is summarized in Figures 4.4 and 4.6, respectively.

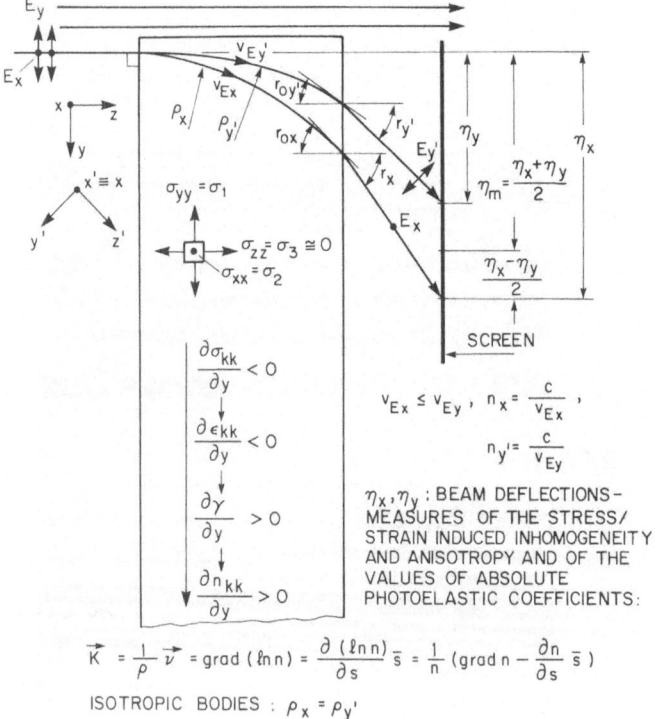

Fig. 4.6. Separation and curvature of the conjugated light beams propagating in the plane of symmetry of a stressed plate, at the normal incidence. Influence of gradients of inhomogeneity and anisotropy caused by stress/strain gradients.

The light beams presented in these figures consist of two conjugated beams represented by electric vectors E_x (normal to the plane of the figure) and E_y (within the plane of the figure). Both conjugated light beams propagating through an inhomogeneous-isotropic body are curved but stay collinear, whereas they are curved and separated when traveling through an inhomogeneous and anisotropic body. Such effects are caused by the anisotropy and inhomogeneity gradients of the stress/strain/density fields produced in a body by external and internal forces, or existing naturally in a body.

It must be noted that about 30 years ago Acloque *et al.* inferred from birefringence observations that the path of light entering glass plates under oblique incidence must be curved when particular residual or induced stress states exist in glass plates. They called this phenomenon 'mirage' and used it to develop new methods of stress analysis and quality control [2, 3].

The phenomenon of light-beam curvature inside stressed bodies can be of significant consequence in theories which relate patterns of light propagation within the bodies to quantities characterizing the mechanical state of these bodies. The most representative examples of such theories are those of transmission photoelasticity, shadow methods (including Schlieren), so-called caustic methods, integrated photoelasticity methods and the optical tomography methods based on phase measurement. As mentioned at the beginning of this section, the classical phenomenological theories of methods of photoelasticity and of caustics are based on the assumption that the light paths within stressed bodies are rectilinear, and by definition, both conjugated light beams are collinear. However, it has been recently observed [30] that light beams propagating through a photoelastic specimen may deviate from a rectilinear path by about 4 mm at the distance of 100 mm and can separate by the same amount. In contrast, the physical theories of the shadow (Schlieren) methods and of the phase-measurement optical tomography are based on the phenomenon of bending of the light path.

Summarizing, in initially homogeneous and isotropic bodies, the optical inhomogeneity and anisotropy which generally are caused by the deformation/ stress states result in the above-presented effect which is called strain-gradient effect. There is a need to develop a simple, qualitatively sufficiently reliable and quantitatively sufficiently accurate, mathematical model relating certain parameters of the stress field to chosen parameters which describe the patterns of light propagation in deformed bodies. It may be noted that whereas the qualitative reliability is rather imperative, the requirements regarding the quantitative accuracy depend on the purpose of a particular task.

Having such a model, it should be possible to answer some basic questions related to the reliability and accuracy of all optical methods in mechanics which allow gathering of information on quantities such as stress, strain, elastic waves, etc., using light propagating through the body under investigation, under the assumption that light propagates rectilinearly.

It should also be possible to provide a more general basis for the advanced photoelastic methods based on the mirage phenomenon which have been developed by Acloque and Guillemet [3]. Following this path, it should also be possible to relate the strain-gradient effect to the theories and techniques of shadow methods, and, in particular, to the modern optical tomography methods. An example of such a model, developed for studies of local effects, is presented in the next section.

4.2.4.1 Strain Gradient Method of Stress Analysis. Sufficiently comprehensive local state relations between components of the strain/stress tensors and refractive indices are presented in Section 4.2.1. To utilize these relations for a qualitative and quantitative global description of the light path in stressed/strained bodies, it is necessary to incorporate sufficiently general relations between the indices of refraction and the direction of light path into the model being developed. For this purpose Pindera and Hecker [35] used the concept of the geometrical wave surface or a geometrical wave front, S, propagating in an isotropic inhomogeneous body, where $S(\bar{r})$, a kind of optical path, is a real scalar function of position in an isotropic nonhomogeneous body [6]. Accepting the fact that such a concept is based on the assumption that the wave number is sufficiently high to disregard some involved phenomena like diffraction, the relation between the wave front and the refractive index can be presented as

$$\text{grad} \cdot \text{grad}\, S = \nabla \cdot \nabla S = n^2, \tag{4.58}$$

where

$$\nabla = \frac{\partial}{\partial x}\, \bar{i} + \frac{\partial}{\partial y}\, \bar{j} + \frac{\partial}{\partial z}\, \bar{k},$$

or

$$\left(\frac{\partial S}{\partial x} \right)^2 + \left(\frac{\partial S}{\partial y} \right)^2 + \left(\frac{\partial S}{\partial z} \right)^2 = n^2(x, y, z). \tag{4.59}$$

Equation (4.59) is known as the eikonal equation and the function S is known as the eikonal.

The surfaces $S(\bar{r}) = $ const. are called geometric wave fronts (surfaces), and the vector

$$\bar{s} = \frac{\text{grad}\, S}{|\text{grad}\, S|} = \frac{\text{grad}\, S}{n}, \tag{4.60}$$

is the unit vector of the surface S, related to the position vector $\bar{r}$ by

$$\frac{d\bar{r}}{ds} = \bar{s}, \tag{4.61}$$

where ds is the distance between the surfaces S and $(S + dS)$ along the ray s, defined by $\bar{r}(s)$. Thus equation (4.61) can be presented as

$$n\, \frac{d\bar{r}}{ds} = \text{grad}\, S = n\, \bar{s}. \tag{4.62}$$

In equation (4.62) the light rays are the orthogonal trajectories to the geometrical wave front $S(x, y, z)$, $\bar{r}$ is the position vector of a point on a ray, and s is the length of the ray measured from a certain point on it.

Following the approach presented in Ref. [6] and differentiating equation (4.62) along the light rays, one obtains

$$\frac{d}{ds}\left(n\,\frac{d\bar{r}}{ds}\right) = \text{grad } n \qquad (4.63a)$$

where n, being a scalar function of spatial coordinates, also depends on the wavelength, $n = n(x, y, z, \lambda)$. This leads to the equation

$$\bar{K} = \frac{1}{\rho}\,\bar{\nu} = \text{grad } \ln n - \frac{\partial}{\partial s}(\ln n)\,\bar{s}$$

$$= \frac{1}{n}\left(\text{grad } n - \frac{\partial n}{\partial s}\,\bar{s}\right) \qquad (4.63b)$$

where $\bar{K}$ and ρ denote (curvature and radius of curvature of) light ray, respectively. Because of reasons stated above, the considerations presented below are limited to the cases when light propagates very closely to the axes of anisotropy represented by the principal directions of the stress/strain/index tensors, whose principal directions are constant along the paths of light rays.

As in all cases of methods using induced optical anisotropy, it is accepted here that light can propagate through dielectric and nonabsorbing bodies in the form of linearly polarized radiation, whose directions of polarization are collinear with the principal directions of strain/stress tensors normal to the direction of propagation. One must note that this assumption is not acceptable in cases when the influence of rotational birefringence is noticeable.

Consequently, the accepted model of the phenomenon under consideration is represented by the following features: light propagates through the birefringent body in the form of linearly polarized rays whose velocities are given by the Ramachandran–Ramaseshan model. When the direction $i = 3$ is the direction of light propagation, then the direction of polarization of the ray described by equation (4.25) for $i = 1$ is 1, and for $i = 2$ is 2. Thus the propagation of both related rays is treated as propagation of light in two isotropic and nonhomogeneous bodies, which are characterized by principal directions denoted by $(1, 2, 3)$; the light rays propagate in the direction 3 and are polarized in planes $(1, 3)$ or $(2, 3)$. Choosing z as an independent variable and using standard notation, the position vector $\bar{r}$ and the ray length s defined in equation (4.61) can be presented in the forms

$$\bar{r} = u^r(z)\bar{i} + v^r(z)\bar{j} + z\bar{k} \qquad (4.64)$$

$$s = \int_0^z ds = \int_0^z \left[\left(\frac{du^r}{dz}\right)^2 + \left(\frac{dv^r}{dz}\right)^2 + 1\right]^{1/2} dz \qquad (4.65)$$

where $u^r(z)$ and $v^r(z)$ are components of the position vector $\bar{r}$.

It is shown by Pindera and Hecker, [30] that relations (4.63), (4.25), (4.61) and (4.62) can yield simple linear relations between the geometry of both conjugated light rays and the particular functions of stress components. Since the problem has been reduced to a linear one, the relative displacements Δu^r and Δv^r of both rays and their average values, u_m^r and v_m^r in an arbitrary plane $z = z_0$, can only be linear combinations of the corresponding displacement components:

$$\Delta u^r = u_1^r - u_2^r, \tag{4.66a}$$

and

$$\Delta v^r = v_1^r - v_2^r, \tag{4.66b}$$

or

$$u_m^r = 1/2(u_1^r + u_2^r), \tag{4.67a}$$

and

$$v_m^r = 1/2(v_1^r + v_2^r). \tag{4.67b}$$

Thus

$$n_0 \frac{d^2}{dz^2}(\Delta u^r) = C_\sigma \frac{\partial}{\partial x}(\sigma_1 - \sigma_2), \tag{4.68a}$$

$$n_0 \frac{d^2}{dz^2}(\Delta v^r) = C_\sigma \frac{\partial}{\partial y}(\sigma_1 - \sigma_2). \tag{4.68b}$$

Analogously,

$$n_0 \frac{d^2 u_m^r}{dz^2} = 1/2 \left[C_b \frac{\partial}{\partial x} I_\sigma - C_\sigma \frac{\partial}{\partial x} \sigma_3 \right], \tag{4.69a}$$

and

$$n_0 \frac{d^2 v_m^r}{dz^2} = 1/2 \left[C_b \frac{\partial}{\partial y} I_\sigma - C_\sigma \frac{\partial}{\partial y} \sigma_3 \right], \tag{4.69b}$$

where I_σ denotes the first invariant of the stress tensor, C_σ denotes the difference of the absolute photoelastic coefficients, $(C_\sigma = C_{1\sigma} - C_{2\sigma})$, called photoelastic stress coefficient, and C_b denotes the sum of the absolute photoelastic coefficients, $(C_b = C_{1\sigma} + C_{2\sigma})$, called photoelastic bulk coefficient.

Equations (4.68) and (4.69) show that two simple functions of partial displacements of both conjugated light rays in the u and v directions, namely the relative displacement and the average displacement, are functions of the corresponding components of the gradient of maximal shear stresses, of the corresponding components of the gradient of volumetric stress or volumetric deformation, and of the photoelastic stress and density coefficients.

The final equations (4.68) and (4.69) relate the displacement of the light rays to the gradient of principal stresses or strains. Because of this, the term "strain gradient method" or 'gradient photoelasticity' is introduced to denote all methods based on equations (4.68) and (4.69).

The following case of the asymmetry of a stress field has a significant practical importance, namely the stress state in a plate where the stress components in the direction of the plate thickness are much smaller than the stress components in planes parallel to the plate faces. In such cases the following relation is satisfied.

$$\left(\frac{\partial}{\partial x} + \frac{\partial}{\partial y}\right) \sigma_3 \ll \left(\frac{\partial}{\partial x} + \frac{\partial}{\partial y}\right)(\sigma_{xx} + \sigma_{yy}),$$
(4.70)

where the principal stress component σ_3 is almost collinear with the z direction: $\sigma_3 \cong \sigma_{zz}$. When the refractive index n is a weak function of the coordinate z collinear with the incident light ray, one more condition exists:

$$\frac{\partial}{\partial z} \sigma_3 \ll \left(\frac{\partial}{\partial x} + \frac{\partial}{\partial y}\right)(\sigma_{xx} + \sigma_{yy}).$$
(4.71)

In such a case, the equations for the differences of displacements, equations (4.68), remain the same, but the equations for the average displacements (4.69) simplify to

$$n_0 \frac{d^2}{dz^2}(u'm) = \frac{1}{2} C_b \frac{\partial}{\partial x}(\sigma_{xx} + \sigma_{yy}),$$
(4.72a)

and

$$n_0 \frac{d^2}{dz^2}(v'_m) = \frac{1}{2} C_b \frac{\partial}{\partial x}(\sigma_{xx} + \sigma_{yy}).$$
(4.72b)

Equations (4.68) and (4.72) describe light propagation through bodies subjected to stress states close to two-dimensional, including the limiting case of two-dimensional stress states.

In the case of local symmetry of stress state about a certain line, say y or $(x = x_0)$, defined by the condition

$$\left[\frac{\partial}{\partial x}(\sigma_{xx})\right]_{x = x_0} = \left[\frac{\partial}{\partial x}(\sigma_{yy})\right]_{x = x_0} = 0,$$
(4.73)

equations (4.68) and (4.72) yield

$$n_0 \frac{d^2 \Delta u'}{dz^2} = n_0 \frac{d^2 u_m}{dz^2} = 0,$$
(4.74)

$$n_0 \frac{d^2 \Delta v'}{dz^2} = C_\sigma \frac{\partial}{\partial y}(\sigma_{xx} - \sigma_{yy}),$$
(4.75a)

$$n_0 \frac{d^2 v'_m}{dz^2} = \frac{1}{2} C_b \frac{\partial}{\partial y}(\sigma_{xx} + \sigma_{yy}),$$
(4.75b)

where σ_{xx} and σ_{yy} are principal stresses σ_1 and σ_2, respectively. The above implies that both rays deflect in the plane containing the line of symmetry and the direction of incident light ray, and that the difference and the average value of deflections depend on the gradients of the difference and of the sum of the principal stresses, respectively.

Equation (4.75) can be used to determine the values of the photoelastic coefficients $C_\sigma(\lambda)$ and $C_b(\lambda)$, and consequently the values of the absolute photoelastic stress coefficients $C_1(\lambda)$ and $C_2(\lambda)$.

$$C_{1\sigma} = \frac{1}{2}(C_b + C_\sigma) \tag{4.76a}$$

and

$$C_{2\sigma} = \frac{1}{2}(C_b - C_\sigma) \tag{4.76b}$$

The conditions underlying the developed theory are satisfactorily fulfilled in the case of the so-called engineering two-dimensional stress states. Since such cases are the most likely ones to be of practical interest, they are taken as the basic condition in what follows. However, it must be noted, as discussed earlier, that the engineering concept of a two-dimensional stress state is not always applicable to the so-called plate problems, particularly in regions of stress concentrations and contact stresses where high local stress/strain gradients exist.

Using presently available techniques, it is difficult in practical applications to measure the values of the second derivatives of the light-ray displacements d^2u^r/dz^2 or d^2v^r/dz^2, which are closely related to the curvature of the light path. However, it is easy to measure the slope of the light ray emerging from the specimen. Without limiting the generality of the treatment of the subject, the following presentation is limited to the case of the local symmetry.

Accepting the physical model of experiment as presented in Figure (4.7) and the specimen in the shape of a plate of thickness b, the angles r_x and r_y can be determined by integrating equations (4.75),

$$r_x = \frac{dv_x^r}{dz}\bigg|_{z=b} = \frac{b}{2n_0}\left[C_b \frac{\partial}{\partial y}(\sigma_{xx} + \sigma_{yy}) + C_\sigma \frac{\partial}{\partial y}(\sigma_{xx} - \sigma_{yy})\right] \tag{4.77a}$$

and

$$r_y = \frac{dv_y^r}{dz}\bigg|_{z=b} = \frac{b}{2n_0}\left[C_b \frac{\partial}{\partial y}(\sigma_{xx} + \sigma_{yy}) - C_\sigma \frac{\partial}{\partial y}(\sigma_{xx} - \sigma_{yy})\right] \tag{4.77b}$$

Analogous relations can be derived for the deflections of rays in orthogonal planes of symmetry.

Equations (4.77) have been chosen as the principle of measurements in the strain gradient methods. Consequently, Figure 4.7 represents a scheme of a strain gradient recorder. Immersion liquid is introduced to quantitatively separate the light deflections caused by the gradient effect from the light deflections caused by the rotations of the faces of the specimen as a result of the stress-induced strain state.

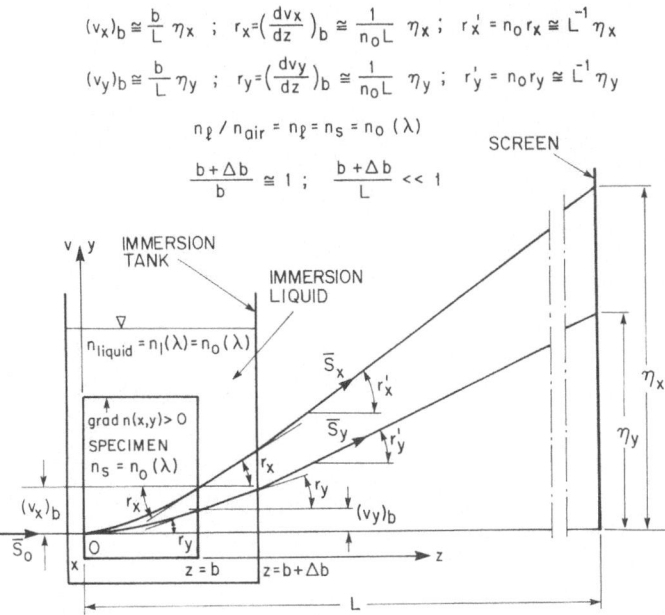

$$(v_x)_b \cong \frac{b}{L}\,\eta_x \;\;;\;\; r_x = \left(\frac{dv_x}{dz}\right)_b \cong \frac{1}{n_0 L}\,\eta_x \;\;;\;\; r_x' = n_0 r_x \cong L^{-1}\eta_x$$

$$(v_y)_b \cong \frac{b}{L}\,\eta_y \;\;;\;\; r_y = \left(\frac{dv_y}{dz}\right)_b \cong \frac{1}{n_0 L}\,\eta_y \;\;;\;\; r_y' = n_0 r_y \cong L^{-1}\eta_y$$

$$n_\ell / n_{air} = n_\ell = n_s = n_0 \ (\lambda)$$

$$\frac{b+\Delta b}{b} \cong 1 \;\;;\;\; \frac{b+\Delta b}{L} << 1$$

Fig. 4.7. Scheme of the physical model and a simplified mathematical model of the experiment for determination of deflection of light rays which is caused by the strain gradient effect and by the rotation of the faces of specimens.

The deflection η_x and η_y can be readily determined with an accuracy of a few percent. When the distance L between the front face of the plate and the screen is sufficiently large, the slopes in equations (4.77) are given by

$$\left.\frac{dv_x'}{dz}\right|_{z=b} \cong \frac{1}{n_0 L}\,\eta_x \tag{4.78a}$$

and

$$\left.\frac{dv_y'}{dz}\right|_{z=b} \cong \frac{1}{n_0 L}\,\eta_y \tag{4.78b}$$

where n_0 denotes the isotropic index of refraction of the specimen, or, by definition, of the immersion liquid. The value of the refractive index of the container for the immersion liquid is immaterial in the specified conditions of the experiment. Some information on which to base the choice of a suitable immersion liquid for a chosen wavelength of radiation is given in [30].

To allow assessment of the usefulness and efficacy of strain gradient methods, some typical results are given in Figure 4.8. The results were selected to show that all materials used in model mechanics are applicable in gradient methods, including Plexiglas, and that the strain gradient method of stress analysis readily yields the actual values of all stress-optic coefficients in the entire spectral range.

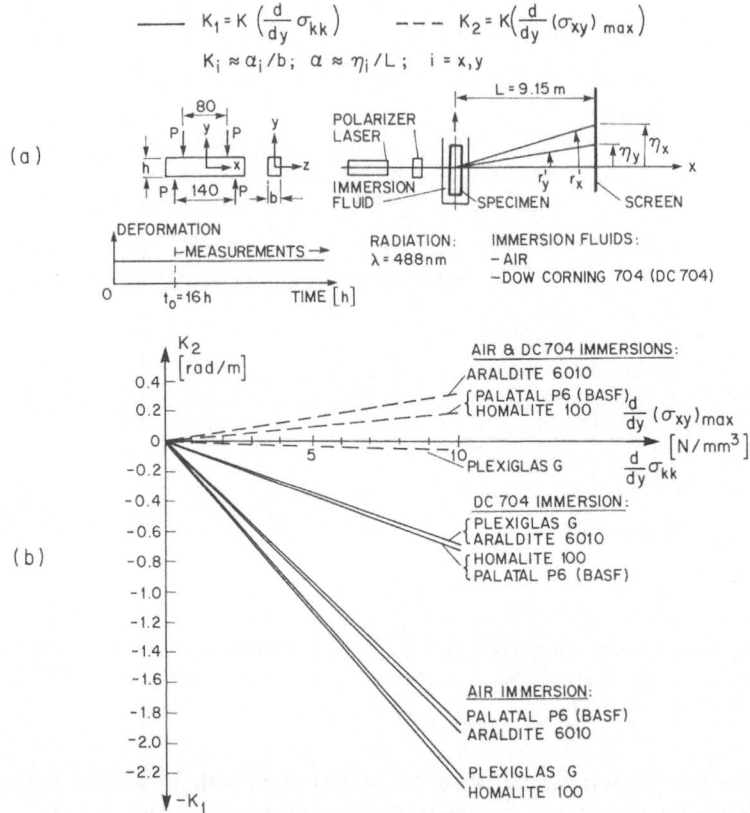

Fig. 4.8. Light propagation through solid bodies at normal incidence. Specimens: beams in pure bending. Volumetric and deviatoric normalized gradient deflections and bending of light beams for four different materials and two different immersion conditions.

(a) Data on experiments. (b) Deflections of light beams caused by bending and separation of light rays and by rotation of the place faces.

4.2.4.2 Summary. Differences between the common speculative phenomenological models of interaction between radiation and matter and the pertinent physical-theoretical models are essential. Acceptance of a physically inadmissible model of interaction may invalidate the reliability of the pertinent theory of stress/strain/deformation analysis. Figure 4.9 illustrates this problem.

4.3 Transmission photoelasticity

4.3.1 General comments

Classical methods of transmission photoelasticity [8, 11, 14, 16, 31, 33] utilize relations which represent a very simplified version of the relations presented in

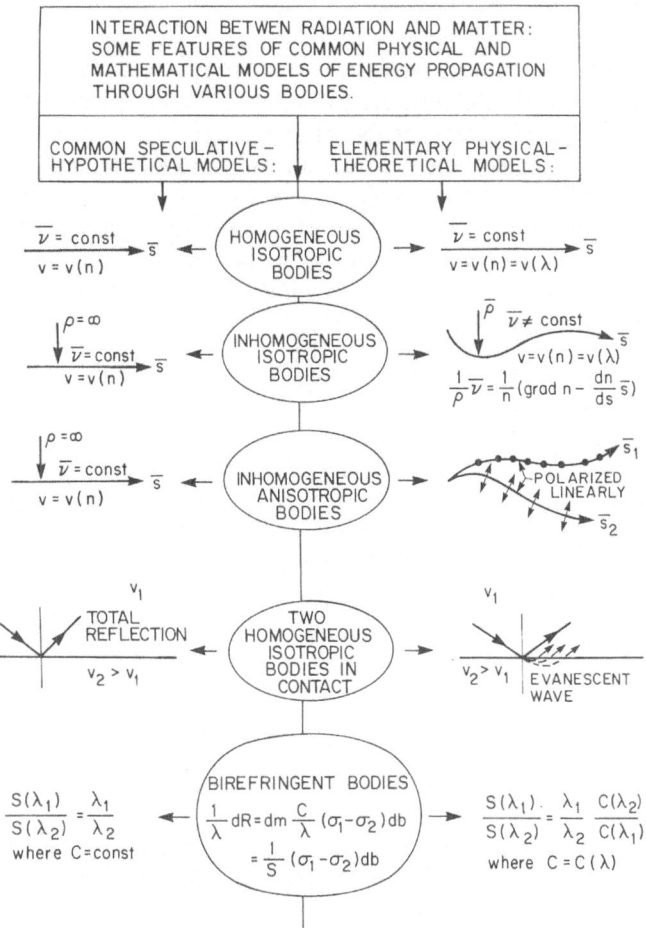

Fig. 4.9. Two incompatible sets of models of the flow of radiant energy through a nonconducting and nonabsorbing bodies, which are used concurrently in experimental mechanics.

Section 4.2.1. It should be remembered that the relations presented in Section 4.2.1 pertain to the local state with all the limitations following thereof. The uncritical extension of those relations to the global state made on the basis of far-reaching, tacitly accepted and unspecified, assumptions may result in development of physically inadmissible relations whose range of applicability cannot be analytically assessed and only can be tested experimentally.

To preserve coherence of presentation of the subject, some of the major features of Section 4.2.1 are summarized using a more explicit notation. This allows us to present the theoretical and analytical foundations of the transmission photoelasticity in a more modern form, compatible with the format of advanced research required by high technology, and makes the presented relations amenable to theoretical analysis and testing, and to experimental testing.

4.3.2 Basic models

The term "transmission photoelasticity" used in this text denotes the photo-elastic techniques which are applied to determine quasi-plane stress fields in plates. The quasi-plane stress fields, characterized by low values of stress gradients and by negligible values of the stress components in the direction of plate thickness, exist in plates loaded at the boundary surfaces by distributed tractions whose resultants are coplanar with the middle plane of the plate. In such plates the stress state is constant along the thickness coordinate.

The basic phenomenological-physical model of the common techniques of transmission photoelasticity consists of:

1. the model of an initially homogeneous, isotropic, continuous, linear elastic, athermal body — a Hooke's body which does not exhibit temperature alteration during elastic deformation;
2. the Ramachandran—Ramaseshan (R—R) model [34] which consists of:
   (a) the assumption that in homogeneously deformed solids all relations which describe the propagation of light derived for homogeneous anisotropic solids are valid. The effect of deformation is only to alter the parameters contained in these relations,
   (b) within elastic limits, the variation of an optical parameter of the solid caused by deformation can be expressed as a homogeneous linear function of the six stress or strain components;
3. the generalized Maxwell's relation between the light velocity and the components of the dielectric tensor;
4. the assumption that light propagates rectilinearly in a stressed solid, regard-less of the strain-induced inhomogeneity;
5. the assumption that both conjugated ordinary and extraordinary rays propagate collinearly, regardless of the magnitude of the stress-induced optical anisotropy, that is, regardless of the gradients of the optical inhomo-geneity and regardless of the values of the angles between the direction of propagation of light and the three principal directions.
6. the assumption that the engineering elastic coefficients, Young's modulus and Poisson's ratio, remain scalar quantities, regardless of the amount of the stress-induced optical anisotropy and inhomogeneity.
7. the assumption that the alteration of the values of the principal stress components with the distance from the plate surface is negligible, even when the gradients of principal stresses at the plate surface are not only mea-surable, but strongly pronounced;
8. the assumption that the alteration of all three principal directions with the distance from the plate surface is negligible, even when the gradients of principal stresses at the plate surface are pronounced.

The influence of the assumption (1) on the results of experiments may be negligible when the creep and relaxation phenomena either are negligible or remain within the linear viscoelastic range, or do not produce orientational optical effects.

The assumption (2) is acceptable when the distortional optical effect is at least two orders of magnitude higher than the eventual orientational optical effects.

The assumption (3) is acceptable when the electromagnetic materials coefficients, the specific conductivity and the magnetic permeability, are close to zero or to one, respectively.

The assumptions (4) and (5) may result in unacceptable errors, well above the accepted 3 percent in the routine, nondemanding photoelastic measurements. An example pertaining to this issue is given at the end of this chapter.

The influence of the assumption (6) depends on the level of sophistication of mathematical models used for determination of dynamic stress states such as those that are the object of studies in acoustoelasticity.

The consequences of the assumption (7) are far-reaching. In regions of local effects where the simplification of the fundamental relation of photoelasticity to the relation of engineering photoelasticity is not permitted, the phenomenological evaluation of photoelastic measurements results in errors in the determination of extremal stress values up to about 30 percent on the unsafe side.

The assumption (8) results in small errors, since the observed rotations of principal directions in regions of local effects do not exceed a few degrees.

The first assumption of the R—R model contains information that the optically anisotropic solid permits two monochromatic plane waves with two different mutually perpendicular linear polarizations and two different velocities to propagate in any chosen direction. When the chosen coordinate system (x, y, z) is collinear with the principal axes of the stress tensor, and when the deformed solid is initially isotropic, it follows from the R—R model that:

$$a_{xx} - \frac{1}{\varepsilon} = q_{11}\sigma_{xx} + q_{12}\sigma_{yy} + q_{13}\sigma_{zz},$$

$$a_{yy} - \frac{1}{\varepsilon} = q_{21}\sigma_{xx} + q_{22}\sigma_{yy} + q_{23}\sigma_{zz},$$

$$a_{zz} - \frac{1}{\varepsilon} = q_{31}\sigma_{xx} + q_{32}\sigma_{yy} + q_{33}\sigma_{zz},$$

$$a_{yz} = a_{zx} = a_{xy} = 0, \tag{4.79a–d}$$

where a_{ij} denote components of the index tensor, ε denotes scalar dielectric coefficient (zero strain-stress state), q_{ij} denotes stress-optic coefficients, and σ_{ii} denotes principal stresses.

For isotropic materials,

$$q_{11} = q_{22} = q_{33}.$$

$$q_{12} = q_{21} = q_{23} = q_{32} = q_{31} = q_{13}. \tag{4.80a, b}$$

The basic Maxwell's model

$$v = \frac{c}{\sqrt{\varepsilon\mu}}, \tag{4.81a}$$

relates the light velocities in vacuum and in an isotropic body, c and v respectively, to the dielectric constant (permittivity) ε, and the magnetic permeability μ, when ε and μ are normalized with respect to the ε and μ of the vacuum.

It is convenient to present this relation in a reciprocal normalized form, which for nonmagnetic materials can be presented as

$$\frac{c}{v} = n = \sqrt{\varepsilon}, \tag{4.81b}$$

where n denotes the index of refraction, which depends on the wavelength of the light

$$n = n(\lambda), \tag{4.81c}$$

because the dielectric coefficient ε is a function of the frequency of the electric field, f,

$$\varepsilon = e(f). \tag{4.81d}$$

The relations of crystal optics presented in Section 4.2.1 yield information that when light propagates through a body represented by the relations (4.79), in direction z perpendicular to plane (x, y), only two linearly polarized light rays are permitted: the ray-vibrating in the plane (zx) and propagating with the velocity v_1, and the ray-vibrating in the plane (zy) and propagating with the velocity v_2. It follows from (4.81b) that velocities v_1 and v_2 represented by the corresponding indices of refraction n_1 and n_2 are related to the components of the index tensor, and that the index of refraction of a non-deformed body, n_0, is related to the original scalar dielectric coefficient, ε:

$$n_1 = \frac{1}{\sqrt{a_{xx}}}, \quad n_2 = \frac{1}{\sqrt{a_{yy}}}, \quad \text{and} \quad n_0 = \sqrt{\varepsilon}. \tag{4.82a, b}$$

where ε is a nondimensional quantity, normalized with respect to the permittivity of space as mentioned earlier. As a rule, it is almost always much easier to measure differences between the values of two identical quantities than their absolute values. Accordingly, presenting relations (4.79) and (4.8) in terms of refractive indices,

$$\frac{1}{n_1^2} - \frac{1}{n_0^2} = q_{11}\sigma_{xx} + q_{12}(\sigma_{yy} + \sigma_{zz}) = q_{11}\sigma_1 + q_{12}(\sigma_2 + \sigma_3), \tag{4.83a}$$

and

$$\frac{1}{n_2^2} - \frac{1}{n_0^2} = q_{11}\sigma_{yy} + q_{12}(\sigma_{zz} + \sigma_{xx}) = q_{11}\sigma_2 + q_{12}(\sigma_3 + \sigma_1), \tag{4.83b}$$

where σ_1, σ_2, σ_3 denote principal stresses σ_{xx}, σ_{yy}, σ_{zz}, respectively, one can obtain:

$$\frac{1}{n_1^2} - \frac{1}{n_2^2} = (q_{11} - q_{12})(\sigma_1 - \sigma_2). \tag{4.84}$$

Because of convenience, it is common to linearize relation (4.89) by replacing the tensorial quantities n_1^{-2} and n_2^{-2} by the algebraic quantities n_1 and n_2,

$$\Delta n_{1,2} = n_1 - n_2 = (C_1 - C_2)(\sigma_1 - \sigma_2) = C_\sigma(\sigma_1 - \sigma_2)$$
$$= \Delta n_{1,2}(\lambda, \sigma_1, \sigma_2), \tag{4.85}$$

where $C_1 = C_1(\lambda)$, $C_2 = C_2(\lambda)$, and $C_\sigma = C_\sigma(\lambda)$ are called the absolute stress photoelastic coefficients and the photoelastic stress coefficient, respectively, and the difference of the principal refractive indices, $\Delta n_{1,2}$, is called birefringence. Obviously, such an analytical simplification results in partial loss of information and introduces numerical error. What is most important, the matrix of refractive indices at a point is not a tensorial quantity.

It should be noted that the dependence of the photoelastic coefficients C_1, C_2 and C_σ on the wavelength of the birefringence detecting radiation, λ, was neglected for a long time, despite the theoretical and empirical evidence that this dependence is significant [7, 20–22]. Recently, however, under the pressure of extensive, easily available empirical evidence this dependence has been recognized as significant with respect to the accuracy of photoelastic measurements. However, the significance of this dependence in analysis of viscoelastic and plastic deformations is still not recognized.

Relation (4.85) predicts that the stress-induced birefringence, $\Delta n_{1,2}$, is caused by the distortional component of the stress tensor, and is proportional to the maximal value of the related shear stress component

$$(\sigma_{xy})_{max} = 0.5(\sigma_1 - \sigma_2).$$

This dependence justifies the use of the term "distortional birefringence" to denote the stress-induced birefringence related to linear elastic deformation. This term is also correct regarding the elementary, very simplified physical theory of the stress-induced birefringence. Relation (4.85) is the basis of the classical photoelastic methods and techniques. To determine the indirectly measurable quantity $\Delta n_{1,2} = n_1 - n_2$, it is necessary to convert it into another quantity which, using a convenient measurement system, can be related to an easily measurable quantity, such as the light intensity.

From the utilitarian, descriptive point of view, in classical photoelasticity it is sufficient, and more convenient, to consider light propagation as a harmonic wave phenomenon within the framework of the electromagnetic theory of light, rather than as a corpuscular phenomenon within the framework of the quantum theory which gives a deeper insight into the basic physical processes. The related notions of the wavefronts, and of the wavefront separation expressed in terms of the linear relative retardation R or in terms of the phase relative retardation ϕ, allow the establishment of simple relations for the stress-induced birefringence when a monochromatic randomly polarized wave front impinges normally on a plate of thickness b, which is subjected to a homogeneous plane stress field, $(\sigma_1 = \text{const}, \sigma_2 = \text{const}, \sigma_3 = 0)$, Figure 4.10:

$$dR_{1,2} = \lambda \frac{1}{2\pi} d\phi = (n_1 - n_2) db = \Delta n_{1,2} db = C_\sigma(\sigma_1 - \sigma_2) db$$

$$= C_\varepsilon(\varepsilon_1 - \varepsilon_2) db. \tag{4.86}$$

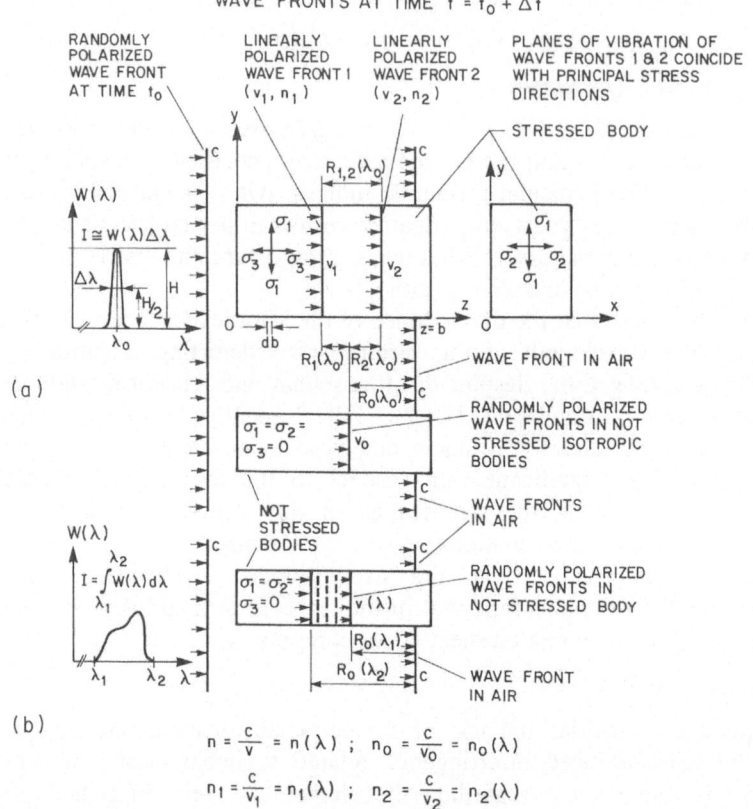

Fig. 4.10. Propagation of monochromatic, randomly polarized wave front through a solid birefringent body. Normal incidence, collinear with one of the principal directions of mechanical and optical anisotropy. A-stressed body exhibiting the strain-induced homogeneous anisotropy. B-not stressed, isotropic and homogeneous body. (a) Propagation of monochromatic wave fronts (b) Propagation of polychromatic wave fronts.

It is convenient to normalize the relative retardation with respect to the wavelength of light, λ, by introducing the dimensionless parameter m representing the normalized relative retardation,

$$\mathrm{d}R = \lambda\,\mathrm{d}m = \lambda\,\frac{1}{2\pi}\,\mathrm{d}\phi. \tag{4.87}$$

Thus:

$$\frac{\mathrm{d}R}{\lambda} = \mathrm{d}m = \frac{1}{2\pi}\,\mathrm{d}\phi = (n_1 - n_2)\,\frac{\mathrm{d}b}{\lambda}$$

$$= \frac{C_\sigma}{\lambda}\,(\sigma_1 - \sigma_2)\,\mathrm{d}b = S_\sigma^{-1}(\sigma_1 - \sigma_2)\,\mathrm{d}b, \tag{4.88}$$

where the wavelength dependent coefficient S_σ is denoted as the unit stress-optical coefficient, or the stress-optical coefficient,

$$S_\sigma = S_\sigma(\lambda).$$

When the stress state in a plate of thickness b is two-dimensional, and the angle of incidence is normal, the total normalized relative retardation between two conjugated wave fronts is given by

$$m = \frac{R}{\lambda} = \frac{\phi}{2\pi} = \int_0^b S_\sigma^{-1}(\sigma_1 - \sigma_2)\,db$$

$$= S_\sigma^{-1}\int_0^b (\sigma_1 - \sigma_2)\,db = S_\varepsilon^{-1}\int_0^b (\varepsilon_1 - \varepsilon_2)\,db \tag{4.89}$$

When the principal stresses σ_1 and σ_2 are constant across the thickness of a plate, and the stress state is homogeneous (stress gradients are equal to zero), the integral in (4.89) can be replaced by a product of the involved quantities

$$m = \frac{R}{\lambda} = \frac{\phi}{2\pi} = \frac{b}{S_\sigma}(\sigma_1 - \sigma_2). \tag{4.90}$$

4.3.3 Measurements of stress-induced birefringence using interference techniques

It is easy to represent the parameter m in equation (4.90) by the modulation of light intensity using the system depicted in Figure 4.11. Considering that within the wave theory of light the intensity of electro-magnetic radiation

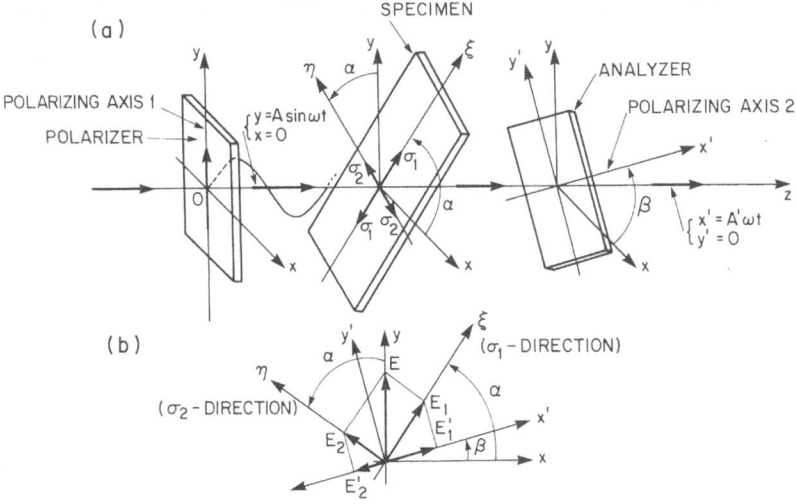

Fig. 4.11. Transmission polariscope. Geometry of the polarizing subsystem.

(radiant power per unit area), I, can be presented as an integral of the radiant power per unit cross-sectional area per unit band width $W(\lambda)$, across the band width $\Delta\lambda = (\lambda_2 - \lambda_1)$, Figure 4.12,

$$I = \int_{\lambda_1}^{\lambda_2} W(\lambda)\, d\lambda, \tag{4.91}$$

the transfer function of the polarizing subsystem given in Figure 4.17 can be presented as [16, 21, 31]:

$$F_T = \frac{I}{kI_0}$$

$$= (kI_0)^{-1} \int_{\lambda_1}^{\lambda_2} \left[\sin^2 \beta + \sin 2\alpha \cdot \sin 2(\alpha - \beta) \cdot \right.$$

$$\left. \cdot \sin^2 \left(\pi \frac{R(\lambda)}{\lambda} W(\lambda) \right) \right] d\lambda, \tag{4.92}$$

where the coefficient k accounts for the losses in the system, the angle $(90 - \beta)$ is the angle between the polarizing axes of the polarizer and analyzer, and the parameter α denotes the angle between the chosen principal direction and the chosen reference direction which is usually collinear with the polarizer's axis.

Usually, it is more convenient to perform measurements when the axes of the polarizer and analyzer are crossed, $\beta = 0$, and when the angle α does not depend on the wavelength of radiation,

$$F_T = (kI_0)^{-1} \sin^2 2\alpha \int_{\lambda_1}^{\lambda_2} \sin^2[\pi m(\lambda)]\, W(\lambda)\, d\lambda, \tag{4.93a}$$

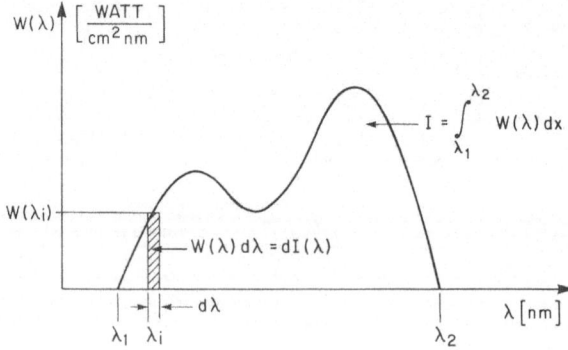

Fig. 4.12. The total spectral power of a particular band of electromagnetic radiation I, and the unit spectral power $W(\lambda)$. Definitions.

or

$$F_T = F(\alpha) \cdot F(\sigma_1 - \sigma_2, \lambda) = F(\alpha) \cdot F(\sigma, \lambda). \qquad (4.93b)$$

Equation (4.93) is a satisfactorily rigorous analytical solution, obtained within the framework of the assumed physical model, for the relations between a plane, homogeneous plane stress state in a plate and the related birefringence-induced modulation of light intensity propagating through a polarizing sub-system given in Figure 4.11. It can be utilized in measurements using the visible and infrared bands of electromagnetic radiation [7].

Customarily, this rigorous relation is extrapolated, usually intuitively and tacitly, for nonhomogeneous quasi-plane stress states which exist in plates of finite thickness which have an arbitrary geometry and are loaded by arbitrary distributed boundary tractions. The extrapolated relation (4.93) is presented in the form of an approximate relation, pertaining to a general case of a nonhomo-geneous plane stress field

$$F(x, y) = (kI_0)^{-1} \sin^2 2\alpha \int_{\lambda_1}^{\lambda_2} \sin^2[\pi m(x, y, \lambda)]\, W(\lambda)\, \mathrm{d}\lambda, \qquad (4.94a)$$

where

$$m = m(x, y, \lambda) = \frac{R(x, y, \lambda)}{\lambda} = [S_\sigma(\lambda)]^{-1} b[\sigma_1(x, y) - \sigma_2(x, y)]. \qquad (4.94b)$$

The reliability of relation (4.94) depends on the values of the gradients of stresses at the plate surface since the gradients of the surface stresses are directly related to the degree of the actual three-dimensionality of the stress state in a plate. As discussed earlier, in such three-dimensional stress states whose presence is indicated by the non-negligible gradients of the surface stresses, the values of the stress components alter with the distance from the plate surface, and exhibit extreme values usually at the middle plane of a plate. In other words, the actual state of stress in a plate whose surface is subjected to a two-dimensional by assumption, and actually a nonhomogeneous stress state, is always three-dimensional. The actual state of stress can be weakly or pronounceably three dimensional, depending on the values of the gradients of the surface stresses. All stress components in such a case alter with the distance from the surface of a plate. The gradients of the surface stresses are pronounced in the regions of the local effects. This correspondence facilitates recognition of the local effects which may cause unexpected structural failures.

Consequently, the extrapolation and transformation of relation (4.93) into relation (4.94) is tantamount to the introduction of the four components of the overall physical model of photoelastic measurements, listed earlier under the numbers (5)—(8).

In analogy with relation (4.93b), relation (4.94a) can be presented conven-iently as a product of two functions [13, 21, 31],

$$F_T(x, y) = F_\alpha(x, y) \cdot F_{\Delta\sigma}(x, y, \lambda). \qquad (4.94c)$$

When the chosen spectral band $\Delta\lambda = \lambda_2 - \lambda_1$ is sufficiently wide, the radiant power $W(\lambda)$ relatively constant within the chosen spectral band and the isochromatic order m high, then the function $F_{\Delta\sigma}$ approaches a constant value and the intensity of light (radiant power) transmitted through a polarizing subsystem, Figure 4.11, depends essentially on the values of α,

$$F_T(x, y) \simeq F_\alpha(x, y). \tag{4.95}$$

Obviously, the function

$$F_\alpha(x, y) = \text{const}, \tag{4.96a}$$

or

$$\alpha = \text{const}, \tag{4.96b}$$

represents isoclinics, one of the characteristic lines of the plane stress fields, known also in the theory of differential equations.

In other words, to determine isoclinics which represent geometric loci of points at which principal directions are constant, $\alpha = \text{const}$, it is useful to use a wide band of radiation.

The photoelastic isoclinics are related to plane stress fields, and can be presented as a particular function of the components of the plane stress tensor

$$\text{tg}\, 2\alpha = \frac{2\sigma_{xy}}{\sigma_{xx} - \sigma_{yy}} = \text{const}, \tag{4.97}$$

where α is the parameter of an isoclinic.

Relations (4.92) and (4.93a) can be simplified if the radiation band is infinitesimal, $d\lambda$, centered around the dominant wavelength, λ_0,

$$F_T(x, y, \lambda) = \frac{W(\lambda)\, d\lambda}{W_0(\lambda)} = \sin^2\beta + \sin^2\alpha \sin^2(\alpha - \beta) \sin^2 \pi m \tag{4.98a}$$

and

$$F_T(x, y, \lambda_0) = \frac{W(\lambda_0)\, d\lambda}{W_0(\lambda_0)\, d\lambda} = \sin^2 2\alpha \cdot \sin^2 \pi m(x, y, \lambda_0)$$

$$= F_\alpha(x, y) \cdot F_{\Delta\sigma}(x, y, \lambda_0). \tag{4.98b}$$

It can be shown that in a general case relation (4.98) is satisfactorily reliable when the spectral band is less than one nanometer. This condition is satisfied by many commercial lasers.

When linear polarizer and analyzer are replaced by circular ones, the directional effects are eliminated, $F_\alpha(x, y) = 1$, and the relation (4.98) simplifies to

$$F_T(x, y, \lambda_0) = \sin^2[\pi m(x, y, \lambda_0)] = \sin^2 \pi \frac{\sigma_1(x, y) - \sigma_2(x, y)}{S_\sigma(\lambda_0)}\, b, \tag{4.99a}$$

or

$$F_T(x, y, \lambda_0) = \cos^2[\pi m(x, y, \lambda_0)] = \cos^2 \pi \frac{\sigma_1(x, y) - \sigma_2(x, y)}{S_\sigma(\lambda_0)}\, b. \tag{4.99b}$$

depending on whether or not the signs of circular polarization produced by the polarizer and analyzer are the same or opposite.

The transfer function $F_T(x, y, \lambda_0)$ represents the light intensity distribution produced by the polarizing subsystem, which is related to the field of the stress-induced birefringence in a plate subjected to a plane stress field.

The lines of constant light intensity

$$F_T(x, y, \lambda_0) = \text{const,} \tag{4.100a}$$

identical with

$$m(x, y, \lambda_0) = \text{const,} \tag{4.100b}$$

and with

$$\sigma_1(x, y) - \sigma_2(x, y) = \text{const,} \tag{4.100c}$$

are called — historically — isochromatics, and the parameter m is called the order of isochromatic. It is useful to distinguish between the analytical elastic isochromatics related to the Airy stress function of a plane stress field, $\Phi(x, y)$, and defined rigorously,

$$\left(\frac{\partial^2}{\partial y^2} \Phi \right)_{\text{max}} - \left(\frac{\partial^2}{\partial x^2} \Phi \right)_{\text{min}} = (\sigma_{xx})_{\text{max}} - (\sigma_{yy})_{\text{min}}$$

$$= \sigma_1 - \sigma_2 = \text{const.} \tag{4.101}$$

and the photoelastic isochromatics, which are as reliable as the underlying physical and mathematical models are reliable.

The reliability of particular photoelastic measurements can be assessed by checking the reliability of the particular components of the accepted physical and mathematical models. This issue has been discussed in Chapters 1 and 2 with respect to the actual stress/strain states in plates loaded by tractions acting on the boundary surfaces, and with respect to the actual patterns of propagation of energy through nonhomogeneous and anisotropic bodies.

A related factor is the actual transfer function of the image-producing system. This issue is illustrated in Figure 4.13 [13, 17]. Depending on whether or not photoelastic photographic recordings are made using collimated or diffused light beam, in regions of high stress/strain gradients photoelastic isochromatics either shrink or disappear depending on the aperture, or do not represent elastic isochromatics. The actual error in the determination of maximal values of boundary stresses using photoelastic transmission isochromatics can be up to about 30 percent on the unsafe side.

The fact that both conjugated wave fronts carrying information on the stress induced birefringence are warped, Figure 4.14, imposes specific conditions upon the theory and technique of recording photoelastic interference fringes, that is, upon the design of the transfer function of the fringe-recording optical system [21, 31, 32]. For example, it is easy to show that the recorded photoelastic interference fringes related to the regions of high stress/strain gradients

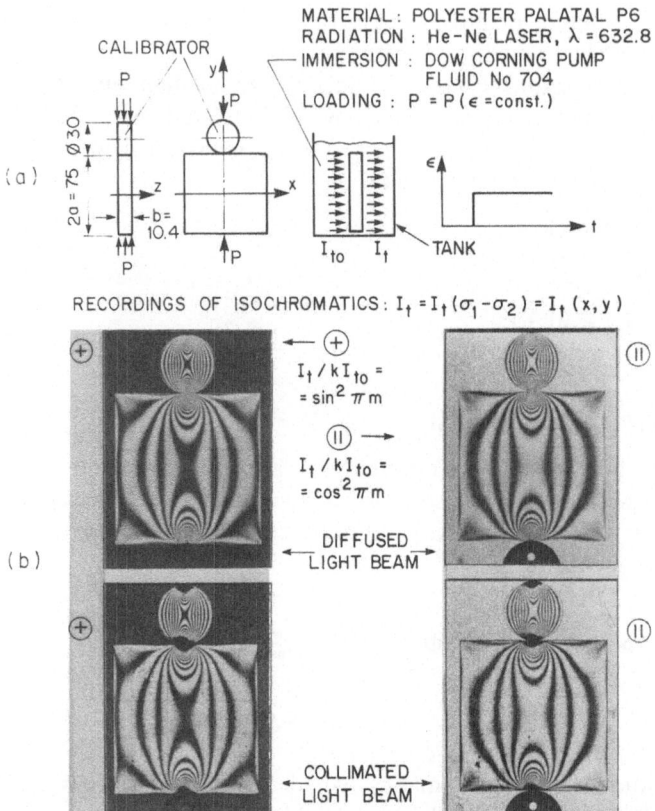

Fig. 4.13. An example of one of the particular features of the transfer functions of the image-producing subsystems of a transmission polariscope: influence of the state of collimation of the birefringence-detecting light beam on the isochromatic fields in regions of high strain gradients in a square plate loaded axially.

(a) Data on experiment. (b) Recordings of isochromatics light intensity modulation. Top recordings: photoelastic isochromatics in diffused light are recorded in the contact regions, however, they do not represent analytical isochromatics. Bottom recordings: photoelastic isochromatics in collimated light are obscured in the contact regions by the gradient effect.

depend strongly on the distance between the photoelastic specimen and the recording camera, even when the influence of local warping of the plate faces is eliminated by the use of a matching immersion liquid [1].

It should be noted that it is difficult to record correctly the light intensity distribution which carries information on the values of difference of principal stresses and/or on the isoclinic parameters presented in relations (4.98) and (4.99). Most typical recordings are made using either photographic emulsion or couple charged device. Both processes cannot be described in terms of the wave theory of light, but can be understood in terms of the quantum theory of light assuming the existence of light photons. The photographic emulsion, being an energy-storing device, converts radiant energy into density of the photographic

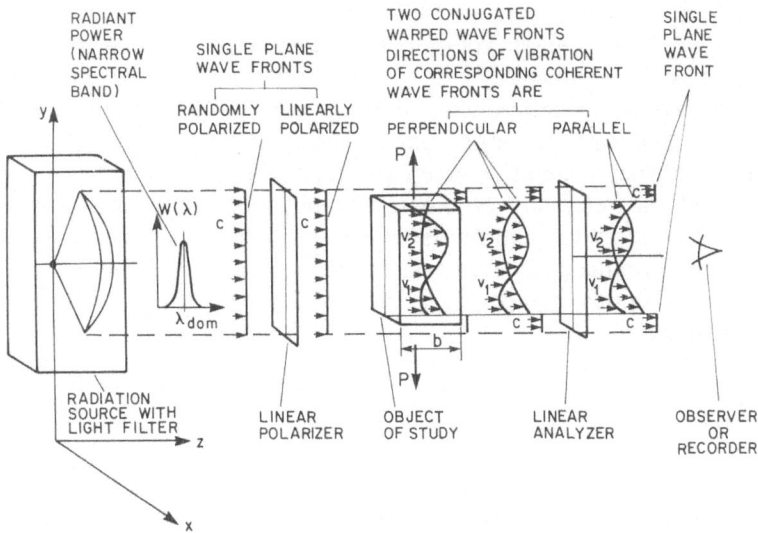

Fig. 4.14. Schematic presentation of propagation of the light wavefront through a transmission polariscope. Warping of the wavefronts.

emulsion, and modulates its level according to the characteristic curve of the emulsion and to the related quantity called gamma value. The couple charged device, being a radiant power sensor, converts radiant power into electric charge according to a specific transfer function which is commonly represented by the gamma value. The gamma value usually depends either on radiant energy in the case of the energy-storing process of the photographic emulsion, or on radiant power when the photoelectric effect is used as the principle of measurement. In both cases the gamma value depends on the energy of light quanta corresponding to the spectral band of radiation.

Summarizing, it is rather easy to record sharp, clear photoelastic fringes which look very convincing. However, the reliability of the result obtained by means of photoelastic techniques depends on the level of sophistication of the theory used to retrieve desired information from the sharply recorded and precisely determined photoelastic fringes. It follows from relations (4.98) and (4.99) that one loses information when photoelastic fringes are recorded using high gamma-value emulsion or a CCD system which produces sharp fringes.

When the photoelastic specimen is a geometric, reduced model of an actual structure, and when the constitutive material relations for materials of both the structure and photoelastic specimen satisfy certain specific conditions discussed in pertinent references [11, 14, 15, 20, 22, 24], the results are transferable from the photoelastic specimen to the original structure. Specifically, a photoelastic specimen represents a photoelastic model only when the so-called stress similarity conditions are satisfied [15].

With regard to the well-known simplified formulation of the stress similarity conditions, two particular issues deserve particular attention. At first, it follows from the compatibility conditions for the strains and rotations that in order to satisfy the condition of similarity of stress fields, the strains in a prototype and its model must be identical. Secondly, with some very infrequent exceptions, photoelastic models of components and structures, which are made of metals whose behaviour can be approximated by Hooke's model, are made of viscoelastic polymers exhibiting creep and relaxation phenomena. To allow the assessment of the reliability of the results obtained using photoelastic techniques, it is necessary to quantify the effect of the excessive deformation which is commonly produced in photoelastic models, and the influence of the mechanical and optical creep, relaxation, and recovery which inherently occur in solid polymers.

4.4 References

[1]   Aben, H., Krasnowski, B. R., and Pindera, J. T., "Nonrectilinear Light Propagation in Integrated Photoelasticity of Axisymmetric Bodies", *Transactions of the CSME* **8**(4), 1984, pp. 195–200.

[2]   Acloque, P. and Guillemet, G., "Method for the Photoelastic Measurement of Stresses in Equilibrium in the Thickness of a Plate (Particular Cases of Toughened Glass and Bent Glass)", Selected Papers on Stress Analysis, Chapman and Hall Ltd., London, 1961, pp. 71–76.

[3]   Acloque, P., "Methods for Local Determination of Both Principal Stresses in a Plate, by Using the Oblique Incidence and the Index of Total Reflection" (in German), in: Gotthart Haberland (Ed.), *Proc. Intern. Symp. Photoelasticity*, Berlin, April, 1961, Akademie Verlag, Berlin, 1962, pp. 10–15.

[4]   Baldwin, G. C., *An Introduction to Nonlinear Optics*, Plenum Publishing Corporation, New York, 1969.

[5]   Bokshtein, M. F., "On the Resolving Power of the Polarizing System for Stress Investigations" (in Russian), *Zhurnal Tekhnicheskoi Fiziki* **19**, 1949, pp. 1103–1106.

[6]   Born, Max and Wolf, Emil, *Principles of Optics*, Pergamon Press, Toronto, 1975.

[7]   Cloud, G. L. and Pindera, J. T., "Techniques in Infrared Photoelasticity", *Experimental Mechanics* **8**(5), 1968, pp. 193–201.

[8]   Coker, E. G. and Filon, L. N. G., *A Treatise on Photo-Elasticity*, Cambridge At the University Press, Cambridge, 1931 (1957).

[9]   Crosignani, B., DiPorto, P., and Bertolotti, M., *Statistical Properties of Scattered Light*, Academic Press, New York, 1975.

[10] Fowles, G. R., *Introduction to Modern Optics*, Holt, Rinehart and Winston, Inc., New York, 1968.

[11] Frocht, M. M., *Photoelasticity*, Vol. 1 and 2, John Wiley & Sons, New York, 1941 and 1948.

[12] Kerker, M., *The Scattering of Light*, Academic Press, New York and London, 1969.

[13] Mazurkiewicz, S. B. and Pindera, J. T., "Integrated Plane Photoelastic Method — Application of Photoelastic Isodynes", *Experimental Mechanics* **19**(7), 1979, pp. 225–234.

[14] Mesmer, G., *Spannungsoptik*, Springer-Verlag, 1939.

[15] Mönch, E., "Similarity and Model Laws in Photoelastic Experiments", *Experimentai Mechanics* **4**(5), 1964, pp. 141–150.

[16] Pindera, J. T., *Outline of Photoelasticity* (in Polish), Panstwowe Wydawnictwa Techniczne, Warszawa, 1953.

[17] Pindera, J. T., *Techniques of Photoelastic Investigations of Two-Dimensional Stress Problems* (in Polish), Engineering Transactions, (Rozprawy Inzynierskie) Polish Acad. Sciences, Vol. III (1), 1955, pp. 109–176.

[18] Pindera, J. T. and Cloud, G., "On Dispersion of Birefringence of Photoelastic Materials", *Experimental Mechanics* **6**(9), 1966, pp. 470—480.

[19] Pindera, J. T. and Straka, P., "Responses of the Integrated Polariscope", *Journal of Strain Analysis* **8**(1), 1973, pp.65—76.

[20] Pindera, J. T. and Straka, P., "On Physical Measures of Rheological Responses of Some Materials in Wide Ranges of Temperature and Spectral Frequency", *Rheologica Acta* **13**(3), 1974, pp. 338—351.

[21] Pindera, J. T., "Response of Photoelastic Systems", *Transactions of The CSME* **2**(1), 1973—74, pp. 21—30.

[22] Pindera, J. T., "On Physical Basis of Modern Photoelasticity Techniques", *Beiträge zur Spannungs- und Dehnungsanalyse*, Vol. V, Akademie-Verlag, Berlin, 1968, pp. 103—130.

[23] Pindera, J. T. and Mazurkiewicz, S. B., "Photoelastic Isodynes: A New Type of Stress Modulated Light Intensity Distribution", *Mechanics Research Communications* **4**, 1977, pp. 247—252.

[24] Pindera, Jerzy, T., "Foundations of Experimental Mechanics: Principles of Modelling, Observation and Experimentation", in: Pindera, J. T. (Ed.), *New Physical Trends in Experimental Mechanics*, Springer-Verlag, Wien 1981, pp. 188—326.

[25] Pindera, J. T. and Mazurkiewicz, S. B., "Studies of Contact Problems Using Photoelastic Isodynes", *Experimental Mechanics* **21**(12), 1981, pp. 448—455.

[26] Pindera, J. T., "New Development in Photoelastic Studies: Isodyne and Gradient Photoelasticity", in: Chiang, F. P. (Ed.), Incoherent Optical Techniques in Experimental Mechanics, *Optical Engineering* **21**(4), 1982, pp. 672—678.

[27] Pindera, J. T., Hecker, F. W., and Krasnowski, B. R., "Gradient Photoelasticity", *Mechanics Research Communications* **9**(3), 1982, pp. 197—204.

[28] Pindera, J. T. and Krasnowski, B. R., *Theory of Elastic and Photoelastic Isodynes*, SMD-Paper No. 184, IEM-Paper No. 1. Solid Mechanics Division, University of Waterloo, October 1983, pp. 1—127.

[29] Pindera, J. T., "Isodyne Photoelasticity and Gradient Photoelasticity: Physical and Mathematical Models, Efficacy, Applications", *Mechanika Teoretyczna i Stosowana* (Journal of Theoretical and Applied Mechanics) **22**(1/2), 1984, pp. 53—68.

[30] Pindera, J. T. and Hecker, F. W., "Basic Theories and Experimental Techniques of the Strain-Gradient Method", *Experimental Mechanics* **27**(3), 1987, pp. 314—327.

[31] Pindera, J. T. and Sinha, N. K., "Determination of Dominant Radiation in Polariscope by Means of Babinet Compensator", *Experimental Mechanics* **12**(1), 1972, pp. 1—5.

[32] Post, D., "Optical Analysis of Photoelastic Polariscope", *Experimental Mechanics* **10**(1), 1970, pp. 15—23.

[33] Pirard, A., *La Photoélasticité*, Dunod, Paris, 1947.

[34] Ramachandran, G. N. and Ramaseshan, S., "Crystal Optics", in: Flugge, S. (Ed.), *Encyclopedia of Physics*, Vol. 25/1, Springer-Verlag, Berlin, pp. 1—217, 1961.

[35] Weller, R., Three-Dimensional Photoelasticity Using Scattered Light, *J. Appl. Phys.* **12**, 1941, 610—616.

[36] Young, M., *Optics and Lasers*, Springer-Verlag, New York, 1977.

Theory of optical isodynes

5.1 Concept of plane optical isodynes

A sufficiently general physical model of phenomena of importance to the theory of optical isodynes should contain one more component in addition to those listed in Section 4.3. This component would replace assumptions (4) and (5), and describe adequately the actually non-rectilinear flow of radiant energy through a stressed dielectric, using, for instance, the concept of the ray surface. However, the resulting relationships would be very cumbersome and would unnecessarily obscure the theory of the actual experiment in which, in contrast to some thought experiments, all physical quantities have finite, limited values.

Consequently, it is methodologically quite correct to analyse some particular cases of the general physical model, and then to discuss the limits of applicability of the derived relationships, as determined by the simplifying assumptions. The final relations derived on the basis of simplified assumptions will also be rigorous within the specified limits which depend on the required accuracy.

The theory of optical isodynes utilizes all the relations presented in Section 4.2 and 4.3. However, information on the relative retardation is collected in a manner different than in the case of the normal incidence transmission photoelasticity because, in this case, it is necessary to obtain information on the values of relative retardation at each point of the optical path of the light beams. This information can be obtained in several ways. At present, the simplest and most convenient method to determine the values of the relative retardation along the path of a light beam is based on the phenomenon of light scattering. The resulting relations are particularly simple when the actual light scattering in the given body can be approximated by the Rayleigh model of scattering. Only basic relations are presented below. More details are given in the pertinent literature [2, 9, 13–20]. The presented relations and techniques pertain only to linear elastic or linear viscoelastic states of deformation, therefore the word "elastic" in terms such as "elastic analytical isodynes" is often omitted.

5.1.1 Basic relations

The classical presentation of basic photoelastic relations between stresses and birefringence, equations (4.86)–(4.90), is not sufficiently explicit to clearly present the elementary model of optical isodynes. For the sake of clarity those relations are repeated here, in a more explicit form.

It is sufficient for the purpose of this work to represent the energy flow through a birefringent body, which is used to detect the birefringence and to carry information on the detected birefringence parameters, in terms of the model of a single frequency harmonic transversal wave represented by the classical electromagnetic theory of light, or even by the Fresnel wave model. As in classical transmission photoelasticity, the simplest techniques of optical isodynes use the dependence of the relative retardations between two linearly polarized wavefronts represented by the electric vectors $\bar{E}_1$ and $\bar{E}_2$ upon the related components of the stress tensor. These electric vectors are assumed to be collinear and to propagate rectilinearly, and to coincide with the principal directions.

It is known that the separation of harmonic wavefronts of radiation travelling through two bodies with the velocities $v_1 = c/n_1$ and $v_2 = c/n_2$ can be presented in terms of the difference between the optical paths. Thus,

$$dR_{1,2} = (n_1 - n_2)\, db, \tag{5.1}$$

where dR is linear separation of wavefronts propagating with velocities v_1 and v_2 produced on the interval db. The quantity R is denoted relative linear retardation.

Using the wave theory of light, the relative retardation also can be presented in terms of the phase angle ϕ:

$$dR_{1,2} = \lambda\, \frac{1}{2\pi}\, d\phi, \tag{5.2}$$

where λ is the corresponding wavelength.

Within the linear range of interaction between radiation and matter the principle of simple superposition is applicable so that any harmonic motion can be represented as a sum of partial harmonic motions, spatially and temporarily coherent, as illustrated by Figures 5.1, 5.2, 5.3 and 5.4. It follows from these figures that the state of polarization described by the azimuth of the axis of the ellipse of polarization and by the amount of the phase difference $\Delta\phi$ can be easily determined at each point of the path of a polarized light beam propagating inside a birefringent body if the body scatters the light in a manner which can be approximated by the Rayleigh model of scattering, relations (4.49)–(4.51). The obtained relations are further simplified when the amplitudes of the component radiation, E_x and E_y, at the beginning of the optical path are equal.

In section 4.2.2 the following relations are presented for the components of scattered light predicted by the Rayleigh model, Figure 4.1,

$$I_\Phi = K'E^2 \sin^2 \Phi \tag{4.53}$$
$$I_\theta = K'E^2 \cos^2 \Phi \cos^2 \theta \tag{4.54}$$

where Φ and θ denote the azimuthal angle and observation angle, respectively. It may be noted that the model presented by Weller [23] is not suitable as a theoretical foundation of optical isodynes.

Two major cases are discussed in the subsequent Sections, 5.1.2 and 5.1.3.

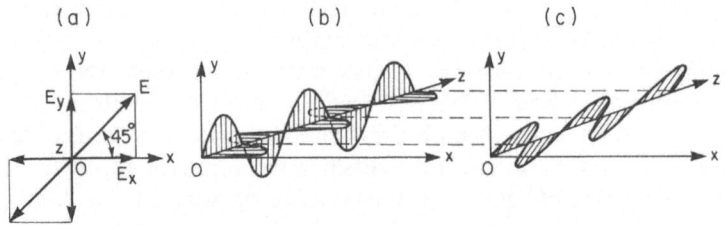

Fig. 5.1. Additive properties of simple harmonic motion (a) E-resultant vector; Ex, Ey-component vectors (b) propagation of component motions represented by vectors Ex and Ey; (c) propagation of resultant motion represented by vector E.

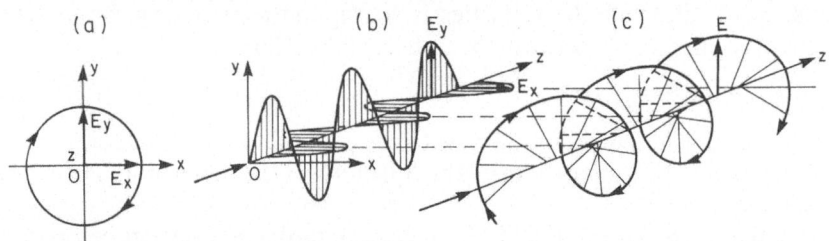

Fig. 5.2. Circularly polarized harmonic motion, which is represented by vectors E_x and E_y having phase difference of $\pi/2$. (a) — path of the vector end in (x, y) plane; (b) — propagation of component motions represented by vectors E_x and E_y; (c) — path of resultant vector in the (x, y, z) space.

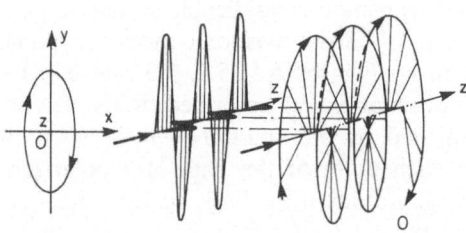

Fig. 5.3. Three equivalent presentations of an elliptically polarized harmonic motion.

5.1.2 *The case of the primary light beam propagating along a straight isostatic of a plane stress field*

In this case the unit vectors of both permitted wave fronts propagating coherently in a birefringent plate are collinear and are characterized by mutually perpendicular electric displacement vectors. Consequently, all the relations derived above can be applied rigorously. The displacement vector and the electric vector are collinear.

$$E_x = E_y \qquad 0 \le \phi \le 2\pi \quad \text{or} \quad 0 \le R \le \lambda$$

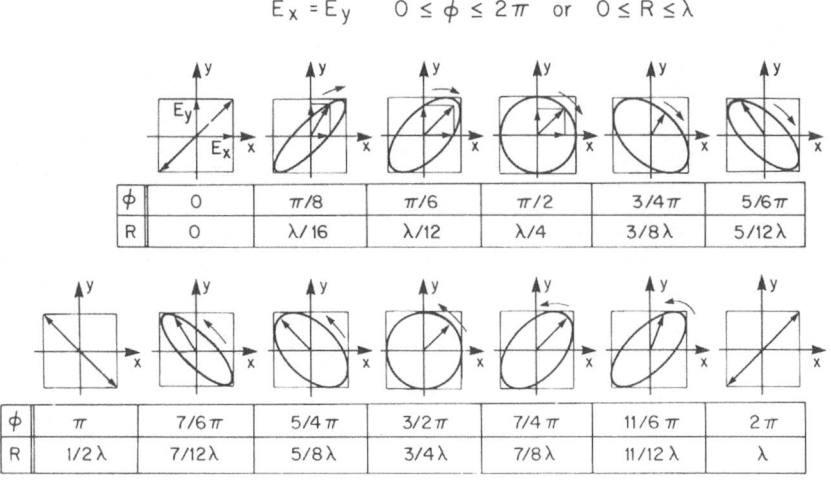

ϕ	0	$\pi/8$	$\pi/6$	$\pi/2$	$3/4\pi$	$5/6\pi$
R	0	$\lambda/16$	$\lambda/12$	$\lambda/4$	$3/8\lambda$	$5/12\lambda$

ϕ	π	$7/6\pi$	$5/4\pi$	$3/2\pi$	$7/4\pi$	$11/6\pi$	2π
R	$1/2\lambda$	$7/12\lambda$	$5/8\lambda$	$3/4\lambda$	$7/8\lambda$	$11/12\lambda$	λ

Fig. 5.4. Polarized harmonic motion resulting from superposition of two coherent and collinear motions represented by vectors E_x and E_y of equal amplitude and shifted in phase by $\Delta\phi$.

The amplitude and the azimuth of the vector $\overline{E}$ (describing the primary beam), or the amplitudes and relative phase difference of the component vectors E_1 and E_2 representing the primary beam as a source of scattered radiation are modulated by the stress components in the manner described by the formulae (4.86). Consequently, the distributions of the scattered-light intensities are modulated by the stress components acting along the path of the primary beam. It is convenient to analyse this modulation by using the same elementary technique [3, 4, 11, 22] which leads to the equations (4.92) and (4.98a, b). The assumed mathematical model, given in Figure 5.5 [9], represents a plate of thickness b subjected to a plane stress field (disk problem). The axis y along which the light beam propagates is an isostatic of the principal stress σ_2.

The physical modulation processes of the radiant energy represented by the mathematical models given in Figures 4.11 and 5.5 are closely related. Both mathematical models yield identical expressions for the output/input ratios of the monochromatic radiant energy if only the I_Φ component of the Rayleigh scattering is considered and when the angle β in Figure 4.1 corresponds formally to the angle of scattering Φ (azimuthal angle), Figure 4.1, since it is easy to show that

$$\frac{W(\lambda)\,d\lambda}{kKW_0(\lambda)\,d\lambda} = \frac{I_\Phi}{kKI_0} = \sin^2\Phi + \sin 2\alpha \sin 2(\alpha - \Phi)\sin^2\frac{\phi}{2}, \qquad (5.3a)$$

where

$$\frac{\phi}{2} = \pi m = \frac{\pi}{S_\sigma}\int_0^y (\sigma_1 - \sigma_3)\,dy, \qquad (5.3b)$$

according to relations (4.86) and (4.88), when the direction of σ_3 is constant.

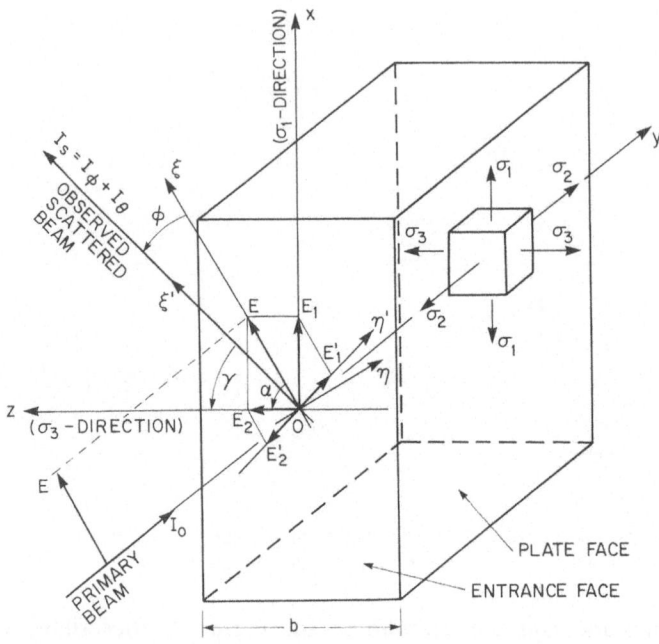

Fig. 5.5. Model of collecting information on the stress state in a plate along the path of the information-collecting light beam using the concept of optical isodynes.

Equation (5.3a), which is formally identical to equation (4.98a) does not contain any terms related to the second component of the scattered light represented by I_θ. This follows from the fact that I_θ is always less than 3 percent of I_Φ if θ is not less than 80°, and that it vanishes when θ is 90°. A generalization of equation (5.3) for an arbitrary value of θ is simple. Equation (5.3) represents a rigorous solution of an assumed mathematical model of Rayleigh scattering for $\theta = 90°$ and for a very narrow spectral band of radiation, such as that emitted by a single-mode helium-neon laser, when the radiant power is not too high.

It is useful to introduce the term "complementary radiation" for the scattered radiation propagating in a scattering plane (the "complementary scattering plane") normal to the scattering plane having the azimuth Φ. This complementary scattering plane has the azimuth $\Phi_c = \Phi + 90°$. The intensity of the complementary scatterd radiation is described by

$$\frac{W(\lambda)_c \, d\lambda}{kKW_0(\lambda) \, d\lambda} = \frac{I_{\Phi c}}{kKI_0} = \cos^2 \Phi - \sin 2\alpha \, \sin 2(\alpha - \Phi) \sin^2 \frac{\phi}{2}. \tag{5.4}$$

Obviously, the sum of the primary scattered radiation for $\theta = 90°$ and of the complementary scattered radiation is constant and independent of the relative retardation, ϕ or R or m:

$$I_{\Phi m} + I_{\Phi c} = kKI_0 \tag{5.5}$$

It follows from equations (5.3) and (5.4) that the intensity of the scattered radiation does not depend on the relative retardation if the scattering angle Φ is equal to the angle α between the plane of vibration of the electric vector of the incident primary beam and the principal direction.

For $\Phi = \alpha$

$$\frac{I_{\Phi m}}{kKI_0} = \sin^2 \Phi, \quad \text{and} \quad \frac{I_{\Phi c}}{kKI_0} = \cos^2 \Phi \tag{5.6}$$

For $\Phi = \alpha = 45°$

$$I_{\Phi m} = I_{\Phi c} = \frac{1}{2} kKI_0. \tag{5.7}$$

In practical applications, it is more convenient to relate the scattering plane to one of the principal directions by introducing the angle γ (Figure 5.5):

$$\Phi = \alpha - \gamma. \tag{5.8}$$

In this case, equations (5.3) and (5.4) become

$$\frac{I_{\Phi m}}{kKI_0} = \sin^2(\alpha - \gamma) + \sin 2\alpha \sin 2\gamma \sin^2 \frac{\phi}{2}, \tag{5.9}$$

and

$$\frac{I_{\Phi c}}{kKI_0} = \cos^2(\alpha - \gamma) - \sin 2\alpha \sin 2\gamma \sin^2 \frac{\phi}{2}. \tag{5.10}$$

Hence, the optical compensation, i.e. the introduction of a known additional retardation ΔR or $\Delta \phi$ between two wave fronts vibrating in the planes determined by the principal directions, can yield additional useful information on the stress state at a chosen point, see Section 5.1.3. Additional retardation $\Delta \phi$ can be introduced either by altering the magnitude of the acting loads or by producing retardation in a compensator located in the path of the primary beam I_0. The corresponding relations are

$$\frac{I_{\Phi m}}{kKI_0} = \sin^2(\alpha - \gamma) + \sin 2\gamma \sin^2 \frac{\phi + \Delta\phi}{2}$$

$$= \frac{1}{2} - \frac{1}{2} \cos 2\alpha \cos 2\gamma - \frac{1}{2} \sin 2\alpha \sin 2\gamma \cos(\phi + \Delta\phi); \tag{5.11}$$

and

$$\frac{I_{\Phi c}}{kKI_0} = \cos^2(\alpha - \gamma) - \sin 2\alpha \sin 2\gamma \sin^2 \frac{\phi + \Delta\phi}{2}$$

$$= \frac{1}{2} + \frac{1}{2} \cos 2\alpha \cos 2\gamma + \frac{1}{2} \sin 2\alpha \sin 2\gamma \cos(\phi + \Delta\phi). \tag{5.12}$$

The signal-to-noise ratio in the system can be optimized by choosing $\alpha = 45°$, so that

$$\frac{I_{\Phi m}}{kKI_0} = \frac{1}{2} - \frac{1}{2} \sin 2\gamma \cos(\phi + \Delta\phi), \tag{5.13}$$

and

$$\frac{I_{\Phi c}}{kKI_0} = \frac{1}{2} + \frac{1}{2} \sin 2\gamma \cos(\phi + \Delta\phi). \tag{5.14}$$

Extrema of these functions occur at $\alpha = 45°$ and $\gamma = 45°$, yielding

$$I^n_{\Phi m} = \frac{I_{\Phi m}}{kKI_0} = \frac{1}{2} - \frac{1}{2} \cos(\phi + \Delta\phi), \tag{5.15}$$

which is called the major scattered beam, and

$$I^n_{\Phi c} = \frac{I_{\Phi c}}{kKI_0} = \frac{1}{2} + \frac{1}{2} \cos(\phi + \Delta\phi) \tag{5.16}$$

which is called the complementary major scattered beam.

It follows from (5.13) and (5.14) that for $\alpha = 45°$ and $\gamma = 0°$

$$I^n_\Phi = \frac{I_\Phi}{kKI_0} = \frac{1}{2}. \tag{5.17}$$

Thus the intensity of light scattered in direction normal to the plate face is constant along the path of the primary beam. This is a very important property of the basic model of optical isodynes.

For a circularly polarized primary beam, equations (5.11) and (5.12) become:

$$\frac{I_{\Phi m}}{kKI_0} = \frac{1}{2} - \frac{1}{2} \sin 2\gamma \cos(\phi + \Delta\phi) \tag{5.18}$$

and

$$\frac{I_{\Phi c}}{kKI_0} = \frac{1}{2} + \frac{1}{2} \sin 2\gamma \cos(\phi + \Delta\phi) \tag{5.19}$$

where

$$\phi = \phi(y) = 2\pi m = 2\pi \frac{C_\sigma}{\lambda} \int_0^y (\sigma_1 - \sigma_3) \, dy, \tag{5.20}$$

and y is the path of propagation of the primary beam. Hence for a circularly polarized primary beam, the optimum signal-to-noise ratio occurs when γ is 45°, as is the case for the linearly polarized primary beam. Consequently,

$$I^n_{\Phi m} = \frac{I_\Phi}{kKI_0} = \frac{1}{2} - \frac{1}{2} \cos(\phi + \Delta\phi), \tag{5.21}$$

and

$$I^n_{\Phi c} = \frac{I_{\Phi c}}{kKI_0} = \frac{1}{2} + \frac{1}{2} \cos(\phi + \Delta\phi). \tag{5.22}$$

Thus from the practical point of view, circularly polarized radiation is equivalent to linearly polarized radiation when $\alpha = 45°$.

All of the above relations are derived for the case where the scattered light beams are not modulated along their path. This occurs within a two-dimensional stress field, as presented in Figure 5.5, for the observation angle $\theta = 90°$

when the plane of vibration of the scattered beam contains two principal directions, along the path of the scattered light between the scattering point and the surface of the birefringent plate. The conditions at the surface, i.e. the mode of transfer of radiant power across the boundary, must be analysed separately.

Summarizing, three scattered light beams, each carrying particular information, can be selected:

— The reference scattered beam, which carries information on the location of the primary beam,

$$\gamma = 0; I_{\Phi r} = \frac{1}{2} kKI_0;$$ (5.23)

— The major scattered beam,

$$\gamma_m = +45°; I_{\Phi m} = \frac{1}{2} kKI_0[1 - \cos(\phi + \Delta\phi)];$$ (5.24a)

— The complementary major scattered beam,

$$\gamma_c = -45°; I_{\Phi c} = \frac{1}{2} kKI_0[1 + \cos(\phi + \Delta\phi)].$$ (5.24b)

All three beams are modulated when they pass the boundaries between different media. This modulation may influence the amplitudes of the related electric vectors and their phases.

It should be mentioned that the relations presented in this section are rigorous only within the framework of a disk problem when the primary light beam propagates along a straight isostatics. In such cases, when the principal stress component normal to the disk surface is equal to zero, $\sigma_3 = 0$, and the additional retardation is equal to zero, $\Delta\phi = 0$, equations (5.15), (5.16) and (5.3b) consequently yield:

$$I_{\Phi m}^n = \frac{I_{\Phi m}}{kKI_0} = \frac{1}{2} - \frac{1}{2} \cos\left(\frac{2\pi}{S_\sigma} \int_0^y \sigma_1 \, dy\right)$$ (5.25a)

and

$$I_{\Phi c}^n = \frac{I_{\Phi c}}{kKI_0} = \frac{1}{2} + \frac{1}{2} \cos\left(\frac{2\pi}{S_\sigma} \int_0^y \sigma_1 \, dy\right)$$ (5.25b)

Relations (5.25) can be used to determine the values of the principal stress component normal to the isostatic at each point along the light path.

Summarizing, three particular sheets of light scattered along the path of the primary light beam propagating through a plate defined above deserve particular attention from the point of view of the analysis of plane stress states. The intensities of those beams are presented by relations normalized with respect to the primary beam I_0.

The reference scattered beam, $\gamma = 0°$:

$$I_{sr}^n = \frac{I_{sr}}{kKI_0} = \frac{I_{\Phi m} + I_{\Phi c}}{kKI_0} = \text{const.} = 1.$$ (5.26)

The major scattered beam, $\gamma = 45°$:

$$I^n_{sm} = \frac{I_{\Phi m}}{kKI_0} = \frac{1}{2}[1 - \cos(\phi + \Delta\phi)] = \sin^2\frac{\phi + \Delta\phi}{2}$$

$$= \sin^2\pi(m_s + \Delta m_s). \tag{5.27}$$

The complementary major scattered beam, $\gamma = -45°$:

$$I^n_{sc} = \frac{I_{\Phi c}}{kKI_0} = \frac{1}{2}[1 + \cos(\phi + \Delta\phi)] = \cos^2\frac{\phi + \Delta\phi}{2}$$

$$= \cos^2\pi(m_s + \Delta m_s), \tag{5.28}$$

where m_s denotes the normalized relative retardation detected using scattered light; Δm, or $\Delta\phi$ denotes an arbitrarily induced addition retardation, or birefringence, which may be useful in practical application of relations (5.22), (5.23), (5.24).

When the primary beam propagates through a plate whose birefringence is caused by a plane homogeneous stress field, and when the direction of propagation of the primary beam is collinear with one of the princpal directions of the stress/strain tensor, Figure 5.5, for instance the y direction, the information carried by the major scattered beams is related to,

$$dR_{1,3} = (n_1 - n_3)\,dy = \lambda\,dm_s = \lambda\,\frac{1}{2\pi}\,d\phi$$

$$= C_\sigma(\sigma_1 - \sigma_3)\,dy = S_\sigma^{-1}\lambda(\sigma_1 - \sigma_3)\,dy$$

$$= C_\varepsilon(\varepsilon_1 - \varepsilon_3)\,dy = S_\varepsilon^{-1}\lambda(\varepsilon_1 - \varepsilon_3)\,dy = \lambda\,dm_s. \tag{5.29}$$

In other words, the intensity of major light sheets scattered along the path of a primary beam travelling in one principal direction is modulated by the difference of the principal stress components acting in the plane normal to the direction of propagation of the primary beam.

It is easy to show that the function

$$m_s = m_s(s) \tag{5.30}$$

where s is one of the principal directions of a plane stress state normal to the directions (1) and (2) of the stress tensor is identical with:

— differential isodyne function in the s-characteristic direction when the stress state is quasi-plane, that is when $\sigma_3 \ll \sigma_1, \sigma_2$, or

$$S_\sigma\frac{dm_s}{ds} = \sigma_1 - \sigma_3; \tag{5.31}$$

— plane isodyne function in the s-characteristic direction, when the stress state is two-dimensional,

$$S_\sigma\frac{dm_s}{ds} = \sigma_1. \tag{5.32}$$

Thus the light intensity distribution of the scattered light sheets I_{sm} and I_{sc} corresponds to the isodyne functions along the characteristic direction given by the direction of propagation of the primary beam, I_0. Accordingly, they are called optical isodyne functions, and are represented by a particular modulation of the intensities of the major scattered light beams or sheets.

To record optical isodyne functions, it is necessary to collimate the scattered light sheets I_{sr}, I_{sm} and I_{sc}. This can be done in various manners using the so-called isodyne collector. One of the possible technical solutions based on the application of the integrating prism is presented in Figure 5.6. Parameters of the corresponding optical transfer functions of the integrating prisms are specified in Figure 5.7. An example of an integrated photoelastic measurement using the isochromatics produced by the transmission photoelasticity and the optical isodyne function along an isostatics is given in Figure 5.8. A sample of

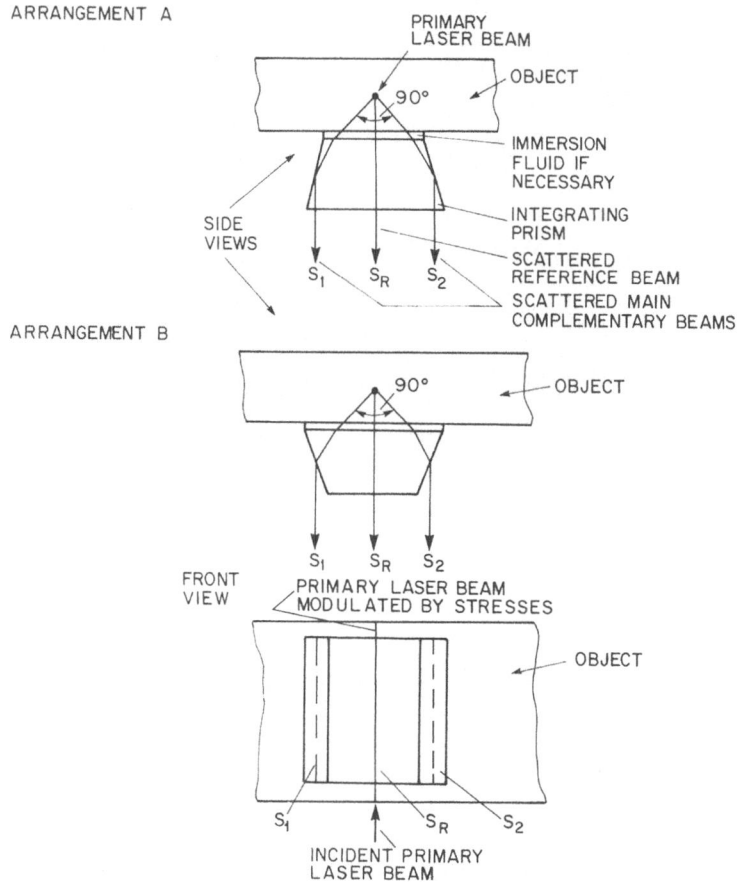

Fig. 5.6. Simple version of the isodyne collector: integrating prism. Arrangement A, and Arrangement B.

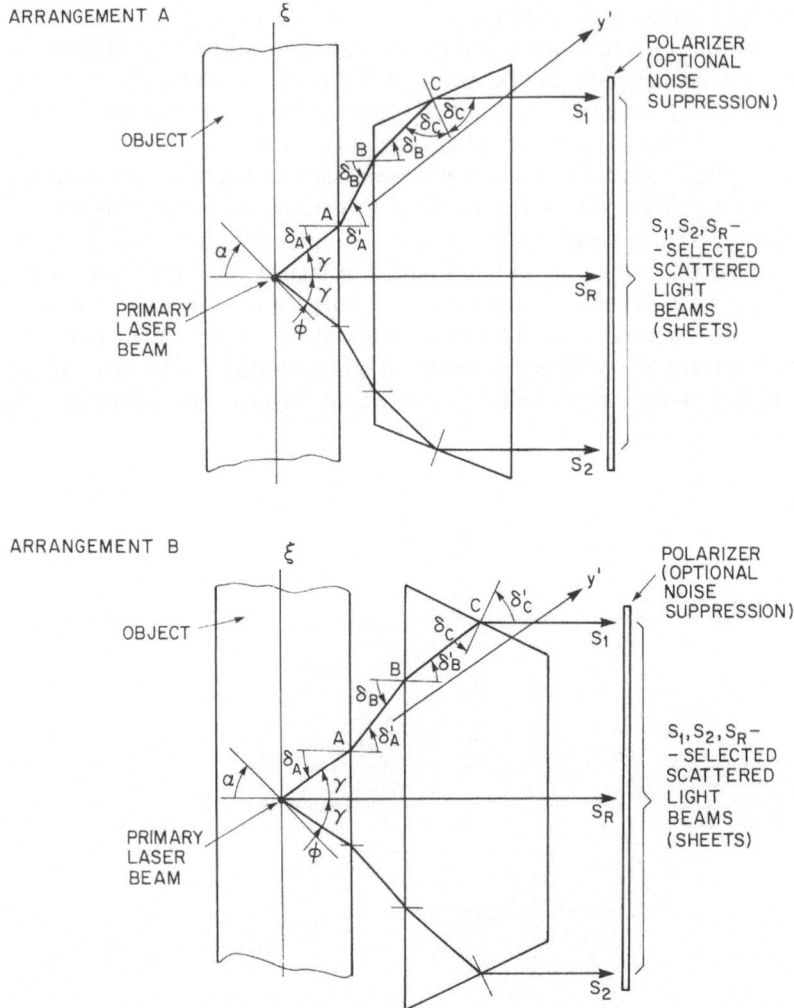

Fig. 5.7. Parameters of the optical transfer functions of integrating prisms of the isodyne collector. Arrangement A, and Arrangement B.

corresponding recordings of the optical isodyne functions is presented in Figure 5.9. A typical example of the integrated photoelastic recording is given in Figure 5.10.

The problem of the reliability limits of optical isodyne functions is illustrated by the example given in Figure 5.11. The stress state in a circular disk in the regions of contact forces is clearly three-dimensional and so the recorded optical isodyne functions represent differential isodyne functions in those regions. In the regions sufficiently remote from the contact forces the stress field is close to a plane stress field and thus the recorded isodyne functions represent plane analytical elastic isodyne functions.

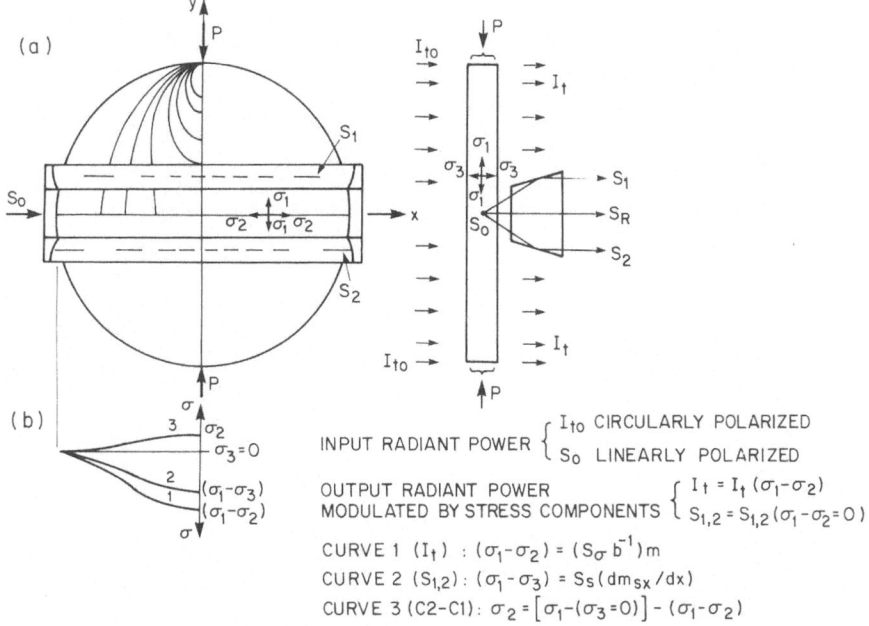

Fig. 5.8. Schematic presentation of evaluation of stress components in a circular disk loaded diametrically by using the integrated photoelastic technique. (a) Sketch of the system. (b) Samples of evaluated stress components along horizontal diameter.

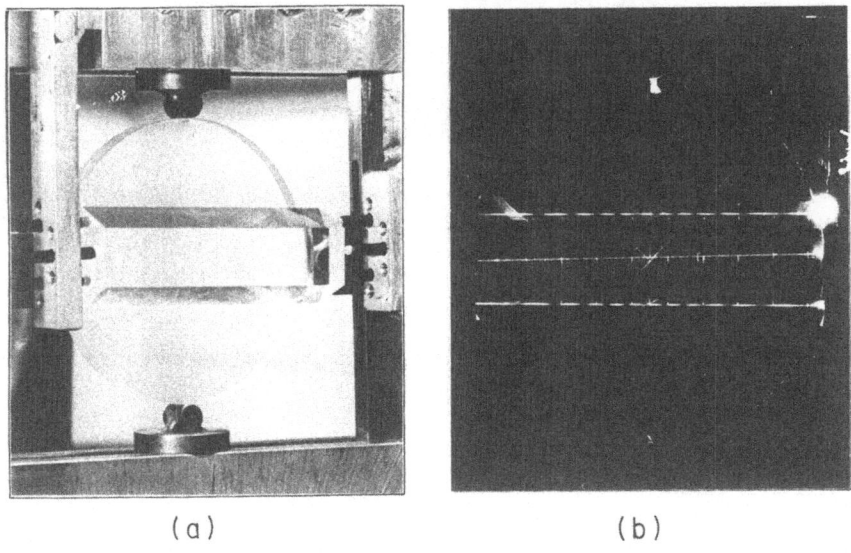

(a)   (b)

Fig. 5.9. Sample of recording of an isodyne function by means of an integrating prism as in Figure 5.8. (a) View of the object with integrating prism. (b) Photographic recordings: the scattered reference light sheet (in the middle) determines the location of the primary beam; the scattered complementary major light sheets carry information on the isodyne function and, consequently, on the normal stress components in the planes normal to the primary beam.

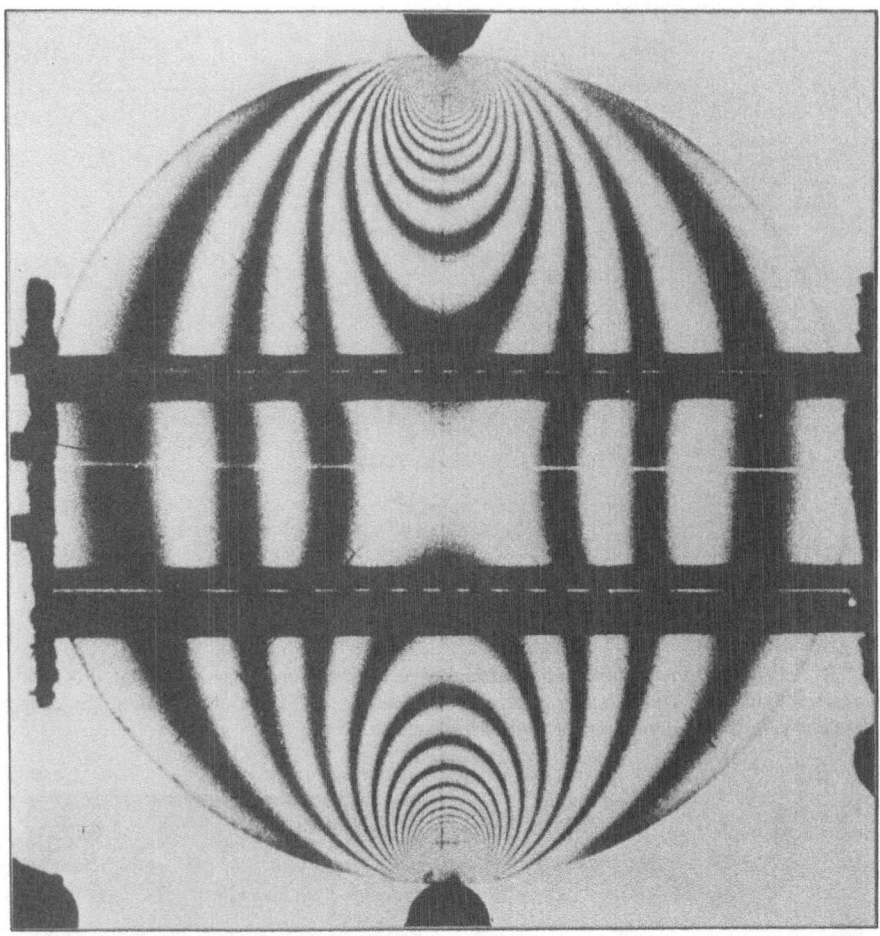

Fig. 5.10. Samples of integrated photoelastic recordings. Circular disk loaded along vertical diameter. Isodyne function is determined in the cross-section along the horizontal diameter of the disk where stress gradients are low.

5.1.3 *The general case of the primary light beam propagating along an arbitrary direction in a disk, parallel to its midplane*

This is a general case of stress analysis of a plane stress field. In this case the unit vectors of both permitted primary wave fronts propagating coherently in a birefringent body are not necessarily collinear, which leads to the spatial separation of the corresponding primary light beams. This separation, however, occurs only within the planes parallel to the middle plane of the disk if the stress state is quasi-plane.

In this general case, two mechanisms influence the propagation of primary light beams:

Fig. 5.11. An example of an integrated photoelastic recording. Circular disk loaded along vertical and horizontal diameters. The isodyne function is determined in a cross-section where stress gradients are high locally.

(a) the double refraction effect caused by the oblique incidence mechanism (the primary beam is generally not collinear with the principal directions of the stress, dielectric, and index tensors) which causes spatial separation of both partial (coherent) primary beams; Figure 4.6.

(b) the strain/stress gradient effect resulting from the inhomogeneity of the strain/stress field, which leads to the bending of both related primary beams; this bending occurs in the plane parallel to the disk's middle plane [21].

Both these effects can be easily observed and measured. Rigorous solution for such a state of birefringence and light interference produced by an unknown plane stress state is practically impossible to obtain using the presently available models and analytical methods. However, according to the obtained empirical data, the influence of these effects on the state of birefringence produced by typical plane stress fields can be neglected if the length of the path of primary beams is less than 300 mm, and when the light beam is characterized by the following parameters: diameter 0.6 mm, divergence 1 mrad, wavelength 632 nm. Within this range of physical parameters the diameter of measurement cylinder (portion of space in which information is collected and averaged) increases from 0.1 mm to 0.3 mm maximum, and the noise level is still acceptable. The noise level itself is an indicator of the degree of separation of both related beams. Both these effects can be utilized to identify regions of high stress gradients and high inhomogeneity of stress field.

Obviously, relation (5.3b) is not rigorously applicable in such a case. To relate the stress induced birefringence to the stress state, use is made of the concept and mathematical model of the secondary principal stresses [1]. According to this model, the maximal and minimal normal stress components occurring in the cross-section normal to the chosen direction y are called secondary principal stresses, σ_1^y and σ_3^y, and the birefringence produced along y is related to the secondary principal stresses in the same manner as described by relation (4.86):

$$(dR_{1,3})_y = \lambda(dm_{1,3})_y = \lambda \, \frac{1}{2\pi} \, (d\phi_{1,3})y = (n_1 - n_3)_y \, dy$$

$$= C_\sigma(\sigma_1^y - \sigma_3^y) \, dy = C_\varepsilon(\varepsilon_1^y - \varepsilon_3^y) \, dy, \text{ etc.} \tag{5.33}$$

Consequently, relation (5.3b) can be generalized as follows:

$$\left(\frac{\phi}{2}\right)_y = \pi m_y = \frac{\pi}{S_\sigma} \int_0^y (\sigma_1^y - \sigma_3^y) \, dy, \tag{5.34}$$

For the case presented in Figure 5.12, where $\sigma_3 = 0$, relations (5.25) can be generalized in the same manner as above:

$$\frac{I_{\Phi m}}{kKI_0} = I_{sy1}^n = \frac{1}{2} - \frac{1}{2} \cos\left(\frac{2\pi}{S_\sigma} \int_0^y \sigma_{xx} \, dy\right), \tag{5.35a}$$

and

$$\frac{I_{\Phi c}}{kKI_0} = I_{sy2}^n = \frac{1}{2} + \frac{1}{2} \cos\left(\frac{2\pi}{S_\sigma} \int_0^y \sigma_{xx} \, dy\right), \tag{5.35b}$$

where I_{s1y}^n and I_{s2y}^n denote normalized light intensities of the major and complementary scattered light sheets along the path of the primary beam I_{oy}.

It is convenient to present relations (5.35b) in terms of normalized linear retardation m_s indicated by the scattered light, which is numerically identical with the normalized linear retardation m,

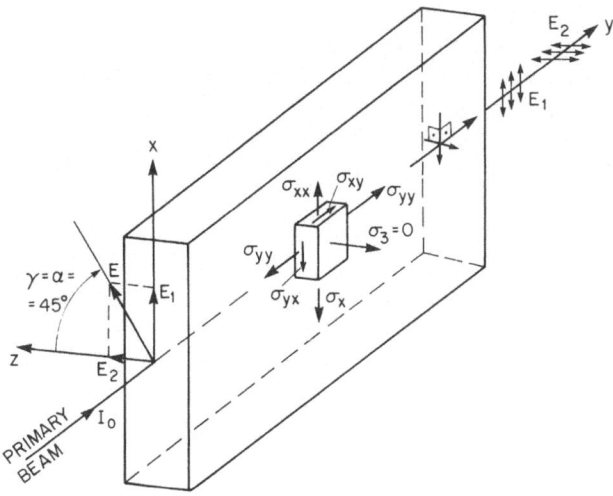

Fig. 5.12. Propagation of a primary beam I_0 in an arbitrary direction through a plane stress field in a plate.

$$I_{syl}^n = \frac{1}{2} - \frac{1}{2} \cos\left(\frac{2\pi}{S_\sigma} \int_0^y \sigma_{xx}\, dy\right) = \frac{1}{2} - \frac{1}{2} \cos\left(2\pi \int_0^m dm_{syy}\right)$$

$$= \frac{1}{2} - \frac{1}{2} \cos(2\pi m_{syy}) = \sin^2 \pi m_{syy}; \tag{5.36a}$$

and

$$I_{sy2}^n = \frac{1}{2} + \frac{1}{2} \cos(2\pi m_{syy}) = \cos^2 \pi m_{syy}, \tag{5.36b}$$

where m_{syy} denotes the normalized linear retardation of the major scattered beam produced by the primary beam I_{0y}, and measured in the y direction, and the subscripts 1 and 2 denote the major and complementary scattered light.

Equations (5.36a, b) represent distributions of the normalized intensity of light scattered along the path of primary beam propagating along a chosen line in an arbitrary direction y, when at each scattering point the observation angle θ is 90° and the aximuthal angle Φ is equal to 0° or 90°, respectively. The arbitrary direction y is called the characteristic direction and the particular lines collinear with the characteristic direction are called the characteristic lines. Thus,

$$m_{syy} = m_{syy}(x = x_0, y), \tag{5.37a}$$

and analogously, for the characteristic direction $y = y_0$,

$$m_{sxx} = m_{sxx}(x, y = y_0). \tag{5.37b}$$

The functions m_{sxx} and m_{syy} unequivocally describe the distribution and values of the normal stress components σ_{yy} and σ_{xx} along the corresponding characteristic lines. The values of m_{sxx} or m_{syy} are proportional to the intensities of total normal forces acting on the corresponding sections of the characteristic lines, since, according to (5.3b), for the y-characteristic direction,

$$\int_0^y \sigma_{xx}\,\mathrm{d}y = S_\sigma m_{syy}(x = x_0, y) = p_x(x = x_0, y), \tag{5.38a}$$

and analogously, for the x-characteristic direction,

$$\int_0^x \sigma_{yy}\,\mathrm{d}x = S_\sigma m_{sxx}(x, y = y_0) = p_y(x, y = y_0), \tag{5.38b}$$

where:

$$\sigma_{xx} = S_\sigma\,\frac{\mathrm{d}}{\mathrm{d}y}\,m_{syy}, \tag{5.39a}$$

and

$$\sigma_{yy} = S_\sigma\,\frac{\mathrm{d}}{\mathrm{d}x}\,m_{sxx}. \tag{5.39b}$$

Thus particular values of m_{sxx} or m_{syy} are proportional to particular values of normal force intensities p_x and p_y.

When the primary light beam propagating in a chosen characteristic direction is displaced normally to this direction and in the plane parallel to the disk's middle plane, the functions

$$I_{sy1}^n = I_{sy1}^n(x = x_0, y), \tag{5.40a}$$

and

$$I_{sy2}^n = I_{sy2}^n(x, y = y_0), \tag{5.40b}$$

become continuous and differentiable functions of two independent variables, x and y,

$$I_{sy1}^n = I_{sy1}^n(x, y), \tag{5.41a}$$

and

$$I_{sy2}^n = I_{sy2}^n(x, y). \tag{5.41b}$$

Thus, it is possible to identify within the region (x, y) bounded by the disk boundary s lines of constant values of the light intensities I_{sx}^n and I_{sy}^n, which are lines of constant values of the normalized relative retardation m_s,

$$I_{sy}^n = I_{sy}^n(x, y) = \text{const}, \tag{5.42a}$$

or

$$m_{sy}(x, y) = \text{const} \tag{5.42b}$$

and

$$I_{sx}^n = I_{sx}^n(x, y) = \text{const}, \tag{5.43a}$$

or

$$m_{sx}(x, y) = \text{const}. \tag{5.43b}$$

According to relations (5.38a, b), these lines are geometric loci of points at which the intensity of normal forces acting on the corresponding sections of the characteristic lines are constant. Such lines were called plane elastic analytical isodynes in Chapter 3 of this book.

Generally, the functions (5.42) and (5.43) represent surfaces above the plane of measurements which are defined by two directions, the characteristic direction and the scanning direction, and which are bounded by the boundary of the disk. These surfaces, each of which is related to a particular direction, are called isodyne surfaces. The isodyne surfaces represent fields of total normal force intensities acting on the corresponding characteristic sections. The isodyne surfaces can be spanned over the surfaces:

$$(x, y, z_i)\text{-planes, where } z_{min} \leqslant z_i < z_{max}; \tag{5.44a}$$

$$(x, z, y_i)\text{-planes, where } y_{min} \leqslant y_i \leqslant y_{max}; \tag{5.44b}$$

$$(y, z, x_i)\text{-planes, where } x_{min} \leqslant x_i \leqslant x_{max}. \tag{5.44c}$$

Consequently, cross-sections through isodyne surfaces with the planes normal to measurement planes and parallel to characteristic directions give isodyne functions in the chosen cross-section. Thus relations (5.37a, b) represent such cross-sections through isodyne surfaces.

The cross-sections through the isodyne surfaces with the planes parallel to the characteristic plane give isodynes $m_{sx}(x, y) = \text{const.}$, (5.42b), and isodynes $m_{sy}(x, y) = \text{const.}$, (5.43b). Projection of the isodynes m_{sx} and m_{sy} on the characteristic plane yields isodyne lines.

It should be noted that the values of isodynes, m_s, represent the values of corresponding total normal force intensities normalized with respect to the value of the stress optic coefficient S_σ:

$$m_{sx} = \frac{p_x}{S_\sigma} = p_{xn} \tag{5.45a}$$

$$m_{sy} = \frac{p_y}{S_\sigma} = p_{yn} \tag{5.45b}$$

These values will be called orders of isodynes.

The normalized light intensity functions (5.36a, b) are cyclic functions of m_s, vary from 0 to 1 and are extremal when m_s or $2m_s$ are integers, respectively. Consequently, it often is convenient to present the isodyne orders m_s (m_{sx} or m_{sy}) in the form,

$$m_s = M_s + \Delta M_s \tag{5.46}$$

where M_s is an integer.

Such isodyne surfaces as those described above are called optical isodyne surfaces. To preserve the consistency of the analytical description of the properties of isodyne surfaces and related functions, the symbols m_{sx} and m_{sy} denote the isodyne orders with respect to the isodyne surfaces, whereas the symbols m_{sxx} and m_{syy} denote the isodyne functions in the characteristic directions. Thus m_{sxx} denotes the cross-sections of the x-isodyne surface with the planes (x, z) for an arbitrary y.

It should be noted that the normalized light intensity functions (5.42a, b) only yield direct information on the absolute values of ΔM_s, and information whether or not two adjacent isodynes are of the same order or differ by one.

5.1.4 Characteristic points of optical isodyne fields: negative and positive sources, zero-order points

It has been mentioned in Section 5.1 that the introduction of a known additional retardation between two related wavefronts (see relations (5.3b), (5.11) and (5.12)) can yield useful information on the stress state when the primary beam is collinear with an isostatics. It follows from (5.15) and (5.16) that an addition retardation ΔR or $\Delta \phi$ causes the displacement of isodynes in a very particular fashion: the isodynes are displaced in such a manner that at certain particular points isodynes emerge or disappear. It is convenient to call such points sources, and to distinguish between positive sources where the isodynes emerge, and negative sources where the isodynes disappear, for a positive ΔR or $\Delta \phi$.

Additional retardation ΔR or $\Delta \phi$ can be produced either by introducing it optically, before the entrance point of the primary beam, or mechanically by changing the loading forces. In the first case ΔR is constant along the path of the beam and in the second case ΔR depends on the stress state, $\Delta R = \Delta R(x, y)$.

It can easily be shown that relations (5.15) and (5.16) are applicable for arbitrary characteristic directions under the same constraints as those given in Section 5.13. That is, relation (5.3b) can be replaced by relation (5.34). Thus, relations (5.36) can be presented conveniently in a general form valid for any characteristic direction:

$$I_{sy1}^n = \frac{1}{2} [1 - \cos(\phi + \Delta\phi)] = \frac{1}{2} \left[1 - \cos \frac{2\pi}{\lambda} (R + \Delta R) \right]$$

$$= \frac{1}{2} [1 - \cos 2\pi(m + \Delta m)] = \sin^2 \pi(m_s + \Delta m_s), \tag{5.47a}$$

$$I_{sy2}^n = \frac{1}{2} [1 + \cos(\phi + \Delta\phi)] = \frac{1}{2} \left[1 + \cos \frac{2\pi}{\lambda} (R + \Delta R) \right]$$

$$= \frac{1}{2} [1 + \cos 2\pi(m + \Delta m)] = \cos^2 \pi(m_s + \Delta m_s). \tag{5.47b}$$

When the additional retardation is optically (externally) produced, that is when

$$\Delta\phi = \frac{2\pi}{\lambda} \Delta R = \text{const.},$$

relations (5.47) lead to the expressions:

$$I_{sy1}^{n} = \frac{1}{2} \left\{ 1 - \cos \left[\frac{2\pi}{S_\sigma} \int_0^y \sigma_{xx} \, dy + \Delta\phi \right] \right\}, \tag{5.48a}$$

and

$$I_{sy2}^{n} = \frac{1}{2} \left\{ 1 + \cos \left[\frac{2\pi}{S_\sigma} \int_0^y \sigma_{xx} \, dy + \Delta\phi \right] \right\}. \tag{5.48b}$$

When the additional retardation is mechanically (internally) produced by change of the load level,

$$\Delta\phi = \Delta\phi(x, y) = \frac{2\pi}{\lambda} \Delta R(x, y),$$

relations (5.47) yield:

$$I_{sy1}^{n} = \frac{1}{2} \left\{ 1 - \cos \left[\frac{2\pi}{S_\sigma} \int_0^y (\sigma_{xx} + \Delta\sigma_{xx}) \, dy \right] \right\}, \tag{5.49a}$$

and

$$I_{sy2}^{n} = \frac{1}{2} \left\{ 1 + \cos \left[\frac{2\pi}{S_\sigma} \int_0^y (\sigma_{xx} + \Delta\sigma_{xx}) \, dy \right] \right\}. \tag{5.49b}$$

Obviously, when at a particular point (x_0, y_0) the normal stress σ_{xx} is equal to zero, $\Delta\sigma_{xx}$ is equal to zero at this point too.

In addition, for the zero-order isodyne:

$$\int_0^y \sigma_{xx} \, dy = \int_0^y (\sigma_{xx} + \Delta\sigma_{xx}) \, dy = 0. \tag{5.50}$$

Relation (5.50) illustrates the method of identifying zero-order isodynes.

Summarizing (see Figure 5.13):

Using the external retardation by changing the additional retardation, it is easy to determine the location of positive and negative sources: isodynes emerge from positive sources and disappear in negative sources.

Using the additional internal retardation by changing the load level, it is easy to determine the location of the sources: isodynes emerge from the sources with the increasing load. In addition it is easy to determine the location of the zero-order isodyne: this isodyne does not move when the load changes.

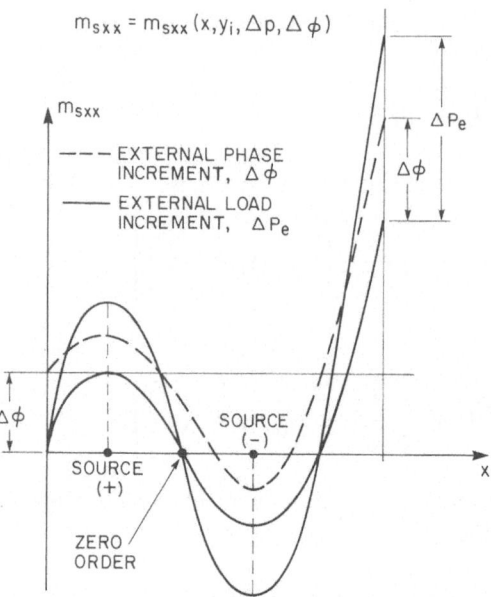

Fig. 5.13. Two techniques for determination of the characteristic points of the plane optical isodyne fields such as positive sources, negative sources, zero-order points: the technique of the initial phase increment $\Delta\phi$, and the technique of the external load increment ΔPe.

5.1.5 Relations between plane elastic analytical isodynes and plane elastic optical isodynes

It has been shown that, in general, an elastic isodyne field is represented by a surface called here the elastic isodyne surface. This surface is spanned over the reference plane (x, y) called characteristic plane, it is bounded by the plate boundary s, and, in general, its ordinates at the boundary are not equal to zero. Isodynes.are the lines on the isodyne surface having the same distance from the characteristic plane. The distance between the characteristic plane and the isodyne surface is the order of the isodyne.

The elastic analytical and optical isodyne surfaces can easily be presented within the reference plane (x, y) by the normal projection of the isodyne lines,

$$m_s = m_s(x, y) = \text{const.} = (m_s)_c$$

where the constant $(m_s)_c$ is conveniently represented by natural numbers and is called the isodyne order.

Since the function $m = m(x, y)$ is a single-valued function, it can be unequivocally represented by the field of $m(x, y) = \text{const.}$ in a plane parallel to the (x, y)-plane. Such an elastic analytical isodyne field is called the isodyne plane. The elastic isodyne lines within the elastic isodyne plane are geometrically identical with the elastic isodyne lines on the elastic isodyne surface. The order of the elastic isodyne lines within the elastic isodyne plane is unequivocally determined by the positive and negative sources and ridges, and by the isodyne order at the entrance boundary.

It follows from the relations presented in Sections 3.4.1 and 3.4.2, and in Sections 5.1.3 and 5.1.4 that the plane elastic analytical and plane elastic optical isodynes are similar. Taking as criteria of identity the identity of the characteristic points of plane elastic isodyne fields and the identity of the results yielded by isodynes with regard to normal stress components, one can readily show that the plane elastic analytical isodynes and optical isodynes are, in fact, identical. Consequently, all the conclusions following from the theory of the analytical plane elastic isodynes are applicable when evaluating the information carried by optical plane elastic isodynes, for instance regarding determination of the shear stress component (3.10a, b) of the stress and strain tensors (3.51) etc.

5.1.6 Discussion of models. Range of validity of derived relations. Criteria of validity

The relations for analytical isodyne fields are rigorous within the framework of a linear theory applied to a plate of an infinitesimal thickness when all deformations are infinitesimal. On the other hand the relations for optical isodynes were derived for plates of finite thickness under several assumptions involving:

— applicability of plane linear elasticity solutions to plates of finite thickness;
— applicability of the concept of the secondary principal stresses;
— assumption that light propagates rectilinearly in stressed bodies;
— assumption that the wave front normals of both conjugated components of the light beam, ordinary and extraordinary, are collinear;
— assumption that the phenomenological model of stress-induced birefringence given by Ramachandran and Ramaseshan is applicable, including the common simplifications.

It is known that these assumptions are arbitrary to some as yet unspecified extent, since no rational analysis of the discrepancies between the predictions of the simplified relations and those of rigorous theories is possible at the present time. This is due to the lack of pertinent theories based on actual physical reality.

In this situation it is necessary to establish criteria of validity of derived relations when these relations are applied to the determination of stress/strain components in real plates of finite thickness, loaded by real forces, under real boundary conditions, and at real finite deformations. The influence of the following real phenomena appear to be the most important when plate problems are considered:

— Actual three dimensional stress states existing in regions of notches, cracks, concentrated loads, etc., which are commonly denoted by the term "local effects".
— Actual, geometric nonlinearities caused by high strain gradients.

- Actual local and global large deformations (finite elasticity problem).
- Creep and creep recovery, and relaxation and relaxation recovery phenomena.
- Eventual local physical nonlinearities.
- Curvature of light beam caused by stress gradients.
- Refraction and angular separation of both coherent components of the primary light beam caused by the stress-induced optical anisotropy and inhomogeneity.

Obviously, all these phenomena are interrelated; hence it is necessary and sufficient to experimentally determine those regions where the actual physical processes noticeably deviate from the simplified and/or linearized presentations, and to assess rationally the magnitude of the deviation.

It can be shown that the existence of the third stress component normal to the plate surface can easily be detected, since in such regions the optical isodyne fields are strongly influenced by the value of the integral

$$\int_0^y \sigma_3 \, dy \neq 0, \tag{5.51}$$

according to relation (5.34)

The curvature of the primary light beam caused by large strain/stress gradients and geometric nonlinearities, and the related separation of the coherent light beam components are easily detectable by means of the strain-gradient effects, Section 4.2.4 [5]. The rotation of the third principal direction with the distance from the plate surface is also easily observable.

As one of the numerical measures of the deviation from the assumption of plane stress state, the value of the integral (5.51) can conveniently be used, when normalized with respect to the corresponding integrals of σ_1 and σ_2.

Disregarding the actual visco-elastic responses of polymeric materials may lead to serious errors and mistakes. Knowledge of the actual time-dependent responses of the material of the object under investigation and of the object itself is imperative [5—8, 10, 12, 14].

5.2 Concept of differential optical isodynes

As stated earlier, stress states in the regions of local effects in plates are clearly three-dimensional: the stress components in the direction of plate thickness, σ_{zz} or σ_3, are not negligibly small; the values of the normal and shear stress components are not constant in the direction of plate thickness but vary noticeably with the thickness coordinate. However, the stress components in the direction of plate thickness z are usually not large, their magnitude may be about up to 10% of the magnitude of the normal stress components acting in planes parallel to the middle plane of the plate (x, y), so the secondary principal directions in the planes normal to the middle plane of the plate do not rotate significantly. Consequently, the differences between the normal stress components σ_{xx} and σ_{yy} and the corresponding secondary principal stresses $\sigma_{x'x'}$ and $\sigma_{y'y'}$ where x' and y' denote the secondary principal directions in the planes (x, z) and (y, z) respectively, are small, usually less than 3%.

In such a situation, the basic assumptions regarding the interaction between the birefringent body and flow of electromagnetic energy, discussed at the beginning of this chapter, still simulate qualitatively the involved processes in a satisfactory manner. The differences are only quantitative within certain ranges of variation of the pertinent parameters, and these ranges can be easily determined empirically, or analytically.

This physical reality allows extrapolation of the concepts of the differential elastic isodynes presented in Section 3.5, and introduction of the concept of the differential optical isodynes represented by relations pertaining to three sets of characteristic planes:

 — (x, y, z_i)-characteristic planes, parallel to the (x, y)-plane;
 — (x, y_i, z)-characteristic planes, parallel to the (x, z)-plane;
 — (x_i, y, z)-characteristic planes, parallel to the (y, z)-plane.

For instance, for the x-characteristic direction within the (x, y, z_i)-characteristic plane the basic relation for the differential optical isodyne surfaces can be given in the form:

$$m_{sx}^d = S_\sigma^{-1} \int (\sigma_{y'y'} - \sigma_{z'z'})\, \mathrm{d}x \cong S_\sigma^{-1} \int (\sigma_{yy} - \sigma_{zz})\, \mathrm{d}x$$

$$= S_\sigma^{-1}(p_y - p_z) = \text{const., etc.} \tag{5.52}$$

It is usually easy to determine regions within which plane optical isodynes transform into differential optical isodynes. In regions of very high stress gradients both conjugated light beams, ordinary and extra-ordinary, clearly separate when the optical anisotropy is high. When the separation is more than the effective light beam diameter, say 0.2 or 0.3 mm, the optical isodynes vanish and the intensity of the scattered light remains constant. It is interesting to note that even in such cases one can obtain clear and sharp isochromatic fringes, which, however, no longer represent the lines of constant maximal shear stresses.

Relations (4.153) representing the differential optical isodynes are useful for determination of all stress components in regions of notches, crack tips, contact loads, heat sources, etc. In such regions, however, the stress gradients are high, the stress states are three-dimensional, as mentioned above, and consequently, the paths of light beams are often curved [3, 5, 17]. The related optical effects identify the regions within the plate where the stress state is clearly three-dimensional. Since the curvatures of light beams are caused by the gradients of the sum and difference of principal stresses, the resulting deflections of light beams can readily be used as an additional piece of information on the stress state [19, 21].

5.3 References

[1] Frocht, M. M., *Photoelasticity*, Vol. 1 and 2, John Wiley & Sons, New York, 1941 and 1948.

[2] Mazurkiewicz, S. B. and Pindera, J. T., "Integrated Plane Photoelastic Method — Application of Photoelastic Isodynes", *Experimental Mechanics* 19(7), 1979, pp. 225—234.

[3] Mesmer, G., *Spannungsoptik*, Springer-Verlag, 1939.

[4] Pindera, J. T., *Outline of Photoelasticity* (in Polish), Panstwowe Wydawnictwa Techniczne, Warszawa, 1953.

[5] Pindera, Jerzy T., *Rheological Properties of Some Polyester Resins*, parts I, II, III (in Polish), Engineering Transactions (Rozprawy Inzynierskie) Polish Acad. Sciences, Vol. VII (3.4), 1959, pp. 361—411, 481—520, 521—540.

[6] Pindera, J. T., "Einige Rheologische Probleme bei spannungsoptischen Unterschungen", in: G. Haberland (Ed.), *Internationales spannungsoptisches Symposium, Berlin, April 10—15*, 1961, Akademie-Verlag, Berlin, 1962, pp. 155—172.

[7] Pindera, J. T., *Rheological Properties of Model Materials* (in Polish), Wydawnictwa Naukowo-Techniczne, Warszawa, 1962.

[8] Pindera, J. T. and Cloud, G., "On Dispersion of Birefringence of Photoelastic Materials", *Experimental Mechanics* 6(9), 1966, pp. 470—480.

[9] Pindera, J. T. and Straka, P., "Responses of the Integrated Polariscope", *Journal of Strain Analysis* 8(1), 1973, pp. 65—76.

[10] Pindera, J. T. and Straka, P., "On Physical Measures of Rheological Responses of Some Materials in Wide Ranges of Temperature and Spectral Frequency", *Rheologica Acta* 13(3), 1974, pp. 338—351.

[11] Pindera, J. T., "Response of Photoelastic Systems", *Transactions of The CSME* 2(1), 1973—74, pp. 21—30.

[12] Pindera, J. T., "On Physical Basis of Modern Photoelasticity Techniques", *Beiträge zur Spannungs- und Dehnungsanalyse*, Vol. V, Akademie-Verlag, Berlin, 1968, pp. 103—130.

[13] Pindera, J. T. and Mazurkiewicz, S. B., "Photoelastic Isodynes: A New Type of Stress Modulated Light Intensity Distribution", *Mechanics Research Communications* 4, 1977, pp. 247—252.

[14] Pindera, Jerzy T., "Foundations of Experimental Mechanics: Principles of Modelling, Observation and Experimentation", in: Pindera, J. T. (Ed.), *New Physical Trends in Experimental Mechanics*, Springer-Verlag, Wien, 1981, pp. 188—326.

[15] Pindera, J. T. and Mazurkiewicz, S. B., "Studies of Contact Problems Using Photoelastic Isodynes", *Experimental Mechanics* 21(12), 1981, pp. 448—455.

[16] Pindera, J. T., "Analytical Foundations of the Isodyne Photoelasticity", *Mechanics Research Communications* 8, 1981, pp. 391—397.

[17] Pindera, J. T., "New Development in Photoelastic Studies: Isodyne and Gradient Photoelasticity", in: Chiang, F. P. (Ed.), Incoherent Optical Techniques in Experimental Mechanics, *Optical Engineering* 21(4), 1982, pp. 672—678.

[18] Pindera, J. T. and Krasnowski, B. R., *Theory of Elastic and Photoelastic Isodynes*, SMD-Paper No. 184, IEM-Paper No. 1. Solid Mechanics Division, University of Waterloo, October 1983, pp. 1—127.

[19] Pindera, J. T., "Isodyne Photoelasticity and Gradient Photoelasticity: Physical and Mathematical Models, Efficacy, Applications", *Mechanika Teoretyczna i Stosowana* (Journal of Theoretical and Applied Mechanics) 22(1/2), 1984, pp. 53—68.

[20] Pindera, J. T., Krasnowski, B. R., and Pindera, M. J., "Theory of Elastic and Photoelastic Isodynes. Samples of Applications in Composite Structures", *Experimental Mechanics* 25(3), 1985, pp. 272—281.

[21] Pindera, J. T. and Hecker, F. W., "Basic Theories and Experimental Methods of the Strain-Gradient Method", *Experimental Mechanics* 27(3), 1987, pp. 314—327.

[22] Pirard, A., *La Photoélasticité*, Dunod, Paris, 1947.

[23] Weller, R., "Three-Dimensional Photoelasticity Using Scattered Light", *J. Appl. Phys.* 12, 1941, 610—616.

6

Theory of isodyne experiments

The ultimate objective of stress analysis, both analytical and experimental, is the determination of the following quantities pertaining to the responses of real bodies:

- — stresses, which represent forces acting inside a body;
- — strains, which represent local deformations, both linear and angular;
- — displacements, which represent movements of sets of particles of a body such as points, lines, squares, referred to a chosen coordinate system which is independent of the body.

Stresses, strains and displacements within a body which is not damaged in the process of deformation must satisfy the conditions imposed by:

- — principle of conservation of energy, which, after simplifications, yields the equilibrium conditions presented in the form of relations between forces, stresses, and spatial derivatives of stresses;
- — requirement of geometric compatability which results in kinematic relations;
- — material constitutive relations which represent relations between stresses, strains, temperature, time, electric and magnetic fields, etc., pertaining to responses of a hypothetical or a real body.

The relations between stresses, strain, deformations, and their spatial and temporal derivatives are defined within chosen mathematical models which may or may not satisfy the condition of a rigorous mathematical admissibility. An "objective" definition of stresses and strains, outside of the pertinent mathematical models, does not exist.

Stresses, strains, and displacements, defined by selected mathematical models, are experimentally determined by means of real physical systems which impose specific deformations upon the tested body and measure the alteration of the flow of energy used to detect alterations of a particular parameter of the response of the tested body. The relation between the signal of interest and the indicated signal is described by the transfer functions of the testing and measurement systems.

This is the general theoretical framework of the theory of experiments which is adopted in the isodyne stress analysis. The pertinent issues have been discussed in the preceeding Chapters. In this Chapter, two basic issues previously not explicitly discussed are shortly outlined. These are:

— Actual constitutive relations for viscoelastic materials used in isodyne measurements;
— An example of the development of the transfer function of a component of an optical measurement system.

The reason for concentrating on these two topics is that the other components of the theory of measurements are usually well known and understood and are correctly applied, whereas the above selected issues are often misunderstood, thus leading to unnecessary errors and mistakes.

6.1 Actual constitutive relations for viscoelastic materials used in isodyne measurements

6.1.1 General remarks

Mechanical and optical responses of materials such as inorganic glasses, or ceramics in general, can be sufficiently accurately described in terms of Hooke's model and Ramachandran—Ramaseshan model. Hooke's model is still applicable with respect to the mechanical responses of birefringent materials such as silicon, but the stress-optical responses of silicon are outside of the descriptive and predictive capacity of the Ramachandran—Ramaseshan model.

The mechanical and optical responses of high polymers, more commonly known as "polymeric materials", are more complicated. These responses belong to the class of rheological responses, that is time-dependent responses, the extreme cases of which are creep with creep recovery, and relaxation with relaxation recovery.

The major problem of stress analysis of homogeneous structures made of viscoelastic materials — original structures or their experimental models — is whether or not the states of stress, strain, and displacement of geometrically identical structures made of linearly elastic materials simulated by Hooke's body and of viscoelastic materials are similar. Fortunately, it was shown that when the viscoelastic response of material is linear in the whole range of stresses and strains, when the amount of creep or relaxation is small, when the boundary conditions are given by tractions, the so-called stress/strain correspondence principle is applicable: the stresses in the linear viscoelastic body are the same as the stresses in Hooke's body, and the strains and displacements which depend on time, are proportional to the ratio of $[D(t)]^{-1}$ over E, where $D(t)$ denotes creep compliance function and E is Young's modulus. For multi-connected homogeneous bodies the additional requirement is that the influence of Poisson's ratio on the stress state be negligible.

Regarding the optical responses of viscoelastic material, the situation is much more complicated because the birefringence response of real viscoelastic materials is theoretically and practically a multi-mechanism response, as illustrated by Figure 6.1. Thus all the speculative-phenomenological models of the time-dependent birefringence responses lack solid theoretical-scientific foundation and in the best case, only have a descriptive power within a specified range of variation of basic physical parameters. Such models by definition lack the explanatory and predictive powers which are major features of modern constitutive materials equations.

The theoretical-scientific physical and mathematical models of any real

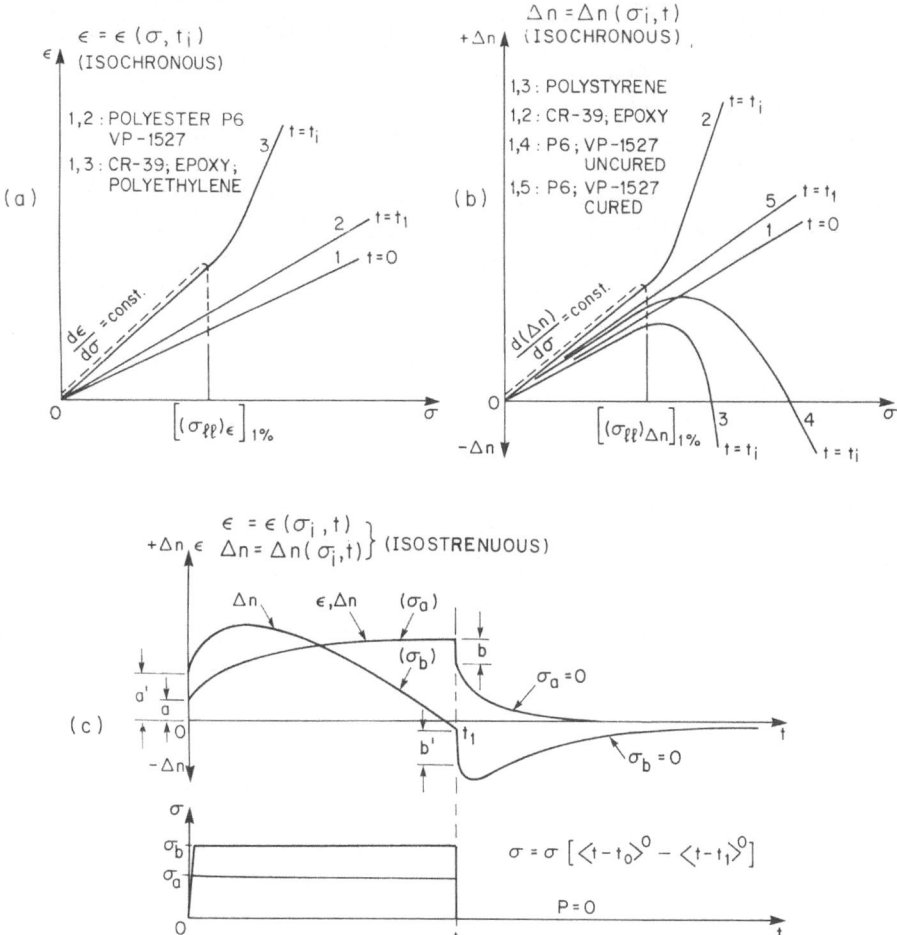

Fig. 6.1. Typical patterns of the rheological responses to uniaxial loads of several high polymers. Indication of several mechanisms of linear viscoelastic responses. (a) Isochronous stress-strain relations. (b) Isochronous stress-birefringence relations, spectral frequency dependent. (c) Isostress relations: two typical creep and creep recovery relations for stress and birefringence at limited unit step load functions; the birefringence response shows existence of at least three mechanisms of birefringence, and is spectral frequency dependent.

phenomenon are characterized by three major features: explanatory power, descriptive power, and predictive power. To develop such a model of the viscoelastic responses of materials, it would be necessary to utilize the known data on the actual structure of a given material and on the patterns of interaction between radiation and matter. In other words, it would be necessary to develop a reliable physical model of the mechanical and optical viscoelastic responses. Figure 6.2 presents a component of such a model. Data presented in Figure 6.2 confirm the conclusion which follows from Figure 6.1(c) that the mechanism of birefringence of high polymers is a multiparameter mechanism, and that there exists a system of coupled mechanisms.

Regarding the requirements imposed by stress analysis procedures, the most important information on the constitutive material relation is whether or not the material response is linearly viscoelastic, mechanically and optically, and what are the ranges of the linear viscoelastic behavior with regard to both the stresses and birefringence. This requirement imposes specific conditions on the methodology of describing and testing viscoelastic materials. Thus it is recognized here that the basic relations between stresses, strains, birefringence and time are isochronous relations which are evaluated using the results of creep and creep recovery tests, and of relaxation and relaxation recovery tests. Because of this, the notions of the creep compliance function obtained from a uniaxial test, $D(t)$, and of the relaxation and relaxation recovery modulus $E(t)$, are adapted here. Such a frame of reference excludes the ambiguities and drastic conceptual simplifications of tests performed at constant stress or strain rates, and imposes no limitations on the physical definitions of the $D(t)$ and $E(t)$ functions. In

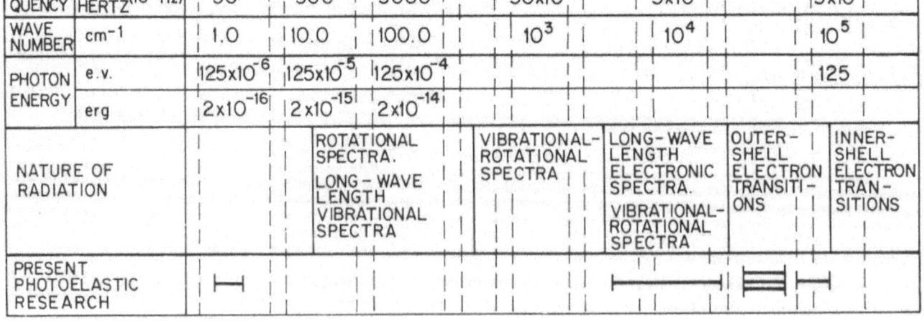

SPECTRAL REGION / QUANTITY	SPECTRAL REGIONS OF ELECTROMAGNETIC RADIATION						
	MICROWAVES	FAR INFRARED	INFRARED	NEAR INFRARED	VISIBLE	ULTRA VIOLET	X-RAYS
WAVE LENGTH — MILLIMETERS	10.0	1.0	0.1	0.05 / 0.01	2×10^{-3} / 1×10^{-3}		
WAVE LENGTH — MICROMETERS	10,000 / 1000	100	50 / 10	2 / 1	0.74–0.38	0.1	
WAVE LENGTH — NANOMETERS				2000 / 1000	740–380	100	
FREQUENCY GIGA-HERTZ (10^9 Hz)	30	300	3000	30×10^3	3×10^5	3×10^6	
WAVE NUMBER cm^{-1}	1.0	10.0	100.0	10^3	10^4	10^5	
PHOTON ENERGY e.v.	125×10^{-6}	125×10^{-5}	125×10^{-4}			125	
PHOTON ENERGY erg	2×10^{-16}	2×10^{-15}	2×10^{-14}				
NATURE OF RADIATION		ROTATIONAL SPECTRA. LONG-WAVE LENGTH VIBRATIONAL SPECTRA	VIBRATIONAL-ROTATIONAL SPECTRA	LONG-WAVE LENGTH ELECTRONIC SPECTRA. VIBRATIONAL-ROTATIONAL SPECTRA	OUTER-SHELL ELECTRON TRANSITIONS		INNER-SHELL ELECTRON TRANSITIONS
PRESENT PHOTOELASTIC RESEARCH	⊢—⊣				⊢——⊣ ⊟ ⊢		

Fig. 6.2. An example of a component of a physical model of the optical viscoelastic responses of solid polymers. Birefringence is generated at the atomic and molecular levels and is detected by radiation having corresponding spectral frequency.

particular, such a frame of reference requires no assumption that one single mechanism of viscoelastic deformation and viscoelastic birefringence exists, and requires no assumption about the spectral dependence of the birefringence. In addition, such a general frame of reference allows the introduction of some new measures of viscoelastic responses which can be used to test the reliability of speculative mathematical models of the photoviscoelastic behavior of high polymers. Those measures are specified in the next section.

Summarizing, due to pragmatic reasons, it is customary to use analytical relations derived for linearly elastic bodies to study and to describe stress/strain states in bodies made of viscoelastic materials. It has been shown that this is justified for linearly viscoelastic materials when the stress-strain viscoelastic correspondence principle is applicable.

With regard to time-dependent values of material parameters, such as the modulus of elasticity or an equivalent parameter, it is common in engineering research to use their average values determined on the basis of tests at constant rates of stress or strain, and to establish very simple relationships between the time-dependent mechanical and optical material responses. However, it has been shown [1—9, 14, 16, 19, 20] that such an approach is too simple, and even too primitive, and leads to unnecessary errors.

6.1.2 Mathematical models of viscoelastic responses.
   New measures for parameters of constitutive relations.
   Examples of actual materials responses

An analysis of experimental data ([1—9; 14—20] and Figure 6.1a, b, c) shows that the traditional measures of rheological responses of materials such as the mechanical creep compliance function $D(t)$, the mechanical relaxation modulus function $E(t)$, and the analogous to the above functions optical creep compliance functions $C_\sigma(t)$ and $C_\varepsilon(t)$, (where the $C_\varepsilon(t)$ function can be presented as a photoelastic relaxation modulus function), and the time-temperature equivalence function are sufficient for an adequate description of the actual rheological responses of polymers used in photoelasticity, isodyne techniques, and in engineering structures. However, they are too elementary for an adequate and commensurate representation of the actual mechanisms of mechanical and optical viscoelastic responses of polymers and of ceramics at high temperatures. Consequently, their utility is restricted to the known range of variation of major physical parameters outside of which no reliable prediction of the viscoelastic responses is possible. Thus far no scientific theory of the viscoelastic responses of materials has been developed as mentioned earlier, so it is acceptable to use the possibly simple and general analytical-empirical functions and to specify the ranges of their reliability and applicability by means of a set of reliable physical measures.

On the basis of information presented in pertinent bibliography, illustrated by Figure 6.1a, b, c, the following quantities appear to be of major importance either as quantities complementary to the set of rheological functions mentioned above, or quantities necessary for a satisfactory design of these functions.

1. The isochronous relations for strain, stress, and birefringence which can be most conveniently and accurately determined using the results of the corresponding sets of creep and relaxation tests conducted at constant stress, or constant strain, respectively.
2. The creep recovery and the relaxation recovery functions, which represent an extension of the creep compliance functions and the relaxation functions into the creep recovery and the relaxation recovery regions, Figure 6.1c.
3. The quantities describing the ranges of linear rheological responses, presented as the linear limit stresses, $\sigma_{\ell\ell}$, and the corresponding linear limit strains, $\varepsilon_{\ell\ell}$. Figure 6.3.
4. The quantity describing the dependence of the magnitude of relative retardation on the spectral frequency of radiation, presented as the normalized spectral birefringence.
5. The quantity describing the rate of change of the spectral birefringence with the radiation frequency, presented as the normalized spectral dispersion of birefringence.
6. The quantity describing the dependence of the dielectric coefficient on the field frequency, presented as the normalized dielectric coefficient.
7. The quantity describing the dependence of the angle between chosen mechanical principal directions and optical principal directions on the frequency of radiation, presented as the spectral dispersion of optical axes.
8. The quantity describing the related interaction between the radiant energy and matter, presented as the spectral transmittance.

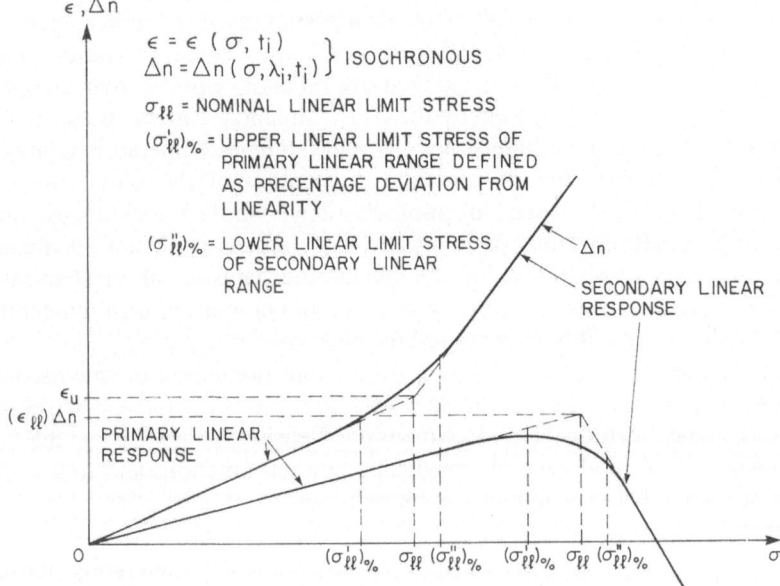

Fig. 6.3. Rheo-mechanical and rheo-optical responses of real solids. Concept and definition of the linear limit stress and the linear limit strain with respect to strain and birefringence.

9. The quantities describing the character of the influence of temperature on the response of materials, presented as the variation of the thermal expansion coefficient with the temperature and the related variation of mechanical damping.

The mechanical and optical creep recovery functions carry information that is not only supplementary to the information provided by the creep function, but also supply essential information on the mechanism of deformation. Obviously, only the creep and creep recovery functions which are produced by a limited step load function are amenable to simple analysis without additional assumptions.

Experimental results shown in Figure 6.1b, c, Figure 6.3 and Figure 6.7, can satisfactorily be explained in the first approximation by assuming that rheo-birefringence is produced by a set of partial effects. While the particular shape of the creep function can be explained to be a result of two partial effects, one distortional and one orientational, the creep recovery function indicates that there exists at least two orientational birefringence effects produced at different stress levels and related to different time intervals.

Since the creep recovery function carries information pertinent to the understanding of the creep function, it seems necessary to present both these components of the creep response as two components of one function. Experimental results show that the simplest presentation of the optical creep for step loading and unloading in the form:

$$\Delta n(t) = C_o(t - t_1)\sigma_0(\langle t - t_0\rangle^0 - \langle t - t_1\rangle^0),\tag{6.1}$$

where t_0 denotes time of loading and t_1 denotes the time of unloading, is not satisfactory, neither quantitatively nor qualitatively and only represents a rough but albeit useful, approximation.

Mathematical models of rheological response of several high polymers such as represented by equation (6.1) are not commensurate with the real response of several major groups of polymers, such as polyester resins or epoxy resins, in both the linear and nonlinear ranges. In a general case the creep response consists of the active creep and creep recovery [15]. The creep compliance functions designed to represent the active creep part of creep response to uniaxial stress in the form

$$\varepsilon(t) = \int_0^t D(t - \tau)\dot\sigma \, d\tau,\tag{6.2}$$

$$\Delta n(t) = \int_0^t C_o(t - \tau)\dot\sigma \, d\tau,\tag{6.3}$$

or

$$\Delta n(t) = \int_0^t C_\varepsilon(t - \tau)\dot\varepsilon \, d\tau,\tag{6.4}$$

for the linear range, and

$$\varepsilon(t) = \int_0^t D_1(t - t_1)\dot{\sigma}\,dt$$

$$+ \int_0^t \int_0^t D_1(t - \tau_1, t - \tau_2)\sigma(\tau_1)\sigma(\tau_2)\,d\tau_1\,d\tau_2 + \dots, \text{ etc.,} \qquad (6.5)$$

for the nonlinear range, cannot satisfactorily describe the responses of several materials, Figure 6.1a, b, c.

The creep recovery function seems to be of primary importance in constructing general creep functions, in describing both the creep and the creep recovery components of creep response, and in understanding the coupling mechanisms between the mechanical and the optical effects.

Experimental results indicate that the birefringence creep response of several materials could be approximated by a creep compliance function of the form:

$$C(t) = C_0(t) + C_1(t - t_1) + C_2(t - t_2) + \dots, \qquad (6.6)$$

where the time intervals $[t - t_i]$ depend on the strain and the strain rate,

$$t_i = t_i(\varepsilon, \dot{\varepsilon}). \qquad (6.7)$$

The creep recovery curves of several polymeric materials show that the mechanical and the birefringence response to the step load in the time interval $[t_0, t_1]$, where t_0 and t_1 denote the time of loading and of unloading, respectively, can be presented in the form:

$$\varepsilon(t) = [D(t_0) + D(t)]\sigma(\langle t - t_0 \rangle^0 - \langle t - t_1 \rangle^0), \qquad (6.8)$$

and

$$\Delta n(t) = [C_\sigma(t_0) + C_\sigma(t)]\sigma(\langle t - t_0 \rangle^0 - \langle t - t_1 \rangle^0), \qquad (6.9)$$

for

$$t_0 < t < (t_1)^+.$$

The statement represented by equations (6.8) and (6.9), that the instantaneous mechanical and birefringence effects of the step function loading and unloading are equal at loading, $t = t_0^+$, and unloading, $t = t_1^+$, characterizes the response of several polymeric materials within a certain period of loading only. However, it also occurs that the instantaneous mechanical effects are equal, while the instantaneous birefringence recovery effect is greater than the instantaneous load effect.

The isochronous relations between stress, strain and birefringence are linear only in limited ranges of stresses or strains and strongly depend on the time parameter. Response of materials can be linear in one or more stress-strain ranges, as presented schematically in Figures 6.3 and 6.4.

It is useful to introduce reliable measures of the linear ranges of rheological

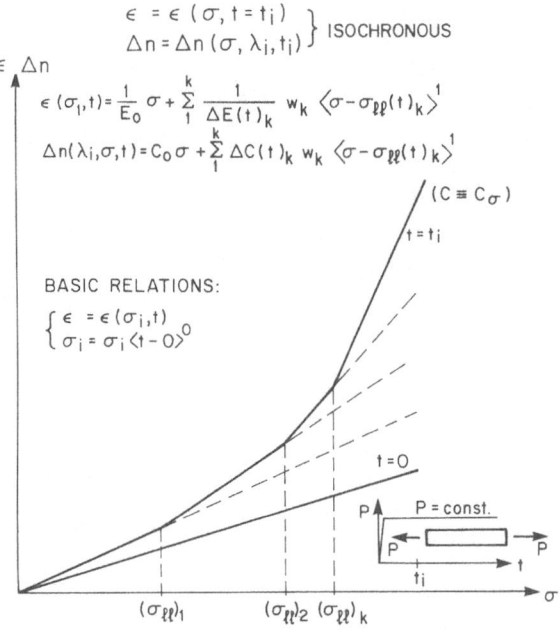

Fig. 6.4. Isochronous presentation of the viscoelastic stress-birefringence-strain responses of a material exhibiting two linear viscoelastic ranges connected by a wide transition region. (Nominal limit stresses are not sharply defined.)

response of materials in the form of the linear limit stresses, $\sigma_{\ell\ell}$, and the linear limit strains, $\varepsilon_{\ell\ell}$ Figure 6.3. Two particular measures have been introduced, the nominal linear limit stress, $(\sigma_{\ell\ell})_{nom}$ or $\sigma_{\ell\ell}$, and the percentual linear limit stress $(\sigma_{\ell\ell})\%$ [1, 6]. Both measures can be related to the primary linear range of response, $\sigma'_{\ell\ell}$ or to the secondary linear range of response, $\sigma''_{\ell\ell}$.

The concept of the linear limit stress is of interest with regard to the mathematical models of material responses. The percentual linear limit stress, which often describes the width of the transition range between two linear ranges, is also of engineering interest.

The linear limit stresses and strains are time-dependent and may be different for the strain and the birefringence, Figure 6.5. The linear limit stresses depend on temperature and decrease to zero in the ranges of transition temperatures.

The linear limit stresses for birefringence may depend also on the wavelength of radiation, hence the pertinent relation should be presented as follows:

for strain: $(\sigma_{\ell\ell})_\varepsilon = \sigma_{\ell\ell}(t, T)$, (6.10)

for birefringence: $(\sigma_{\ell\ell})_{\Delta n} = \sigma_{\ell\ell}(t, T, \lambda)$. (6.11)

The dependence of the linear limit stress on time, temperature and wavelength of radiation supplies information on the mechanisms of deformation.

In accordance with the interpretation of experimental results outlined above,

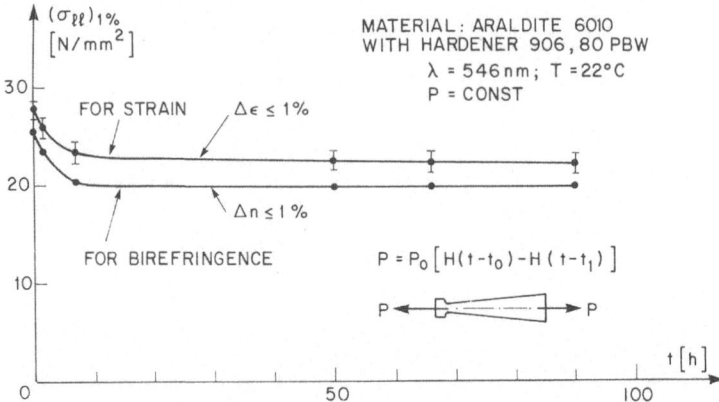

Fig. 6.5. Example of the time dependence of the linear limit stresses for strain and for bire-fringence.

the multi-linear response of materials presented schematically in Figure 6.1, 6.3 and 6.4 can be approximated by the relations, Figure 6.4,

$$\varepsilon(t) = D_0\sigma + \sum_k D_k(t) W_k\langle\sigma - \sigma_{\ell\ell}(t_k)\rangle^1 \tag{6.12}$$

$$\Delta n(t) = C_0\sigma + \sum_k C_k(t) W_k\langle\sigma - \sigma_{\ell\ell}(t)_k\rangle^1 \tag{6.13}$$

where W denotes statistical weight which depends on temperature, D_0 and C_0 denote instantaneous values of the creep compliance functions, D_k and C_k denote partial creep compliance functions representing partial effects, and $(\sigma_{\ell\ell})_k$ denotes the linear limit stress of the kth partial effect [8].

It must be noted that C_σ and C_ε depend, nonmonotonically, on the spectral frequency (wavelength of radiation) Figure 6.6; that is, contrary to the widespread belief,

$$\Delta n(\lambda_1) \neq \Delta n(\lambda_2), \text{ and } \lambda_1\Delta n(\lambda_2) \neq \lambda_2\Delta n(\lambda_1), \tag{6.14a}$$

or

$$C_\sigma(\lambda_1) \neq \frac{\lambda_1}{\lambda_2} C_\sigma(\lambda_2) \tag{6.14b}$$

where C_σ does not depend on the loading time for the practically linear elastic bodies such as inorganic glass.

Obviously, the information on the rheological behavior of viscoelastic materials is incomplete without data on the mechanical and optical relaxation response. Formally, it is possible to treat the optical relaxation and relaxation

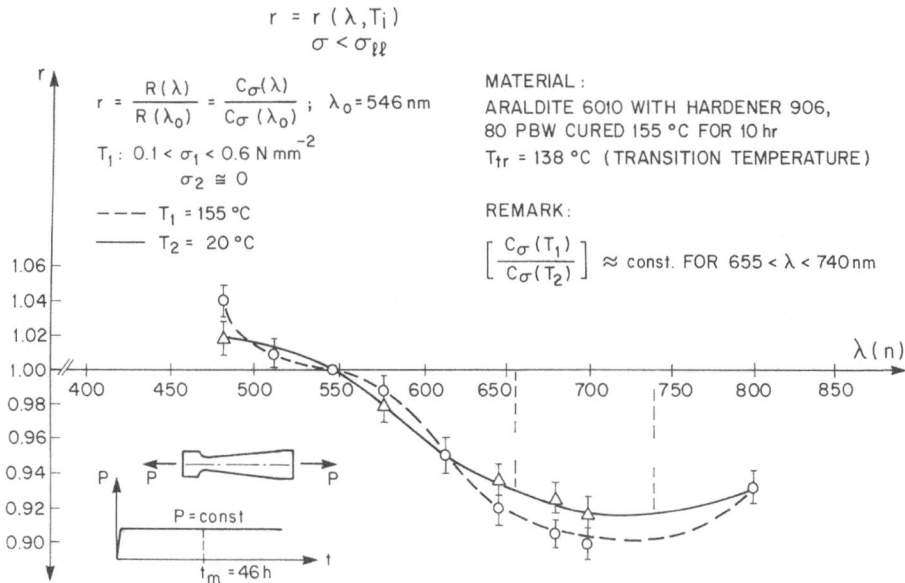

Fig. 6.6. Example of the spectral frequency dependence of the normalized birefringence at temperatures below and above the second order transition temperature. Normalizing wavelength of light: $\lambda_0 = 546$ nm.

recovery functions in a manner analogous to the mechanical relaxation function.

$$\sigma(t) = \int_0^t E(t - \tau)\dot{\varepsilon}\,dt, \tag{6.15}$$

where $E(t)$ denotes mechanical relaxation modulus; that is,

$$\Delta n(t) = \int_0^t C_\varepsilon(t - \tau)\dot{\varepsilon}\,dt. \tag{6.16}$$

However, the known phenomenological relation for the mechanical response,

$$D(t)E(t) \cong 1, \tag{6.17}$$

which is valid in the range where the stresses and strains have unequivocal physical meaning and the deformations are small, is not applicable in the case of the birefringence response, that is,

$$C_\sigma(\lambda, t)C_\varepsilon(\lambda, t) \neq \text{const} \tag{6.18}$$

as it is shown in Figure 6.7a for temperatures below and above the second order transition temperature in the creep and creep recovery ranges.

On the other hand, the information that for a particular material and particular spectral range the value of the strain optic coefficient C_ε does not depend on time, Figure 6.7a,

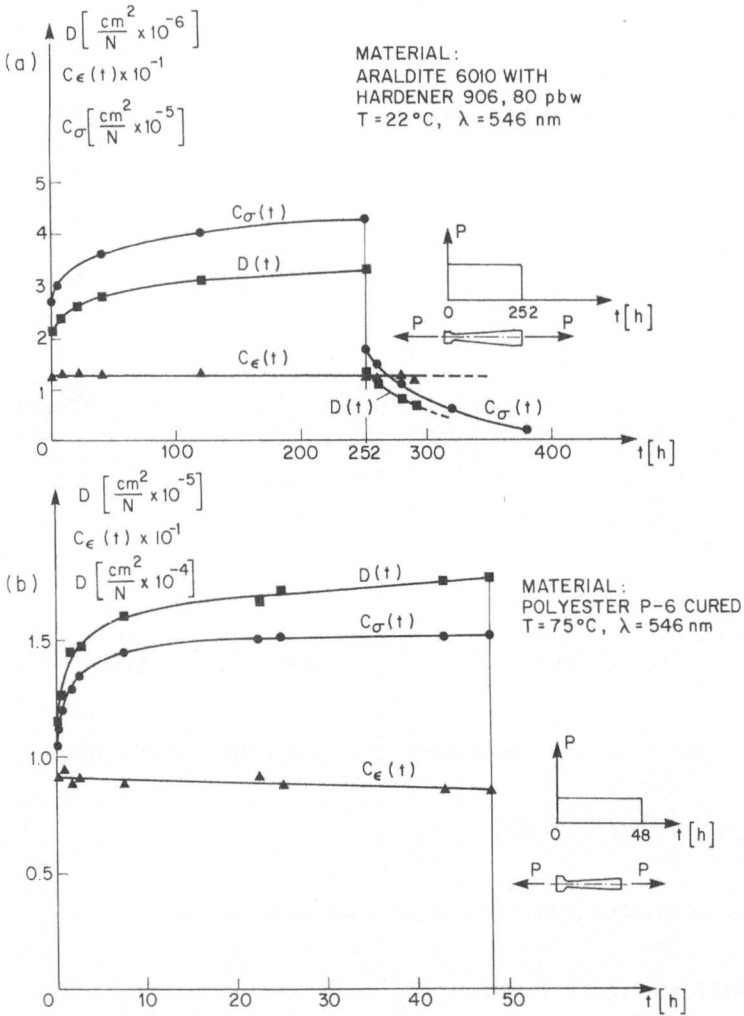

Fig. 6.7. Examples of the mechanical and optical creep and creep recovery functions for some polymers, for stress and strain. (a) Functions for the epoxy resin Araldite 6010 below the transition temperature of 138 °C. (b) Functions for the polyester resin Patatal P6, above the first transition temperature of 68 °C and below the second transition temperature of 115 °C.

$$C_\varepsilon(\lambda_0, t) = \text{const},\tag{6.19}$$

in both loading ranges enormously facilitates the experimental procedure and opens new horizons, particularly regarding the studies of composite structures.

It follows from the relations presented in this section that for bodies made of linearly viscoelastic materials whose rate of creep decreases with time, and which are loaded by a set of forces, the sequence of loading becomes immaterial after a certain period of time. This property of linearly viscoelastic materials allows an easy verification whether or not the responses of a particular polymeric material can be simulated by a simple mathematical viscoelastic model. Figure 6.8 illustrates this issue.

6.1.3 Temperature-dependence of the thermal expansion coefficients

With regard to the validity of stress analysis data and their applicability in design of advanced engineering structures, differences in the values of thermal expansion coefficients of materials used in composite structures and their

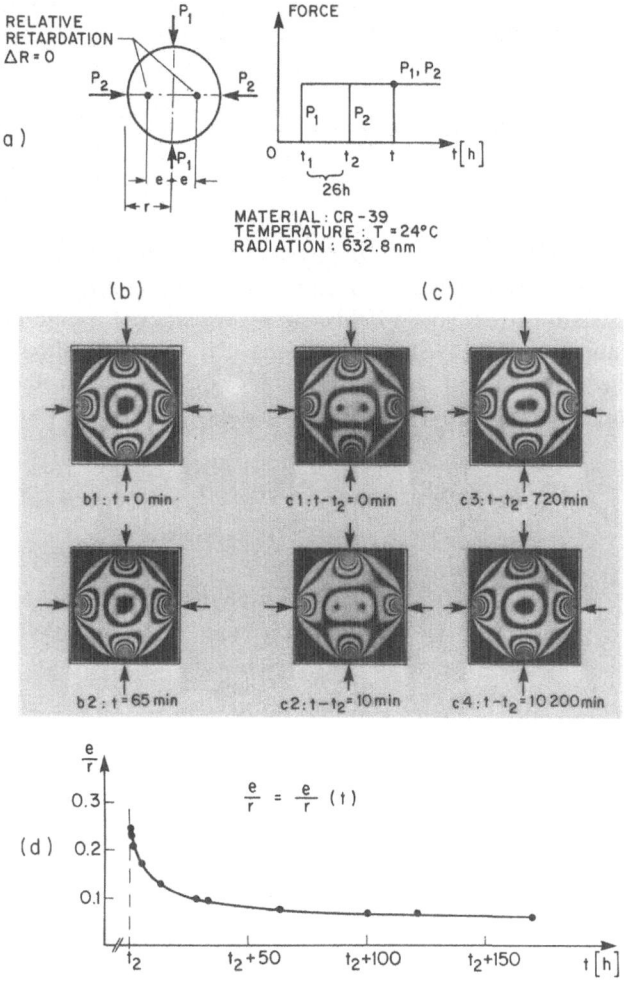

Fig. 6.8. Superposition of loads. Example of a linear viscoelastic response. Time-dependent birefringence in a circular disk, related to two-dimensional stress states caused by two systems of loads applied along the mutually perpendicular diameters. (a) Data on experiment. (b) Isochromatic fields when the loads P_1 and P_2 are applied simultaneously. (c) Isochromatic fields when load P_2 applied with delay of 26 h. (d) Creep of optical singularities characterized by the zero order isochromatics ($m = 0$).

dependence on temperature deserves particular attention. As shown in Figure 6.9, the typical polymeric materials used in manufacturing of composite materials exhibit a more or less sudden alteration of the thermal expansion coefficient in the region of the first and second order transition temperature. This alteration is not negligible — 200—300% on the average — and may cause unexpected structural failures. For some materials, such as polyester resins which are often used as matrix materials of fiber-reinforced composites, the first sudden alteration of the thermal expansion coefficient by about 100% occurs in the temperature range 50 °C—70 °C. Such temperature can easily be produced in a structural component by typical sun exposure. Figure 6.9 presents pertinent data for the most common engineering high polymers.

6.2 Responses of isodyne measurement systems. Transfer functions

6.2.1 General considerations

It is easy to show that, when specific physical conditions are satisfied the optical subsystem of the isodyne recorder used for the determination of static or slowly varying (quasistatic) stress strain fields can be considered a time-invariant, zero-order instrument, [10; 11—13]. Consequently, the simplified mathematical model of the input-output relations, $s_r = s_r(s_e)$, as given by the ordinary differentiation equation with constant coefficients a_j and b_k,

$$\sum_0^n a_j \frac{d^j s_r}{dt^j} = \sum_0^m b_k \frac{d^k s_e}{dt^k} \qquad (6.20)$$

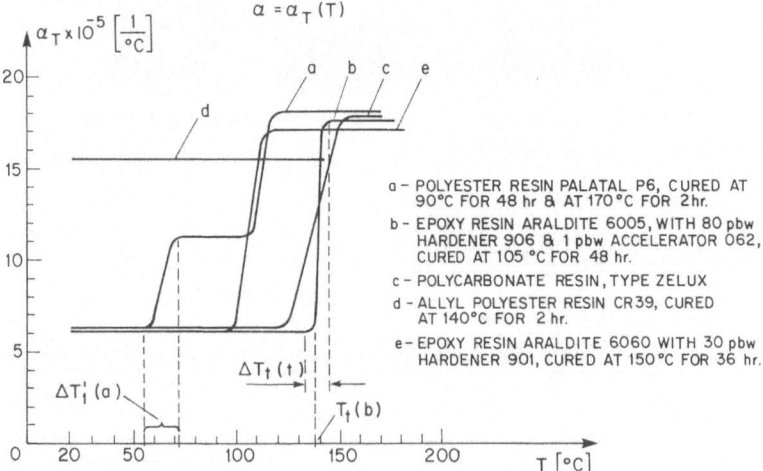

Fig. 6.9. Temperature dependence of the linear thermal expansion coefficient of several polymeric materials. First and second order transition temperatures (phase transformations) are indicated by the significant alterations of the thermal expansion coefficients.

degenerates into a simple algebraic equation for the polariscope components,

$$a_0 s_r = b_0 s_e, \tag{6.21}$$

where the input (entrance) signal $s_e = s_e(f)$, and the output (response) signal $s_r = s_r(f)$, the symbol f denoting the spectral frequency of the information-collecting radiation.

Thus, the component transfer functions relating the output radiant power to the input radiant power are represented by

$$F = \frac{b_0}{a_0}, \tag{6.22}$$

where, of course, $F = F(f)$ and therefore the pertinent components can be considered zero-order instruments only with regard to the time parameter.

The transfer functions of all components of the isodyne recorder modulate the spectral distribution of the transmitted radiant power; this fact requires that the spectral band of radiation used for collection of information of interest should be chosen carefully.

Considering the mutual influence of the component transfer functions, the components of the typical isodyne recorder can be represented by zero-impedance sources and infinite-impedence loads, if the reflection between the surfaces of elements can be practically eliminated and thus neglected. Hence, the overall transfer function of the isodyne polariscope F_{0v}, called the instrument transfer function, can be simplified to:

$$F_{0v} = F_1 F_2 F_3 \ldots F_n, \tag{6.23}$$

where

$$F_i = F_i(x, y, f). \tag{6.24}$$

The major feature of the specific physical conditions mentioned at the beginning of this section is that the system response can be inferred from the local state relations. In practical terms, this means that equation (6.23) is valid only when the thickness of the photoelastic specimen is small and the strain-gradient effects are negligible.

6.2.2 Example of approach

Let us assume that the isodyne collector, presented symbolically in Figure 6.10, is in the form of a prism such as that used by the first author in his previous research [12]. This system is deficient in some respects, however, the analysis of this system is very instructive. The analysis of this system presented below is performed on the basis of Figure 6.10 where the meaning of the basic symbols is obvious.

The modulation of the radiant power when the linearly polarized wave front is resolved into a reflected front and a refracted front on the boundary of two optical media can be conveniently presented by the known Fresnel equations. For the external reflection presented in Figure 6.10 the relations between the amplitudes and phases of the linearly polarized incident electric vector $\bar{E}$, and

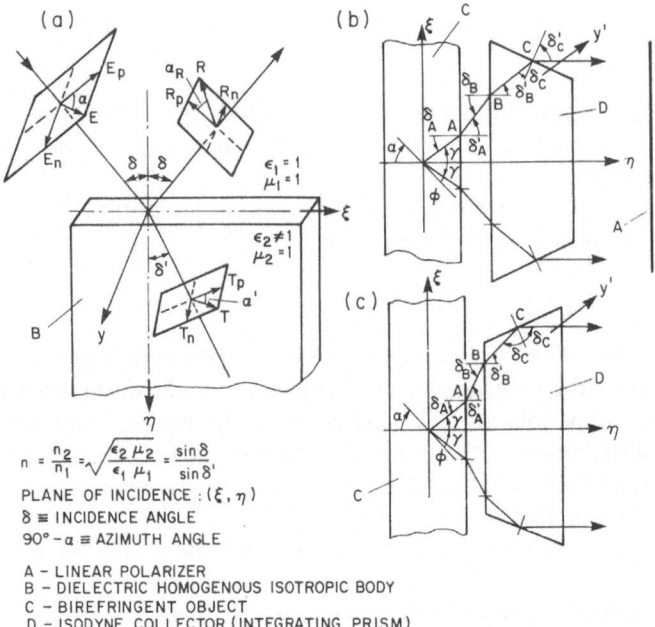

PLANE OF INCIDENCE : (ξ, η)
$\delta \equiv$ INCIDENCE ANGLE
$90° - \alpha \equiv$ AZIMUTH ANGLE

A - LINEAR POLARIZER
B - DIELECTRIC HOMOGENOUS ISOTROPIC BODY
C - BIREFRINGENT OBJECT
D - ISODYNE COLLECTOR (INTEGRATING PRISM)

Fig. 6.10. Modulation of scattered radiant power by the isodyne collector having the form of an integrating prism. (a) Classical relations at refraction and reflection. (b) Optimal arrangement of the isodyne collector in the form of an integrating prism. (c) acceptable arrangement of the integrating prism.

of the reflected and refracted vectors $\bar{R}$ and $\bar{T}$ presented by their components parallel and normal to the plane of incidence can be formulated as follows:

$$\frac{R_p}{E_p} = \frac{\tan(\delta - \delta')}{\tan(\delta + \delta')} = \frac{n_{12} \cos \delta - \cos \delta'}{n_{12} \cos \delta + \cos \sigma'}, \tag{6.25}$$

$$\Delta\phi(R_p) = 0 \quad \text{for} \quad \delta < \delta_p = \arctan n_{1,2},$$

$$\Delta\phi(R_p) = \pi \quad \text{for} \quad \delta > \delta_p, \tag{6.26}$$

$$\frac{R_n}{E_n} = -\frac{\sin(\delta - \delta')}{\sin(\delta + \delta')} = \frac{\cos \delta - n_{12} \cos \delta'}{\cos \delta + n_{12} \cos \delta'}, \tag{6.27}$$

$$\Delta\phi(R_n) = \pi \quad \text{for} \quad 0 < \delta < \pi, \tag{6.28}$$

$$\frac{T_p}{E_p} = \frac{2 \sin \delta' \cos \delta}{\sin(\delta + \delta') \cos(\delta - \delta')} = \frac{2 \cos \delta}{n_{12} \cos \delta + \cos \delta'}, \tag{6.29}$$

$$\Delta\phi(T_p) = 0 \quad \text{for} \quad 0 < \delta < \pi, \tag{6.30}$$

$$\frac{T_n}{E_n} = \frac{2 \sin \delta' \cos \delta}{\sin(\delta + \delta')} = \frac{2 \cos \delta}{\cos \delta + n_{12} \cos \delta'} \tag{6.31}$$

$$\Delta\phi(T_n) = 0 \quad \text{for} \quad 0 < \delta < \pi \tag{6.32}$$

where δ_p denotes the polarizing angle and

$$n_{12} = \frac{n_2}{n_1} = \sqrt{\frac{\varepsilon_2 \mu_2}{\varepsilon_1 \mu_1}} = \frac{\sin \delta}{\sin \delta'} . \qquad (6.33)$$

It follows that for external reflection presented in Figure 6.10 the reflected and the refracted beams are linearly polarized. However, the azimuth of the plane of vibration of the incident beam (the angle between the plane of vibration and the normal to the plane of incidence) is not equal to the azimuths of the planes of vibration of the reflected and of the refracted beams. The azimuths of the reflected and of the refracted linearly polarized beams and the intensity of these beams depend on the angle of incidence and on the relative indices of refraction between both media.

Relations (6.25—6.32) describe the propagation of the scattered light beam from an optically less dense medium to a denser one. When the scattered light beam crosses the boundary between the denser and the less dense medium, i.e. when the internal reflection is considered, relations (6.25—6.32) are only valid when the angle δ is replaced by δ' and vice versa. The phase changes of the polarized parallel components and of the polarized normal components of the internally refracted beams are continuous functions of the angle of incidence; consequently, the internally reflected beam is elliptically polarized. This fact imposes a serious constraint on the technique of collecting information from the scattered beams by means of the integrating prism.

The values of the critical angles for all optical components used in scattered light techniques are important parameters since the signal-to-noise ratio for the arrangements presented in Figure 6.9 depends on the difference between the angles of internal incidence δ_A in both arrangements (b) and (c) and on the angle δ_c in the arrangement (c), and on the critical angles of the total reflection δ_t determined by

$$\sin \delta_t = n_{21} = \frac{n_1}{n_2} . \qquad (6.34)$$

The range of change of the critical angles in the spectral range 400 nm—1000 nm for some materials used for photoelastic models and for the integrating prism is given in Figure 6.11.

It follows from a simple analysis that the relations between the radiant power of all three scattered beams and the values of the stress components along the path of the primary beam can be simplified if the observation angle θ is 90° and if the axis of the isodyne collector (prism) is parallel to the primary beams. Equations (6.25—6.33) then shown that the intensity of the beam scattered at the observation angle θ and emerging from the prism can be described as follows:

— For the arrangement (b), Figure 6.10, without immersion liquid between the

n : REFRACTIVE INDEX, WITH RESPECT TO AIR δ₁ : CRITICAL ANGLE, WITH RESPECT TO AIR						
MATERIAL	λ = 400 nm		λ = 589 nm		λ = 1000 nm	
	n	δ_1	n	δ_1	n	δ_1
ACRYLIC RESIN USED BY APPLIED PRODUCT CORPORATION	1.506	41° 36'	1.489	42° 12'	1.482	42° 30'
SODA–LIME–SILICA GLASS	1.525	40° 55'	1.516	41° 15'	1.519	41° 09'
EPOXY RESIN ARALDITE 502	n = 1.61 – 1.51 δ₁ = 38° 24' – 41° 30'					

Fig. 6.11. Indices of refraction with respect to air and related values of critical angles for various materials, in the visible and in the near infra-red ranges.

specimen and prism, by

$$\frac{I'_\Phi}{k'I_0} = \frac{I_\Phi}{kKI_0} \left(\frac{2 \cos \delta_A}{n_{21} \cos \delta_A + \cos \delta'_A} \right)^2 \times$$

$$\times \left(\frac{2 \cos \delta_B}{n_{13} \cos \delta_B + \cos \delta'_B} \right) \left(\frac{n_{31} \cos \delta_c - \cos \delta'_C}{n_{31} \cos \delta_c + \cos \delta'_C} \right)^2 \times$$

$$\times \frac{\Delta n_{13}}{(n_{13} + 1)^2}, \tag{6.35a}$$

$$I_\theta = 0. \tag{6.35b}$$

— For the arrangement (b), Figure 6.9, with immersion liquid and identical indices of refraction of the specimen (n_{12}) and prism (n_{13}), by

$$\frac{I'_\Phi}{kI_0} = \frac{I_\Phi}{K'I_0} \left(\frac{n_{31} \cos \delta_C - \cos \delta'_C}{n_{31} \cos \delta_C + \cos \delta'_C} \right)^2 \frac{4n_{13}}{(n_{13} + 1)^2}, \tag{6.36a}$$

$$I_\theta = 0. \tag{6.36b}$$

— For the arrangement (c), Figure 6.10, without immersion liquid, by

$$\frac{I'_\Phi}{k'I_0} = \frac{I_\Phi}{kKI_0} \left(\frac{2 \cos \delta_A}{n_{21} \cos \delta_A + \cos \delta'_A} \right)^2 \times$$

$$\times \left(\frac{2 \cos \delta_B}{n_{13} \cos \delta_B + \cos \delta'_B} \right)^2 \left(\frac{2 \cos \delta_C}{n_{31} \cos \delta_C + \cos \delta'_C} \right)^2 \tag{6.37a}$$

$$I'_\theta = 0. \tag{6.37b}$$

— For the arrangement (c), Figure 6.10, with immersion fluid and identical indices of refraction of the specimen and prism by

$$\frac{I_\Phi}{k'I_0} = \frac{I_\Phi}{k'I_0} \left(\frac{2 \cos \delta_C}{n_{31} \cos \delta_C + \cos \delta'_C} \right)^2$$ (6.38a)

$$I'_\theta = 0.$$ (6.38b)

The quantity I_Φ is described by relations (5.6—5.19).

All three parallel scattered beams emerging from the isodyne collector (prism) with its axis parallel to the primary beam, namely, the beam I'_Φ ($\gamma = \gamma_0$, $\theta = 90°$), and the beam I'_Φ ($\gamma = -\gamma_0$, $\theta = 90°$), are linearly polarized. The plane of vibration is perpendicular to the direction of the primary beam. The signal-to-noise ratio of the radiation can be improved when the radiation passes through a linear analyzer with its axis parallel to the plane of vibration of the scattered beams. Such an analyzer does not introduce any change in the state of the scattered radiation other than a small decrease in intensity by the factor $4n_{13}/(n_{13} + 1)^2$ if it is aligned with the scattered radiation. A linear polarizer can be replaced by a circular one. The orientation of the circular polarizer is irrelevant because, regardless of its orientation, the radiant intensity of the scattered beams is reduced by the factor $4n_{13}/(n_{13} + 1)^2$ only.

The limitations of the above-presented relations or the conditions resulting thereof are obvious. For example, these relations require that the residual birefringence of the integrating prisms be within ranges formulated in commercial standards for optical glass.

6.2.3 Transfer functions of the observer or experimenter

In the traditional, mechanistic approach the observer or the experimentalist is not a component of the measurement or experimental system. The observer or the experimentalist is treated in such an approach as an objective entity, performing objective observations or measurements, "as it is". In other words, the observer or experimentalist is considered as a kind of a universal measurement instrument, characterized by a very high input impedance, by the response of the first-order instrument (proportional to any input signal), and responding to all kinds of energy flow in terms of ordinary linear differential equations with constant coefficients.

Such an approach is obsolete, incorrect, and very often leads to incorrect evaluation of accurate experimental data. Regarding the theory of experimentation, the human body and mind can be treated as a kind of measurement subsystem. The influence of the "human transfer function" is already critical at the level of cognition — the perceived reality, represented by constructed physical models, depends on the intellectual, scientific, and professional levels of the observer or experimenter. Not only do the responses of the human mind to reality depend on the level of intellectual sophistication of the particular individual, but in addition the responses of various human senses to various patterns of flow of various kinds of energy are different and depend on the energy form. Figures 6.12 and 6.13 illustrate this issue [19].

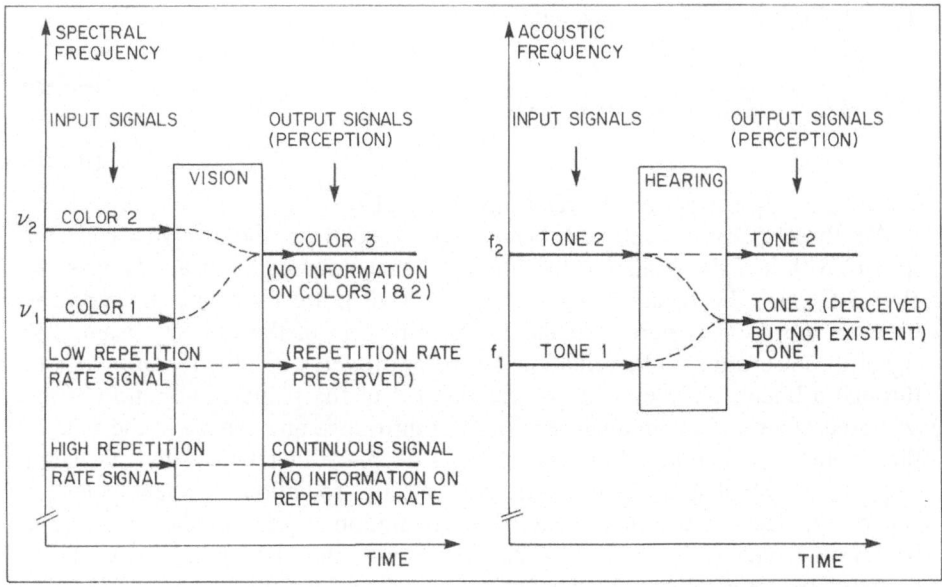

Fig. 6.12. Spectral responses of human vision and human hearing to the flow of radiant power and acoustic power. Human eye responds to photons and their energy. Human hearing responds to density waves (phonons). This is a typical response of various measurement instruments.

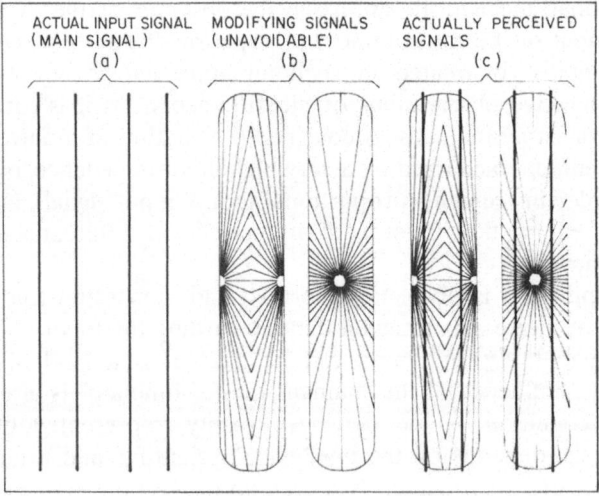

Fig. 6.13. Spatial responses of human vision. (a) — Input (observed) signal. (b) — Modifying signal (usually unavoidable). (c) — Perceived signal. This also is a typical response of various measurement instruments, it is real, thus it is not an illusion.

6.3 References

[1] Kiesling, E. W. and Pindera, J. T., "Linear Limit Stresses of Some Photoelastic and Mechanical Model Materials", *Experimental Mechanics* **9**(8), 1969, pp. 337—347.

[2] Pindera, J. T., *Rheological Photoelastic Properties of Some Polyester Resins* (in Polish), Parts 1, 2, and 3, Engineering Transactions (Rozprawy Inzynierskie), Polish Acad. Sciences, Vol. 7, (3, 4), 1959, pp. 361—411, 481—520, 521—540.

[3] Pindera, J. T., "Some Rheological Problems of Photoelastic Investigations" (in German), in: G. Haberland (Ed.), *Internationales spannungsoptisches Symposium* (Berlin, April 10—15, 1961), Akademie-Verlag, Berlin, 1962, pp. 155—172.

[4] Pindera, J. T., *Rheological Properties of Model Materials* (in Polish), Wydawnictwa Naukowo-Techniczne, Warszawa, 1962.

[5] Pindera, J. T., "Remarks on Properties of Photoviscoelastic Model Materials", *Experimental Mechanics* **7**(6), 1967, pp. 375—380.

[6] Pindera, J. T. and Kiesling, E. W., *On the Linear Range of Behaviour of Photoelastic and Model Materials*, VDI-Berichte No. 102, Experimentelle Spannungsanalyse, VDI-Verlag, Dusseldorf, 1966, pp. 89—94.

[7] Pindera, J. T. and Cloud, G., "On Dispersion of Birefringence of Photoelastic Materials", *Experimental Mechanics* **6**(9), 1966, pp. 470—480.

[8] Pindera, J. T., "On Physical Basis of Modern Photoelasticity Techniques", *Beiträge zur Spannungs- und Dehnungsanalyse*, Vol. V, Akademie-Verlag, Berlin, 1968, pp. 103—130.

[9] Pindera, J. T. and Sinha, N. K., "On the Studies of Residual Stresses in Glass Plates", *Experimental Mechanics* **11**(3), 1971, pp. 113—120.

[10] Pindera, J. T., "On the Transfer Properties of Photoelastic Systems", in: N. J. Prigorovski and H. K. Aben (Eds.), *Proc. of the 7th All-Union Conference on Photoelasticity*, Akademia Nauk Estonskoy SSR, Tallinn, 1971, pp. 48—63.

[11] Pindera, J. T. and Sinha, N. K., "Determination of Dominant Radiation in Polariscope by Means of Babinet Compensator", *Experimental Mechanics* **12**(1), 1972, pp. 1—5.

[12] Pindera, J. T. and Straka, P., "Response of the Integrated Polariscope", *Journal of Strain Analysis* **8**(1), 1973, pp. 65—76.

[13] Pindera, J. T., "Response of Photoelastic Systems", *Trans. of the CSME* **2**, 1973—74, pp. 21—30.

[14] Pindera, J. T. and Straka, P., "Mechanical and Dielectric Rheological Responses of Some High Polymers in Wide Temperature and Spectral Frequency Ranges", in: J. T. Pindera *et al.* (Eds.), *Experimental Mechanics in Research and Development*, Solid Mechanics Division, Study No. 9, 1973, pp. 601—615.

[15] Pindera, J. T. and Straka, P., "On Physical Measures of Rheological Responses of Some Materials in Wide Ranges of Temperature and Spectral Frequency", *Rheologica Acta* **13**(3), 1974, pp. 338—351.

[16] Pindera, J. T., Straka, P., and Krishnamurthy, A. R., "Rheological Responses of Materials Used in Model Mechanics", in: G. Bartolozzi (Ed.), *Proc. of the Fifth International Conference on Experimental Stress Analysis*, CISM, Udine, 1974, pp. 2.85—2.98.

[17] Pindera, J. T. and Krishnamurthy, A. R., "Characteristic Relations of Flow Birefringence, Part 1: Relations in Transmitted Radiation", *Experimental Mechanics* **18**(1), 1978, pp. 1—10.

[18] Pindera, J. T. and Krishnamurthy, A. R., "Characteristic Relations of Flow Birefringence, Part 2: Relations in Scattered Radiation", *Experimental Mechanics* **18**(2), 1978, pp. 41—48.

[19] Pindera, Jerzy T., "Foundations of Experimental Mechanics: Principles of Modelling, Observation and Experimentation", in: Pindera, J. T. (Ed.), *New Physical Trends in Experimental Mechanics*, Springer-Verlag, Wien, 1981, pp. 188—326.

[20] Pindera, J. T. and Krasnowski, B. R., "Rheological Factors in Theory of Isodyne Methods for Determination of Stress/Strain States in Composite Materials and Structures", in: *Proc. of the IX International Congress on Rheology*, Acapulco, Gro., Mexico, October 8—12, 1984. Advances in Rheology. Vol. 3: Polymers, pp. 741—748.

Experimental techniques of isodynes

7.1 Basic techniques

Certain aspects of the methodology and technique of the determination, evaluation and utilization of optical isodynes have already been described in the previous publications of the first co-author and his co-workers [8, 13, 16]. However, the picture presented in these publications is incomplete, partially inconsistent, and partially obsolete. The reason for it is that each new result was opening new horizons and, consequently, forcing generalizations of the developing frame of reference. It appears at the present time that the major features of the concept and consequences of the elastic and optical isodynes already have been developed; this made it possible to design the experiments and related instrumentation in a manner compatible with the underlying theory.

The methodology, technique and instrumentation presented in subsequent sections are disclosed in detail in the following USA patents: the United States Patent No. 4,679, 933 for the "Device for Birefringence Measurements Using Three Selected Sheets of Scattered Light (Isodyne Selector, Isodyne Collector, Isodyne Collimator) and the United States Patent No. 4,703,918 for the "Apparatus for Determination of Elastic Isodynes and of the General State of Birefringence Whole Field-Wise Using the Device for Birefringence Measurement in a Scanning Mode (Isodyne Polariscope) [17, 18].

7.1.1 Principle and scheme of the isodyne measurement system

Two different, but interrelated techniques are described below:

— line-wise measurements,
— plane-wise measurements.

7.1.1.1 Line-wise measurements. The basic model of a device for measuring birefringence along a chosen line within a birefringent body is presented in Figure 7.1, which is self-explanatory. The physical schematic of a device for linear measurements of birefringence, and in particular, of stress/strain bire-

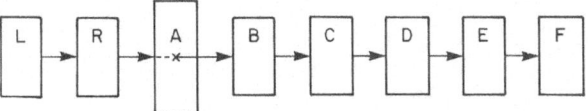

L - SOURCE OF RADIANT ENERGY (e.g. LASER)
R - MEANS TO SUITABLY POLARIZE THE LIGHT BEAM, LINEARLY
    OR CIRCULARLY
A - BIREFRINGENT BODY WHICH INTERNALLY SCATTERS LIGHT
    IN CHOSEN SPECTRAL RANGES
B - MEANS FOR PRIMARY SELECTION OF THREE MAJOR SCATTERED
    SHEETS OF LIGHT MODULATED BY BIREFRINGENCE : ISODYNE
    COLLECTOR , ISODYNE COLLIMATOR
C - MEANS FOR ISOLATION OF A SINGLE SHEET OF LIGHT :
    SHEET SELECTOR
D - MEANS FOR SECONDARY SELECTION AND FOCUSING OF THREE
    MODULATED SCATTERED SHEETS OF LIGHT : ISODYNE SELECTOR
E - MEANS FOR CONDITIONING AND NOISE REDUCTIONS
F - MEANS FOR DISPLAYING AND /OR RECORDING SELECTED SIGNAL

Fig. 7.1. Concept of a device for observation and recording of intensity of light scattered along the path of a primary light beam, in accordance with the theory of optical isodynes: Block schematis of device for birefringence measurements along a chosen line in a body.

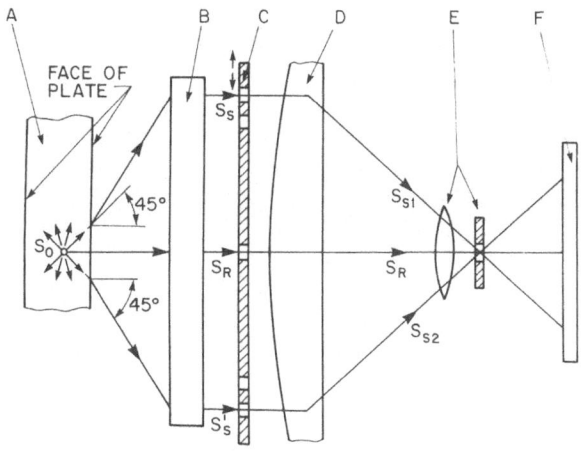

A...F : SEE Fig. 7.1

S_0 – LASER PRIMARY (MAIN) LIGHT BEAM: LINE OF
    MEASUREMENTS

S_R – LASER SCATTERED REFERENCE LIGHT SHEET

S_{S1}, S_{S2} – LASER SCATTERED COMPLEMENTARY LIGHT
    SHEETS OF INTEREST

Fig. 7.2. Practical solution of the optical system for collecting data on the scattered reference light sheet, and the scattered major and complementary light sheets which carry information on the isodyne function along a chosen line (isodyne collector): scheme of the system.

fringence, is given in Figure 7.2. The patterns of flow of radiant energy through
such a system are illustrated in Figure 7.3. The actual design of the component
B depends on several factors. The elementary design of the component *B* used
earlier in connection with the discussion of the transfer function of the system is
deficient in several respects, however it is very simple functionally. The
described system satisfies all the requirements following from the theories of
elastic and optical isodynes which were presented and discussed in the previous
chapters. Thus, the modulation of light intensity along the selected lines which
are obtained using this system corresponds to the cross-sections through the
isodyne surfaces discussed previously.

7.1.1.2 Plane-wise measurements. Plane-wise measurements yield photoelastic
isodynes in the whole field of measurements. The relation between the elastic
isodynes and photoelastic isodynes follow, in each particular case, from the
general theories of elastic and optical isodynes.

The particular system for the determination of optical isodynes, which
satisfies all the theoretical requirements following from the basic theories, is
called the isodyne recorder. The block schematic of the isodyne recorder is
presented in Figure 7.4. The principles of operation of the isodyne recorder are
illustrated by Figure 7.5. A particular technical solution in the form of a
structural scheme of the isodyne recorder is given in Figure 7.6. The obtained
recordings represent plane photoelastic isodyne fields in selected planes of
measurements within the tested specimen.

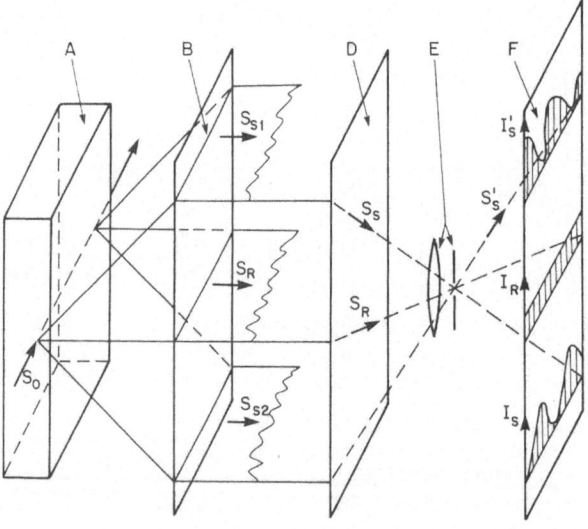

A, B, D, F, S_0, S_R, S_{S1}, S_{S2}: SEE Fig. 7.1

I_S, I_S', I_R : INTENSITIES OF SHEETS S_{S1}, S_{S2}, S_R

Fig. 7.3. Flow of radiant power through the isodyne collector and isodyne linear recorder:
scheme of operation of the system for linear measurements of birefringence.

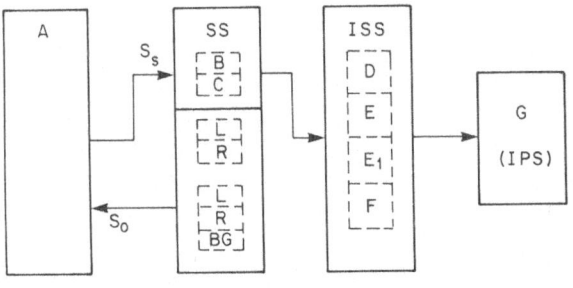

A   – BIREFRINGENT BODY (e.g ISODYNE SPECIMEN OR ISODYNE MODEL)
SS – SCANNING SYSTEM
IBS – IMAGE SELECTING SYSTEM
L   – SOURCE OF RADIANT ENERGY SUPPLYING PRIMARY BEAMS S_0
R   – MEANS TO SUITABLY POLARIZE PRIMARY BEAM S_0
BG – MEANS TO GUIDE THE PRIMARY BEAM
B   – MEANS FOR PRIMARY SELECTION OF THREE MAJOR MODULATED
        SCATTERED SHEETS OF LIGHT
C   – MEANS FOR ISOLATION OF A SINGLE SCATTERED SHEET OF LIGHT
D   – MEANS FOR SECONDARY SELECTION AND FOCUSING OF THREE
        MODULATED SCATTERED SHEETS OF LIGHT
E, E_1 – MEANS FOR CONDITIONING AND NOISE REDUCTION
F   – MEANS FOR IMAGE BUILDING (OPTICAL SYSTEM) AND FOR STORING
        (RECORDING) OF SELECTED SIGNAL
G   – MEANS FOR PROCESSING INFORMATION (ISODYNE PROCESSING
        SYSTEM – IPS)
ISS – IMAGE SELECTING SYSTEM

Fig. 7.4. Conceptual scheme of the isodyne recorder for linear and whole-field recordings (isodyne polariscope).

7.1.2 Immersion liquids

Analysis of the optical transfer function of the isodyne collecting subsystem discussed in Section 6.2 shows that it is desirable and easy to optimize the signal-to-noise ratio by submerging the specimen under investigation in a suitable immersion liquid. Submerging a specimen in an immersion liquid is even imperative when the boundary of the specimen is of an arbitrary shape and when plane-wise measurements are performed.

Two major requirements must be satisfied by the chosen immersion liquid:

— liquid must be safe;
— liquid should not significantly penetrate specimens made of polyester, epoxy and polymethylcrylate resins.

Several popular immersion liquids available on the market are not safe, are dangerous to health, and the use of some of them has been restricted by the law.

The information on the spectral dependence of refractive indices of some safe immersion liquids is given in Figure 7.7. Figure 7.8 presents the optical data on the spectral dependence of refractive indices of three popular resins, the polyester resin Palatal P6, the epozy resin Araldite 6060 and the epoxy

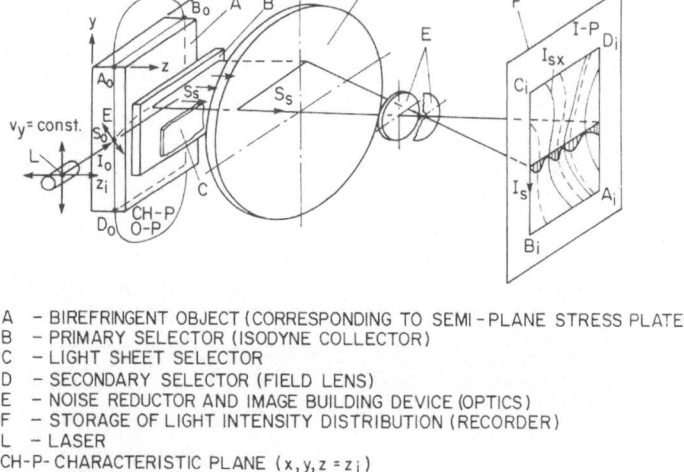

A – BIREFRINGENT OBJECT (CORRESPONDING TO SEMI – PLANE STRESS PLATE)
B – PRIMARY SELECTOR (ISODYNE COLLECTOR)
C – LIGHT SHEET SELECTOR
D – SECONDARY SELECTOR (FIELD LENS)
E – NOISE REDUCTOR AND IMAGE BUILDING DEVICE (OPTICS)
F – STORAGE OF LIGHT INTENSITY DISTRIBUTION (RECORDER)
L – LASER
CH-P- CHARACTERISTIC PLANE $(x, y, z = z_i)$
O-P- OBJECT PLANE
I-P- IMAGE PLANE
x,y,z – CARTESIAN COORDINATES
X – CHARACTERISTIC DIRECTION; DIRECTION OF PRIMARY BEAM S_0
y, v_y – SCANNING DIRECTION AND SCANNING VELOCITY
I_0, I_s – INTENSITIES OF PRIMARY BEAM S_0 AND SCATTERED BEAM S_s
I_{sx} – ISODYNES RELATED TO X – DIRECTION

Fig. 7.5. Flow of radiant power through the isodyne recorder (isodyne polariscope): scheme of the system for linear and whole-field recordings of the birefringence fields representing isodyne fields.

resin Epon 828, and the corresponding data on the matching immersion liquids, Dow Corning Diffusion Pump Fluids No. 704 and No. 705 (silicon polymers).

7.1.3 Recording techniques

The simplest technique of isodyne recording is the photographic technique. Obviously, the overall transfer function of the measurement and recording systems must be well known, otherwise, the produced recordings cannot be accepted as unequivocal. In particular, knowledge of the characteristic curve of photographic emulsion is imperative.

The most efficient recording technique is, at the present time, electronic recording using digital image storage. In this case also knowledge of the transfer function of the overall system is indispensible.

7.2 Particular techniques of isodynes

The theories and techniques of some particular methods developed to date are presented below. They are:

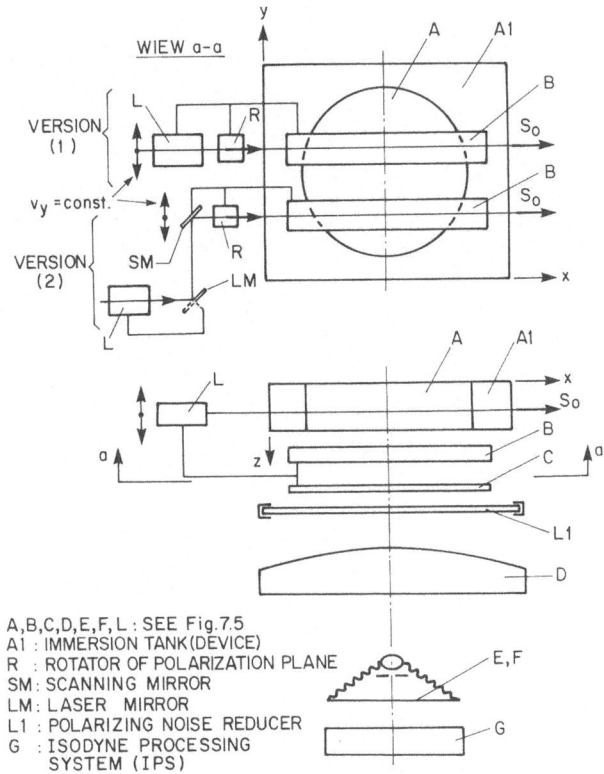

A,B,C,D,E,F,L : SEE Fig.7.5
A1 : IMMERSION TANK (DEVICE)
R : ROTATOR OF POLARIZATION PLANE
SM : SCANNING MIRROR
LM : LASER MIRROR
L1 : POLARIZING NOISE REDUCER
G : ISODYNE PROCESSING
     SYSTEM (IPS)

Fig. 7.6. Functional scheme of the isodyne recorder (polariscope) for linear and whole-field recordings.

— Two basic techniques of experimental determination of isodyne fields;
— Techniques of differentiation of isodyne fields to retrieve information on the normal stress components;
— Technique of isodyne coatings.

To assure reliability and repeatibility of experimentald data obtained using polarized light and polymeric materials, the experiments should follow the guidelines outlined in [6].

7.2.1 Techniques of experimental determination of isodyne fields

Two basic techniques leading to determination of isodyne fields have been developed: the amplitude modulation technique and the spatial frequency modulation technique.

7.2.1.1 Amplitude modulation technique. The amplitude modulation technique produces light intensity modulation which is directly related to the isodyne

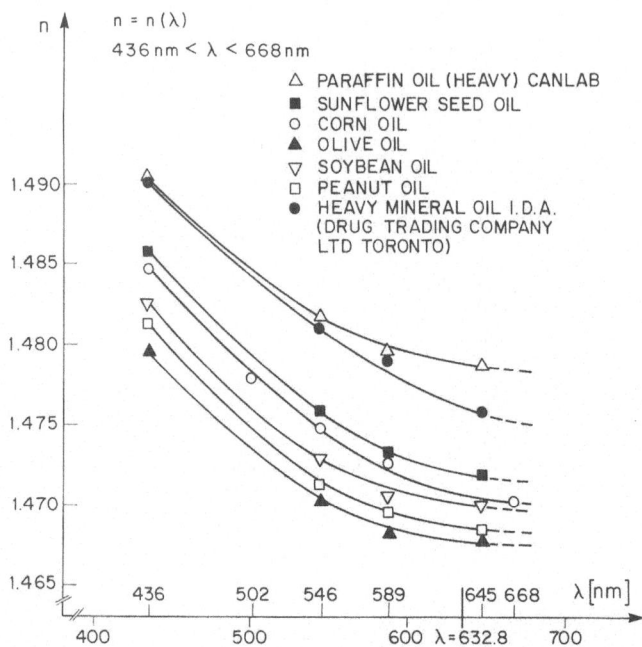

Fig. 7.7. Spectral dependence of the refractive indices of several immersion liquids of interest in isodyne techniques.

fields. This technique is applicable when the residual birefringence is weak. An example of the recordings of isodynes obtained using the amplitude modulation technique is presented in Figure 7.9.

The technique of retrieving information on optical isodynes from amplitude modulation recordings is straightforward. The lines of equal optical density of the photographic recording are the lines of equal radiant energy of the scattered light reaching the photographic plate or any other energy storing device, or of the radiant power reaching a recording device such as the CCD cameras. When the experiment satisfies the requirements which follow from the theory of optical isodynes, the lines of equal radiant energy or radiant power are the optical isodynes.

7.2.1.2 Spatial frequency modulation technique. The concept of the spatial frequency modulation technique is, in principle, identical with the concept of the temporal frequency modulation technique. Accordingly, the temporal carrier frequency is represented by the spatial frequency of a periodic and homogeneous distribution of the residual birefringence, Figure 7.10, left. This residual birefringence produces residual homogeneous isodyne fields for the x-and-y characteristic directions, depicted in Figure 7.10, left. This homogeneous spatial frequency of residual isodynes, analogous to the temporal carrier frequency, is modulated by the isodyne field induced by the applied load, Figure 7.10, right. It follows from the theory of optical isodynes presented

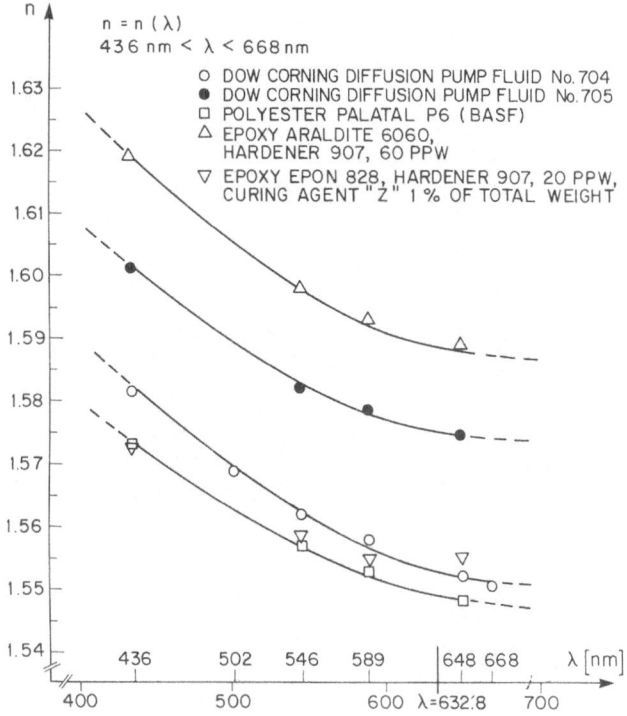

Fig. 7.8. Practical data on spectral refractive indices of some polymers used in isodyne techniques and of suitable immersion liquids.

in Chapter 5 that this is a linear problem, in contrast with the analogous problem of the transmission photoelasticity where the residual birefringence and the load-induced birefringence are not additive quantities. Thus the amount of spatial shift of homogeneously distributed residual isodynes is a measure of the order of load-induced isodynes. An example of application of both techniques is given in [15].

7.2.2 Evaluation of isodyne fields

7.2.2.1 General remarks. Various well-known techniques can be used to retrieve information from the spatial frequency modulation recordings. Here, one particular technique is presented which directly leads to determination of the lines of constant values of the first partial derivatives of an isodyne field, taken with respect to corresponding characteristic direction. Under the conditions specified below, the lines represent lines of constant values of the normal stress components,

$$\sigma_{xx} = \sigma_{xx}(x, y) = \text{const.}, \tag{7.1a}$$

$$\sigma_{yy} = \sigma_{yy}(x, y) = \text{const.} \tag{7.1b}$$

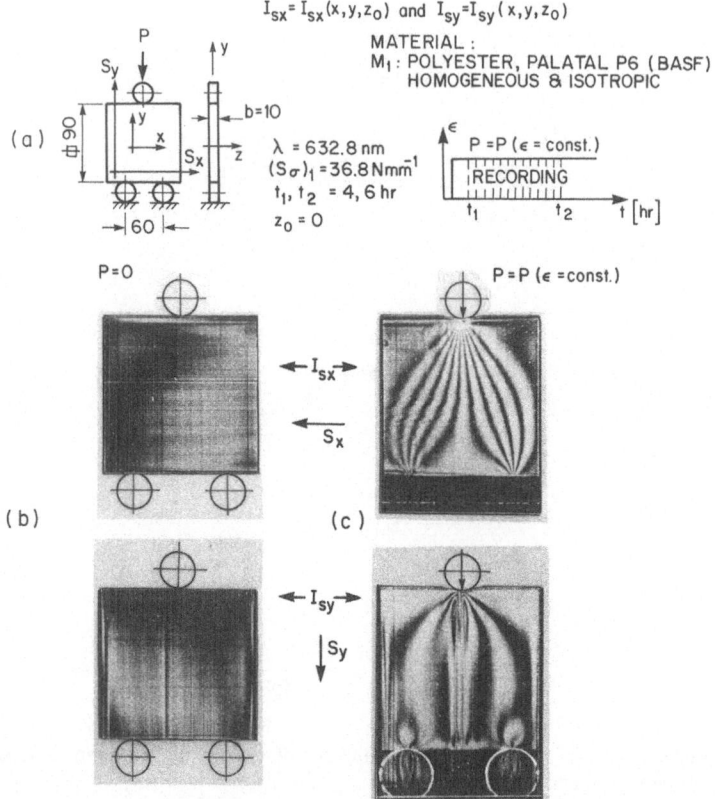

$I_{sx} = I_{sx}(x, y, z_0)$ and $I_{sy} = I_{sy}(x, y, z_0)$

MATERIAL :
M_1 : POLYESTER, PALATAL P6 (BASF)
HOMOGENEOUS & ISOTROPIC

(a)

b=10

λ = 632.8 nm
$(S_\sigma)_1$ = 36.8 Nmm^{-1}
t_1, t_2 = 4, 6 hr
z_0 = 0

$P = P (\epsilon = const.)$

RECORDING

t_1 t_2 t [hr]

P=0

$P = P (\epsilon = const.)$

I_{sx}

S_x

(b)

(c)

I_{sy}

S_y

Fig. 7.9. An example of isodyne recordings using the amplitude modulation technique. The recorded light intensity modulation represents directly the field of optical isodynes. (a) Data on experiment. (b) Zero-load recordings. (c) Load-induced isodynes.

This technique is based on the application of the image shift along the characteristic direction. Such a shift can be produced mechanically, optically, or electronically. The examples presented in Figure 7.11 had been obtained using the mechanical shift.

7.2.2.2 Optical differentiation techniques. The principle and techniques of the differentiation of isodyne fields in directions collinear with, and normal to, characteristic directions have been presented in the preceeding chapters. Such techniques yield values of normal and shear stress components along straight lines collinear with, and normal to, corresponding principal directions. As mentioned above, the technique presented here yields lines of constant values of normal stress components and, under particular circumstances following from the general theory of isodynes, values of shear stress components, as well.

Various methods have been developed to differentiate geometric patterns representing some functional relations. The simplest method is based on

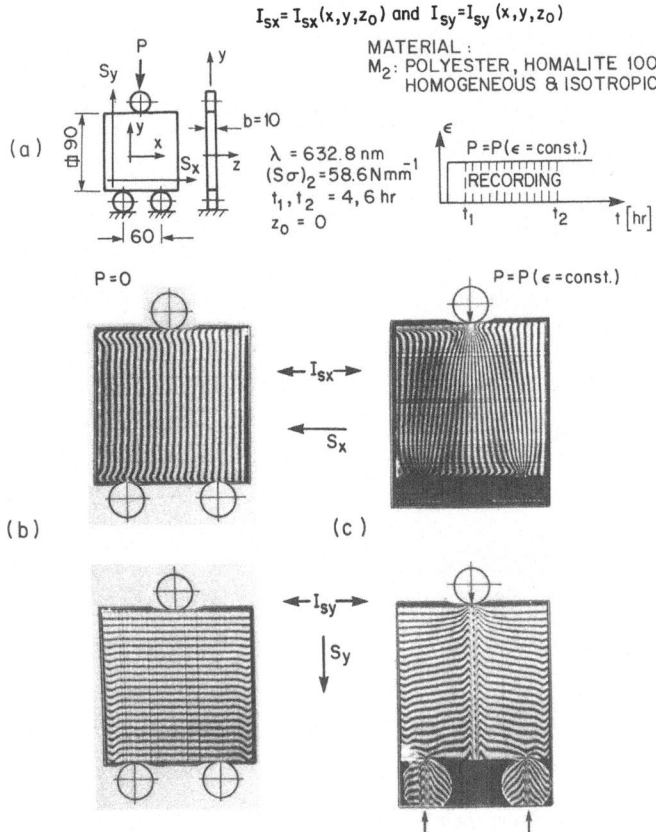

Fig. 7.10. An example of isodyne recordings using the spatial frequency modulation technique. (a) Data on experiment. (b) Zero-load recordings of residual isodyne fields. (c) Load-induced modulation of residual isodyne fields. The residual spatial frequency of light intensity (b) is modulated by the load-induced isodynes (c). The problem is linear, and the linear shift of the residual spatial frequency represent the load-induced optical isodynes. Information on the isodyne field is indirect.

mechanical shifting of identical patterns recorded photographically, as applied by Parks and Durelli [5], Gilbert [3], and others. More versatile methods use spatial filtering, optical image shifting devices, etc., reported by Sciammarella and Chang [23], Chiang [1], Pirodda [20], and others.

Optical differentiation usually requires higher densities of isodynes than those needed for graphical differentiation of a comparable accuracy. This condition can be satisfied by using various graphical or photographic methods. For instance, the Sabattier effect can be applied using the method of isodensitometry reported by Schwieger in 1955 [22]. The resulting isodenses represent isodynes of fractional orders. Such isodenses also can be produced using the known techniques of electronic image processing.

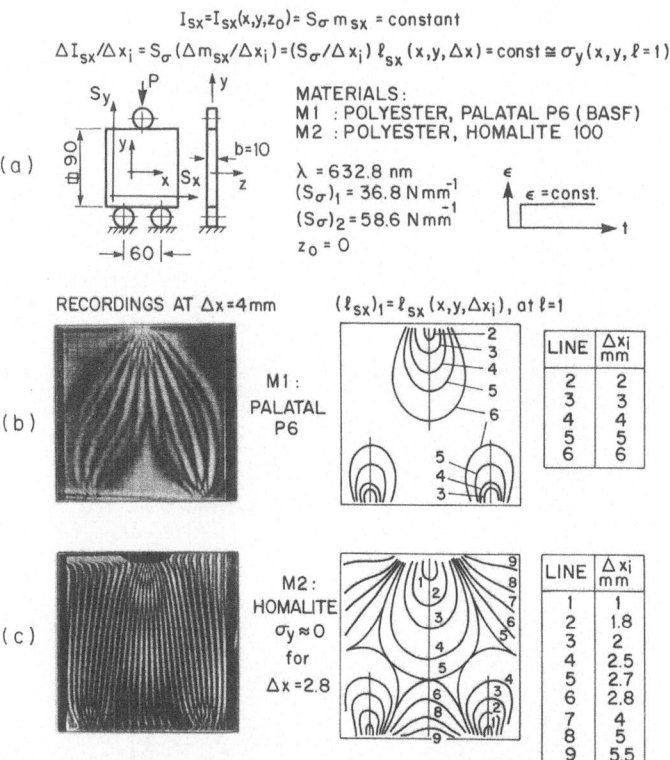

$$I_{sx} = I_{sx}(x,y,z_0) = S_\sigma \, m_{sx} = \text{constant}$$

$$\Delta I_{sx}/\Delta x_i = S_\sigma (\Delta m_{sx}/\Delta x_i) = (S_\sigma/\Delta x_i)\, \ell_{sx}(x,y,\Delta x) = \text{const} \cong \sigma_y(x,y,\ell=1)$$

(a)

MATERIALS:
M1 : POLYESTER, PALATAL P6 (BASF)
M2 : POLYESTER, HOMALITE 100

$b=10$

$\lambda = 632.8$ nm
$(S_\sigma)_1 = 36.8\ \mathrm{N\,mm^{-1}}$
$(S_\sigma)_2 = 58.6\ \mathrm{N\,mm^{-1}}$
$z_0 = 0$

$\epsilon = \text{const.}$

RECORDINGS AT $\Delta x = 4$ mm

$(\ell_{sx})_1 = \ell_{sx}(x,y,\Delta x_i)$, at $\ell=1$

(b)

M1 :
PALATAL
P6

LINE	Δx_i mm
2	2
3	3
4	4
5	5
6	6

(c)

M2 :
HOMALITE
$\sigma_y \approx 0$
for
$\Delta x = 2.8$

LINE	Δx_i mm
1	1
2	1.8
3	2
4	2.5
5	2.7
6	2.8
7	4
8	5
9	5.5

Fig. 7.11. Two examples of the optical differentiation of the isodyne fields by the mechanical shift in the corresponding characteristic direction.
(a) Data on experiment. (b) Differentiation of the amplitude modulation isodyne field. (c) Differentiation of the spatial frequency modulation isodyne field.

A more direct method of increasing the density of isodynes has been developed to avoid the inconvenience of the photographic procedure [12]. This method is based on the modulation of the spatial frequency of a homogeneous residual isodyne field by the induced isodyne field discussed in Section 7.2.1.2. The residual isodyne field of a desired spatial frequency results from a particular homogeneous residual, birefringence field. Only the basic relations are given below. General presentation of the underlying theory of moiré fringes is given by Theocaris [25], and more particular application by Pirodda [20].

The relations describing the fringes which result from a differentiation procedure are described below for one family of isodynes when the differentiation variable is collinear with the characteristic direction. The method of differentiation with respect to other variables is identical, therefore, it is not explicitly analyzed here.

In general, two families of isodynes formed by shifting with respect to the reference isodyne field by the distance $\pm 0.5\,\Delta x$ can be represented by the functions of I_{sx}^n. For simplicity the symbol "n" is deleted, so $I_{sx}^n \equiv I_{sx}$.

$$I_{sx}\left(x - \frac{\Delta x}{2}, y\right) = S_\sigma m'_{sx},\tag{7.2a}$$

$$I_{sx}\left(x + \frac{\Delta x}{2}, y\right) = S_\sigma m''_{sx}.\tag{7.2b}$$

These functions can be presented in the form of the Taylor series:

$$I_{sx}\left(x - \frac{\Delta x}{2}, y\right) = I_{sx}(x, y) - \frac{\partial I_{sx}}{\partial x}\frac{\Delta x}{2} + \frac{1}{2}\frac{\partial^2 I_{sx}}{\partial x^2}\frac{\Delta x^2}{4} -$$

$$- \frac{1}{6}\frac{\partial^3 I_{sx}}{\partial x^3}\frac{\Delta x^3}{8} + \dots,\tag{7.3a}$$

$$I_{sx}\left(x + \frac{\Delta x}{2}, y\right) = I_{sx}(x, y) + \frac{\partial I_{sx}}{\partial x}\frac{\Delta x}{2} + \frac{1}{2}\frac{\partial^2 I_{sx}}{\partial x^2}\frac{\Delta x^2}{4} +$$

$$+ \frac{1}{6}\frac{\partial^3 I_{sx}}{\partial x^3}\frac{\Delta x^3}{8} + \dots\tag{7.3b}$$

According to the theory of moiré fringes [25] the resulting fringes could be described by the equation:

$$I_{sx}\left(x + \frac{\Delta x}{2}, y\right) - I_{sx}\left(x - \frac{\Delta x}{2}, y\right) = (m'_{sx} - m''_{sx})S_\sigma = S_\sigma \ell_{sx}.\tag{7.4}$$

The lines

$$\ell_{sx} = \ell_{sx}(x, y),\tag{7.5}$$

represent a new family of lines produced by the "geometrial interference" of lines given by (7.2a) and (7.2b).

Using relations (7.3a) and (7.3b), one may rewrite equation (7.4) and present it in the form:

$$S_\sigma \ell_{sx} = \frac{\partial I_{sx}}{\partial x}\Delta x + \frac{1}{6}\frac{\partial^3 I_{sx}}{\partial x^3}\frac{\Delta x^3}{4} + \dots,\tag{7.6}$$

or

$$S_\sigma \frac{1}{\Delta x}\ell_{sx} = \frac{\partial I_{sx}}{\partial x} + \frac{1}{6}\frac{\partial^3 I_{sx}}{\partial x^3}\frac{\Delta x^2}{4} + \dots.\tag{7.7}$$

Thus, one can approximate (7.7) with a satisfactory accuracy by:

$$S_\sigma \frac{1}{\Delta x}\ell_{sx} = \frac{\partial I_{sx}}{\partial x},\tag{7.8a}$$

when

$$\frac{\partial I^x}{\partial x} \gg \frac{1}{6} \frac{\partial^3 I_{sx}}{\partial x^3} \frac{\Delta x^3}{4} + \ldots \qquad (7.8b)$$

Thus the lines

$$\ell_s = \ell_{sx}(x, y),$$

represent loci of points where the first partial derivative of the isodyne taken in the characteristic direction is constant under the conditions:

— that the values of higher partial derivatives of isodynes are very small;

or

— that the values of geometric shifts are very small.

The second condition requires that the density of isodynes be sufficiently high relative to the size of the object.

The criterion for "small" given by (7.8b) is optimally approximated when

$$\ell_{sx} = 1. \qquad (7.9)$$

Obviously, the line $\ell_{sx} = 1$ represents the geometric locus of points where

$$\sigma_{yy} = \sigma_{yy}(x, y) = \text{const} = S_\sigma \frac{1}{\Delta x} (\ell_{sx} = 1). \qquad (7.10)$$

Consequently, in order to obtain a set of lines where $\sigma_{yy} = \text{const}$ covering the whole stress field, it is necessary to use a corresponding set of geometric shifts, Δx_i.

It follows from (7.7) and (7.8) that the lines

$$\ell_{sx} = \ell_{sx}(x, y) \text{ for } \ell > 1, \qquad (7.11)$$

are functions of higher order derivatives of isodynes.

One must note that, when using the spatial frequency modulation method described below, one must modify equation (7.8) in the following manner:

$$S_\sigma \frac{1}{\Delta x} \dot{\ell}_{sx} = \frac{\partial I_{sx}}{\partial x} + \left(\frac{\partial I_{sx}}{\partial x} \right)_0, \qquad (7.12)$$

where the term

$$\left(\frac{\partial I_{sx}}{\partial x} \right)_0 = \left[\frac{\partial I_{sx}}{\partial x} (x, y) \right]_0, \qquad (7.13)$$

denotes the influence of the initial modulation.

Presented below are some examples of the initial modulation introduced in the form of initially homogeneous residual modulation (residual isodyne field):

$$\left(\frac{\partial I_{sx}}{\partial x} \right)_0 = \text{const.} \qquad (7.14)$$

The application of the presented method is demonstrated and discussed using, as an example, the stress field in a square plate loaded by three forces, Figures 7.9 and 7.10.

Two techniques of producing the derived fringes are presented:

— amplitude modulation technique using the load induced optical isodyne fields only;

and

— spatial frequency modulation technique using the initially homogeneous isodyne field, modulated by the load induced isodynes.

The employed specimens, boundary and loading conditions, and experimental parameters are given in Figures 7.9—7.11.

The material of the specimen presented in Figure 7.9 was essentially homogeneous and isotropic, both mechanically and optically. The small optical anisotropy phenomenologically represents a noise level which is very low. One may note that for the loading conditions given by $P = P(\varepsilon = \text{const})$, the value of the material stress-optic coefficient for Palatal P6 depends only on the wavelength of light, and is practically independent of the time of loading, according to References [9] and [10]. The values of the material stress optic coefficients were determined using creep calibrators described in [6].

Figures 7.9 and 7.10 present two families of x- and y-isodynes related to x- and y-characteristic directions. For comparison, the isodyne fields in the unloaded state are included in Figure 7.9; the residual isodyne field is very weak and easy to account for.

The experiments presented in Figure 7.10 were carried out under conditions identical with those presented in Figure 7.9, however, the material of the specimen was optically homogeneous, but anisotropic. This particular residual optical anisotropy produces uniform isodyne fields whose lines are straight and normal to the x- and y-characteristic directions, respectively. The recordings of the residual isodyne fields are presented on the left side of Figure 7.10. The results of the load-induced modulation of the residual isodyne fields are visible in the recordings given on the right side of Figure 7.10.

The shift differentiation technique was used to evaluate the isodyne recordings presented in Figures (7.9) and (7.10). The resulting optical fringes are given in Figure 7.11. Both methods, amplitude modulation and spatial frequency modulation, yield qualitatively identical results; the character of the lines

$$\frac{\Delta I_{sx}}{\Delta x} = S_\sigma \frac{1}{\Delta x} \ell_{sx} = \sigma_{yy}(x, y) = \text{const.,} \tag{7.15}$$

where

$$S_\sigma = S_s \tag{7.16}$$

are identical in both cases. However, the quality of recordings and the overall resolution and accuracy of the results is higher when the spatial frequency modulation method is used.

The recordings of the mechanical shift differentiation presented in Figure 7.11 are related to the shift $\Delta x = 4$ mm. These recordings present, in addition to the ℓ-lines of the basic 1st order, $\ell = 1$, ℓ-lines of higher orders which do not directly represent the first partial derivatives of the isodynes, as shown above, (7.8a, b).

It should be mentioned that the method of optical differentiation of the isodyne fields using mechanical shifting, although apparently very attractive, yields reliable results in a limited number of cases only, when the standard technique is applied. However, the reliability and accuracy of this method can be significantly increased when the spatial frequency modulation technique is applied.

The presented methods of optical differentiation allow us to readily determine the character of the normal stress distribution whole field-wise. However, it follows from relation (7.7) that the standard graphical, numerical or electronic differentiation will yield more accurate results, especially in regions where stress gradients are high, e.g. regions of contact, notches, cracks, etc.

7.2.3 Isodyne coatings

Photoelastic coatings introduced by Felix Zandman [26] are widely used in engineering practice because of several practical advantages in comparison with other methods. The theory and technique of the photoelastic coatings method are well developed [27, 28]. One of the limitations of this method is the low level of optical effects. The method of isodyne coatings allows us to increase the efficacy of the photoelastic coatings method by utilizing the technique of optical isodynes.

This Section presents the theory and technique of this particular application of the isodyne method. It is shown that by a suitable choice of materials for the known technique of photoelastic coatings it is possible to obtain fields of optical isodynes carrying information on all stress components at the surface of an actual machine part or engineering structure.

Isodyne coatings differ from typical birefringent coatings in the following ways:

— isodyne coatings do not require a reflective layer;
— isodyne fields in isodyne coatings essentially are not influenced by the thickness of the coatings;

Consequently, the requirements concerning the uniformity of the thickness of photoelastic coatings are not important. No special devices or accessories are needed to determine the stress components at chosen points.

It can be shown that by using an isodyne coating material exhibiting a negligible time dependence of the photoelastic relaxation modulus $C_\varepsilon(t)$ it is possible to eliminate all the problems related to the not-negligible rheological behavior which is exhibited by all photoelastic materials with the exception of glass and semiconductors such as silicon or germanium [2].

As an illustration, analytically and experimentally determined strain fields at the surfaces of a beam in bending are compared and discussed.

The analysis of the elastic interaction between the object and the coating is limited to cases where the influence of the alteration of curvature of the object's surface can be neglected. As a particular case, a plane stress field in an element is considered; however, this fact does not limit the generality of the presented relationships. These relationships apply to objects bounded by curved surfaces too. The derived relations represent transfer functions in a matrix form relating the stress components in a coating to those at the surface of an object.

Two cases of transfer functions are considered:

— coating transfer function for the case when the reinforcing effect of the coating is neglected; and
— coating transfer function with the reinforcing effect taken into account.

In the first case, the following assumptions are made:

— the material of the coating is satisfactorily approximated by Hooke's body;
— the stress field in the coating is two-dimensional;
— the in-plane strain components at the interface between the object and the coating are equal, Figure 7.12.

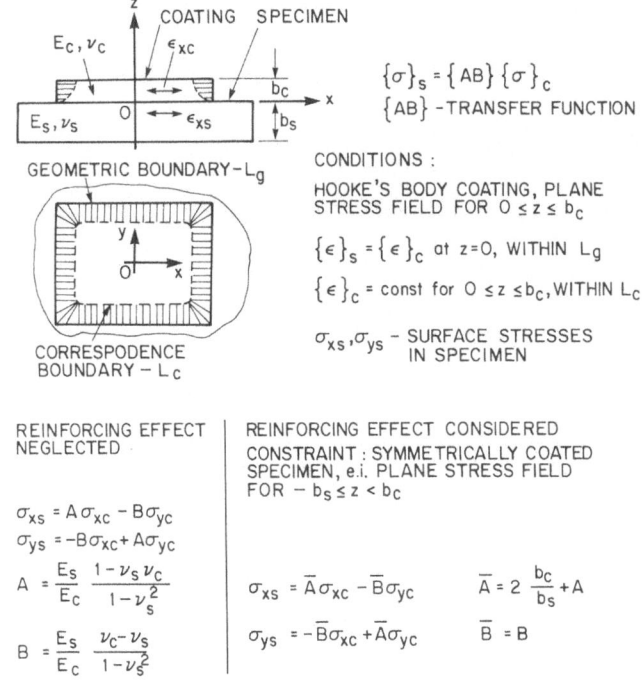

Fig. 7.12. Stress analysis using isodyne coatings. Physical and mathematical models of interaction between an elastic object and an elastic coating. A model of a transfer function. Shaded areas are outside of the predictive power of the presented relations.

This leads to the following formulation of the problem:

— with regard to the in-plane strain components,

$$[\varepsilon]_s = [\varepsilon]_c, \text{ at } z = 0,$$ (7.17a)

$$[\varepsilon]_c = \text{const, at } 0 < z < b_c,$$ (7.17b)

where the subscripts s and c refer to the object, (specimen), and the coating, respectively.

— With regard to the normal stress components at the surface of the specimen,

$$\sigma_{xxs} = \frac{E_s}{1 - v_s^2} (\varepsilon_{xxs} + v_s \varepsilon_{yys}),$$ (7.18a)

$$\sigma_{yys} = \frac{E_s}{1 - v_s^2} (\varepsilon_{yys} + v_s \varepsilon_{xxs}).$$ (7.18b)

— With regard to the normal stress components in the coating,

$$\sigma_{xxc} = \frac{E_c}{1 - v_c^2} (\varepsilon_{xxc} + \varepsilon_{yyc}),$$ (7.19a)

$$\sigma_{yyc} = \frac{E_c}{1 - v_c^2} (\varepsilon_{yyc} + v_c \varepsilon_{xxc}).$$ (7.19b)

The conditions (7.17) and the relations (7.18) and (7.19) yield,

$$[\sigma]_s = \{AB\}[\sigma]_c,$$ (7.20a)

or, in this case,

$$\sigma_{xxs} = A\sigma_{xxc} - B\sigma_{yyc}$$ (7.20b)

$$\sigma_{yys} = -B\sigma_{xxc} + A\sigma_{yyc},$$ (7.20c)

where

$$A = \frac{E_s}{E_c} \frac{1 - v_s v_c}{1 - v_s^2}; \quad B = \frac{E_s}{E_c} \frac{v_c - v_s}{1 - v_s^2}.$$ (7.21)

The matrix $\{AB\}$ represents the transfer function of interest,

$$\{AB\} = \begin{bmatrix} A & -B \\ -B & A \end{bmatrix}.$$ (7.22)

A sample of one of relations (7.20) is presented in Figure 7.13. In the second case, when the reinforcing effect of the coating is considered, two more assumptions are added to those formulated above:

— the stress field in the object is two-dimensional;
— the object is coated symmetrically.

This leads to the following formulation of the problem:

$$\{\sigma\}_s = \{AB\}\,\{\sigma\}_c$$

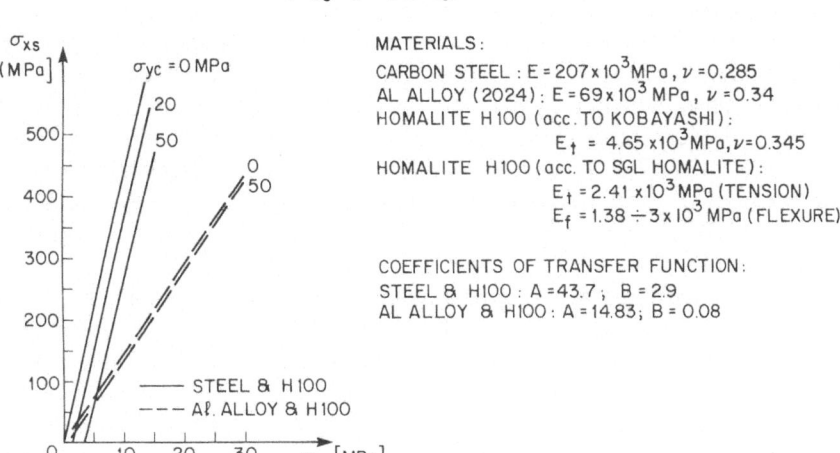

MATERIALS:

CARBON STEEL : E = 207×10^3 MPa, ν = 0.285
AL ALLOY (2024) : E = 69×10^3 MPa, ν = 0.34
HOMALITE H100 (acc. TO KOBAYASHI):
$\qquad E_t = 4.65 \times 10^3$ MPa, ν = 0.345
HOMALITE H100 (acc. TO SGL HOMALITE):
$\qquad E_t = 2.41 \times 10^3$ MPa (TENSION)
$\qquad E_f = 1.38 \div 3 \times 10^3$ MPa (FLEXURE)

COEFFICIENTS OF TRANSFER FUNCTION:
STEEL & H100 : A = 43.7 ; B = 2.9
AL ALLOY & H100 : A = 14.83 ; B = 0.08

Fig. 7.13. Stress analysis using isodyne coatings. An example of the elastic interaction between an object and a coating, represented by the relation $\sigma_{xxs} = \sigma_{xxs}(\sigma_{xxc}, \sigma_{yyc}, \nu_c, \nu_s, E_c/E_s)$ where the subscripts c and s denote the coating and the object surface, respectively: relations for several materials when reinforcing effect is neglected.

$$[\varepsilon]_s = [\varepsilon]_c, \text{at} -(b_c + b_s) < s < b_c. \tag{7.23}$$

The reinforcing effect of both layers of the coating can be determined by equating the forces acting on a three-layered element to the forces acting on the uncoated element, Figure 7.9.

$$b_s \, dy \, \sigma_{xxs} = b_s \, dy \, \sigma'_{xxs} + 2b_c \, dy \, \sigma_{xxc}, \tag{7.24a}$$

$$b_s \, dx \, \sigma_{yys} = b_s \, dy \, \sigma'_{yys} + 2b_c \, dy \, \sigma_{yyc}, \tag{7.24b}$$

where σ'_{xxs} and σ'_{yys} denote stress components in the coated object.

Condition (7.23) together with relations (7.18), (7.19) and (7.24) yields:

$$[\sigma]_s = \{AB\}'[\sigma]_c \tag{7.25}$$

where

$$A' = \frac{2b_c}{b_s} + A; B' = B. \tag{7.26}$$

The matrix $\{AB\}'$ represents the transfer function of interest,

$$\{AB\}' = \begin{bmatrix} A' & -B' \\ -B & A \end{bmatrix}. \tag{7.27}$$

A structural component or a machine part with an applied isodyne coating represents a composite structure consisting of two or three layers. The overall response of such a system depends on whether or not the responses of the individual components are linearly or nonlinearly elastic, or viscoelastic.

The initial mechanical response of a typical model structural component is usually linearly elastic with insignificant thermoelastic effect. Because of known advantages, several high polymers can be used for photoelastic and isodyne coatings. However, as discussed earlier, such materials are inherently visco-elastic [9—11; 14]. Thus, to correctly design an isodyne coating experiment, one has to consider the parameters of the experiment, together with the mechanical and optical viscoelastic responses of photoelastic and isodyne coatings as discussed above.

It was stated earlier, that the optical relaxation function is practically constant with the time of relaxation for some photoelastic materials as some epoxy resins and polyester resins, when the temperature is well below or above the tempera-ture of the second order transition [9, 10]. Even at the temperature very close to the second order transition, the time-dependence of the optical strain coefficient is often weak as illustrated by the bottom graph of Figure 6.7. Nevertheless, the time-dependence of the mechanical relaxation modulus is not negligible and does influence the general reinforcing effect described above. Therefore, the transfer functions described by (7.20), (7.21) and (7.26) are actually time-dependent. At small deformations, this dependence may be approximated by:

$$A = A(t) = \frac{E_s}{E_c(t)} \frac{1 - v_s v_c(t)}{1 - v_s^2} , \qquad (7.28a)$$

$$B = B(t) = \frac{E_s}{E_c(t)} \frac{v_c(t) - v_s}{1 - v_s^2} . \qquad (7.28b)$$

The above-described relationships have been accounted for in the design of the experiments. In particular, the experiments were performed within the linear limit stress and strain ranges [7].

In addition, a supplementary set of experiments has been performed to allow a direct comparison of the responses of isodyne coating beams under two loading conditions, Figure 7.14:

 — coating beam loaded by three forces;
 — coating beam loaded by interface forces.

To increase the resolution of measurement, aluminum alloy has been chosen as the material for the investigated object in conjunction with the technique of the spatial modulation of isodyne fields, see Section 7.2.1.2.

The scheme of the experiments and the recordings of isodyne fields are given in Figure 7.14. The recordings on the left are related to the non-attached coatings; the recordings on the right present isodyne fields in the attached coatings. The top four recordings present residual isodyne fields. The strain-modulated isodyne fields are seen in the four bottom recordings.

The isodyne recordings presented in Figure 7.14 have been used to evaluate the results which are given in Figure 7.15. Two beam cross-sections have been analyzed: cross-section $(x_1 = 5, y)$, and cross-section $(x_2 = 35, y)$. The following

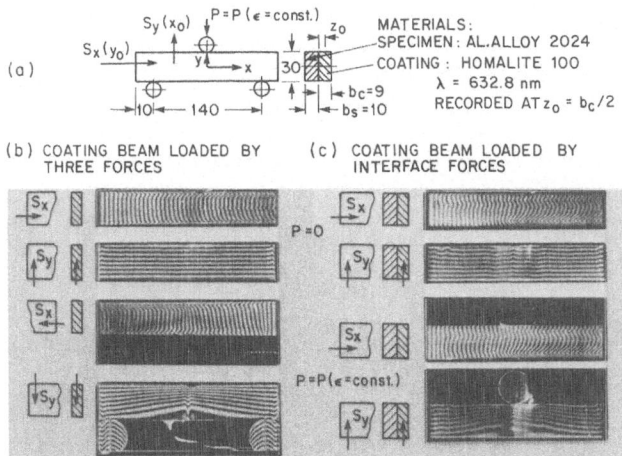

Fig. 7.14. Stress analysis using isodyne coatings and spatial frequency modulation technique. Case of a beam loaded by three forces. Residual and load-modulated isodyne fields in the initially homogeneous and optically anisotropic coatings. (a) Data on experiment. (b) Coating is not attached to the beam under study. (c) Coating is joint with the beam surface.

relations for these cross-sections are presented in Figure 7.15 for both coating beams, non-attached and attached:

   (a) residual isodyne function, $(m_{syy})_0$,
   (b) modulated isodyne function, m_{syy},
   (c) slope of the induced isodyne function, $d/dy(m_{syy})$.

The last function is proportional to the distribution of the stress component normal to the y-direction across the cross-section. Because the value of the coefficient B is very small, Figure 7.13, the values of the x-stress components in the coating are practically proportional to the x-stress components in the object.

Since the basic relations (7.20), (7.21) and (7.26) have been derived for the case where all the components of a composite structure are linearly elastic, whereas one component of the considered structure is actually viscoelastic, relations (7.28) represent a first approximation, acceptable only when the deformations are small.

Calculation of the parameters A and B of the transfer function relating stress fields in the object and the coating, Figure 7.13, is based on the values of material coefficients E and ν reported in [4]. The recordings of the isodynes given in Figure 7.14 show that the actual values of Poisson ratios for materials of the object and the coating (Al. alloy 2024 and Homalite H100) are indeed very close.

One must note that a too optimistic simulation of the actual viscoelastic responses of the coating material by a linear model may lead to serious errors. For instance, values of the E-modulus of polymers depend strongly on the testing technique and may differ by a factor of 3 for the same material as is

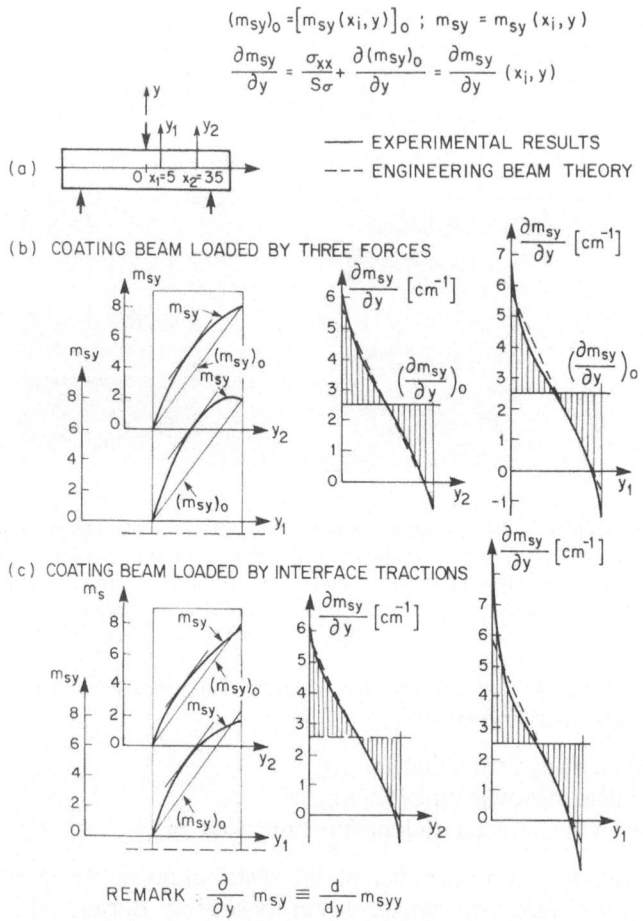

$$(m_{sy})_0 = [m_{sy}(x_i, y)]_0 \; ; \; m_{sy} = m_{sy}(x_i, y)$$

$$\frac{\partial m_{sy}}{\partial y} = \frac{\sigma_{xx}}{S_\sigma} + \frac{\partial (m_{sy})_0}{\partial y} = \frac{\partial m_{sy}}{\partial y}(x_i, y)$$

——— EXPERIMENTAL RESULTS

- - - - ENGINEERING BEAM THEORY

(a)

(b) COATING BEAM LOADED BY THREE FORCES

(c) COATING BEAM LOADED BY INTERFACE TRACTIONS

REMARK : $\dfrac{\partial}{\partial y} m_{sy} \equiv \dfrac{d}{dy} m_{syy}$

Fig. 7.15. Beam loaded by three forces. Stress analysis in selected cross-sections using optical isodyne coatings: evaluation of normal stress components (see Figure 7.14). (a) Data on experiment. (b) Normal stresses in coating beam loaded by three forces. (c) Normal stresses in coating beam loaded by interface tractions when the specimen beam is loaded by three forces.

shown in Figure 7.13 [24]. Such data have a technological significance only, and such values cannot be used as constant parameters of mathematical models to predict behaviour of bodies composed of viscoelastic materials.

It also must be noted that under certain circumstances the stress/strain gradient in the isodyne coating may be high. Such a stress state in the coating may cause a noticeable bending of the light beam and a spatial separation of the ordinary and extraordinary light beams [19]. As a result, both conjugated light beams would not interfere and isodyne fringes would disappear.

7.2.4 Illustrative example

A prismatic, rectangular beam loaded by three forces has been chosen to demonstrate the relationship between the analytical and optical isodynes, Figure 7.16a, b, c, d, in this example.

7.2.4.1 Analytical iscdynes in a beam under a three-point load. The stress state in the beam in Figure 7.16a is approximated by the stress state given in Figure 7.16b, and the related stress function is chosen as

$$\Phi(x, y) = b_2 xy + \frac{d_4}{6} xy^3. \tag{7.29}$$

Thus

$$\sigma_{xx} = \frac{\partial^2 \Phi}{\partial y^2} = d_4 xy, \tag{7.30a}$$

$$\sigma_{yy} = \frac{\partial^2 \Phi}{\partial x^2} = 0, \tag{7.30b}$$

$$\sigma_{xy} = \sigma_{yx} = -\frac{\partial^2 \Phi}{\partial x \partial y} = -b_2 - \frac{d_4}{2} y^2. \tag{7.30c}$$

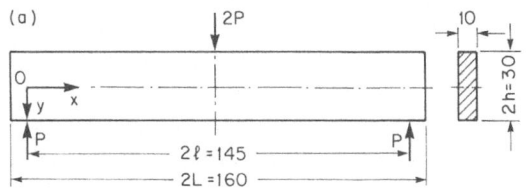

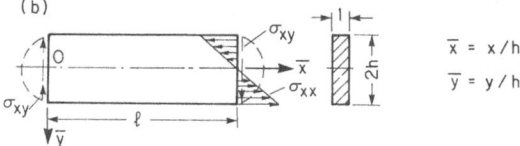

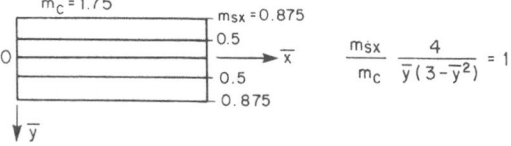

(c) m_{sx} – ANALYTICAL ISODYNES (X – ISODYNES)

$$\frac{m_{sx}}{m_c} \frac{4}{\bar{y}(3-\bar{y}^2)} = 1$$

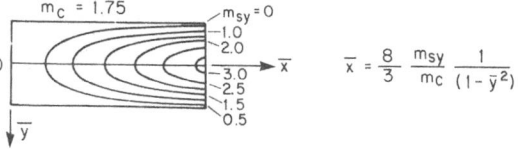

(d) m_{sy} – ANALYTICAL ISODYNES (Y – ISODYNES)

$$\bar{x} = \frac{8}{3} \frac{m_{sy}}{m_c} \frac{1}{(1-\bar{y}^2)}$$

Fig. 7.16. Beam under three point load. Plane analytical elastic isodynes. (a) System under consideration. (b) Boundary conditions. (c) X-isodynes. (d) Y-isodynes.

Boundary and load conditions yield:

$$y = \pm h: \sigma_{xy} = -b_2 - \frac{d_4}{2} h^2 = 0;$$ (7.31)

$$-\int_{-h}^{h} \sigma_{xy} \, dy = P: 2b_2 + \frac{d_4}{3} h^3 = \frac{P}{h};$$ (7.32)

$$b_2 = \frac{3}{4} \frac{P}{h}; \; d_4 = -\frac{3}{2} \frac{P}{h^3}.$$ (7.33)

Consequently:

$$\Phi(x, y) = \frac{3}{4} \frac{P}{h} \left(xy - \frac{1}{3h^2} xy^3 \right);$$ (7.34)

$$\frac{\partial \Phi}{\partial x} = \frac{3}{4} \frac{P}{h} \left(y - \frac{1}{3h^2} y^3 \right) = \frac{3}{4} \frac{P}{h} y \left(1 - \frac{y^2}{3h^2} \right);$$ (7.35a)

$$\frac{\partial \Phi}{\partial y} = \frac{3}{4} \frac{P}{h} \left(x - \frac{1}{h^2} xy^2 \right) = \frac{3}{4} \frac{P}{h} x \left(1 - \frac{y^2}{h^2} \right).$$ (7.35b)

Introducing,

$$\bar{y} = \frac{y}{h}; \; \bar{x} = \frac{x}{h};$$ (7.36)

relations (7.35) can be presented in the form

$$\frac{\partial \Phi}{\partial x} = \frac{3}{4} P\bar{y} \left(1 - \frac{\bar{y}^2}{3} \right);$$ (7.37a)

$$\frac{\partial \Phi}{\partial y} = \frac{3}{4} P\bar{x} \left(1 - \frac{\bar{y}^2}{3} \right).$$ (7.37b)

The integration functions of isodynes (3.18a, b; 3.25a, b; 3.26a, b) are:

$$f_x(y) = -\int_0^s \bar{q} \cdot \bar{j} \, ds = -\int_0^y 1\sigma_{xy} \, dy$$

$$= \frac{3}{4} \frac{P}{b} y \left(1 - \frac{y^2}{3b^2} \right) = \frac{3}{4} P\bar{y} \left(1 - \frac{\bar{y}^2}{3} \right);$$ (7.38a)

and

$$f_y(x) = \int_0^s \bar{q} \cdot \bar{i} \, ds = 0.$$ (7.38b)

Thus the analytical isodynes can be expressed by:

$$S_s m_{sy}(x, y) = \frac{\partial \Phi}{\partial y} = \frac{3}{4} P\bar{x}(1 - \bar{y}^2);$$ (7.39a)

$$S_s m_{sx}(x, y) = \frac{\partial \Phi}{\partial x} + f_x(y) = \frac{1}{2} P\bar{y}(3 - \bar{y}^2)$$ (7.39b)

where

$$P = (1/2)S_s m_c,$$ (7.40)

and m_c is the maximal order of isodyne in calibration disk for the characteristic directions normal to the load direction. Finally,

$$m_{sy}(x, y) = \frac{3}{8} m_c \bar{x}(1 - \bar{y}^2);$$ (7.41a)

$$m_{sx}(x, y) = \frac{1}{4} m_c \bar{y}(3 - \bar{y}^2).$$ (7.41b)

Relations (7.41) represent isodyne fields. Isodynes are lines for wh en the fiues of m_{sx} or m_{sy} are constant:

$$m_{sy} = \frac{3}{8} m_c \bar{x}(1 - \bar{y}^2) = \text{const};$$ (7.42a)

$$m_{sx} = \frac{1}{4} m_c \bar{y}(3 - \bar{y}^2) = \text{const}$$ (7.42b)

Explicitly: y-isodynes are given by:

$$\bar{x} = \frac{8}{3} \frac{m_{sy}}{m_c} \frac{1}{1 - \bar{y}^2}, \quad \text{for} \quad -1 \leqslant \bar{y} \leqslant 1;$$ (7.43a)

and x-isodynes by:

$$4 \frac{m_{sx}}{m_c} \frac{1}{\bar{y}(3 - \bar{y}^2)} = 1 \quad \text{for} \quad -1 \leqslant \bar{y} \leqslant 1.$$ (7.43b)

The isodynes described by (7.43a, b) are shown in Figure 7.16.

7.2.4.2 Optical isodynes. Comparison of results. Optical isodynes for x- and y-characteristic directions are presented in Figure 7.17a, b, c. The comparison between the elastic analytical and optical isodynes is given in Figure 7.18. It follows that in this case the very simple stress function (7.29) satisfactorily describes the isodyne field at sufficient distances from the boundaries where the assumptions regarding the loading and deformation conditions are clearly not satisfied, and where the local, three-dimensional effects are not negligible.

7.2.5 Auxiliary methods

Basically, the optical isodyne methods are self-sustained; the number of obtained empirical data is sufficient to directly check the reliability and accuracy of results. However, in some cases it is worthwhile to obtain independent informa-

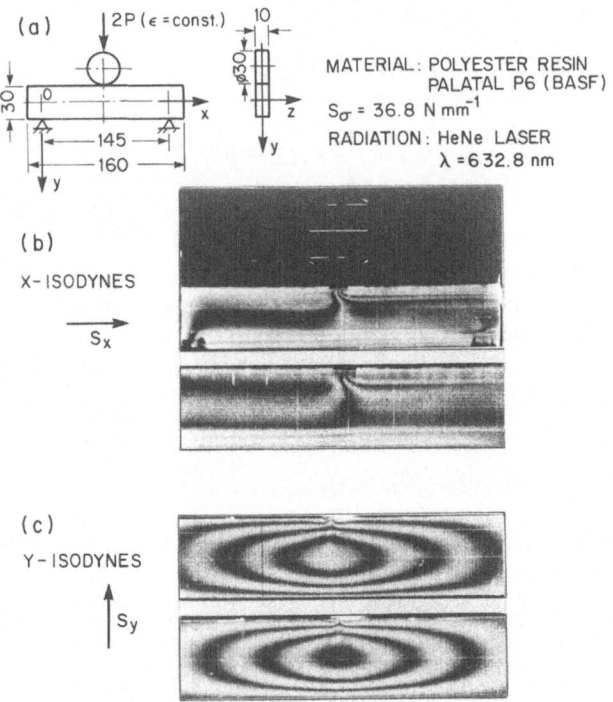

Fig. 7.17. Beam under three point load. Optical isodynes obtained using amplitude modulation technique (see Fig. 7.16). (a) Data on experiment. (b) *X*-isodynes. (c) *Y*-isodynes.

tion when doubt exists whether or not the assumptions are sufficiently well satisfied, or when the differentiation cannot be performed with the usual precision, as in the case, for instance, at the boundary with very low stress gradients. Among the numerous experimental methods, the transmission photoelasticity is one of the most suitable, provided that the results of the transmission photoelasticity. measurements are analyzed within a framework of a sufficiently general theory. For instance, the condition for the applicability of the classical photoelasticity relation between the birefringence and stresses in the form

$$\sigma_1 - \sigma_2 = s_\sigma m, \tag{7.44a}$$

or at free boundary, where $\sigma_2 = 0$:

$$\sigma_1 = s_\sigma m, \tag{7.44b}$$

is, that the effect of the boundary curvature leading to the stress concentration, as expressed by the normalized radius of curvature, ρ/b, is negligible. This condition can be presented in the form

$$\ell/b \geq (\rho/b)^{-1/2}, \tag{7.45}$$

where ℓ denotes the distance from the boundary, with the curvature ρ, of a plate having thickness b.

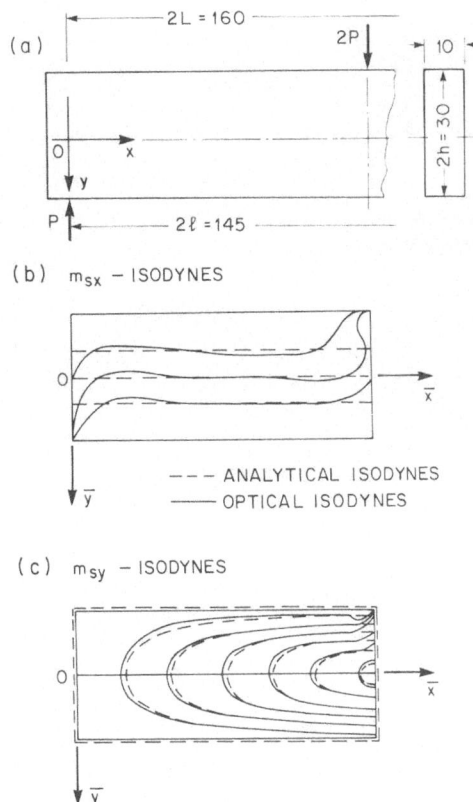

Fig. 7.18. Beam under three point load. Comparison of the plane analytical elastic and the optical isodynes. Optical isodynes are differential isodynes in regions of contact forces (see Figs. 7.16 and 7.17). (a) Data on experiment. (b) X-isodynes. (c) Y-isodynes.

Relation (7.45) is a generalization of the geometric similarity condition for plane stress fields, formulated by one of the authors:

$$b/\rho \ll 1 \tag{7.46}$$

Consequently, the relations (7.44a, b) can be applied safely only when the condition (7.45) or (7.46) is satisfied.

Advanced methods of moiré interferometry, such as those developed by Post [21] are excellent auxiliary methods of the isodyne stress/strain analysis. Those methods are of particular importance when local stresses caused by concentrated boundary loads are investigated. In such cases, the size of the region of contact is of primary interest. Figure 7.19 illustrates the technique of determination of the contact area using the technique of total reflection of the incident light. In the regions of close contact, when the separation of contacting surfaces is less than a half wavelength of the incident light, no reflection occurs because of the influence of the evanescent wave. Thus the resolution of this technique is 0.2– 0.3 μm when the visible light is used.

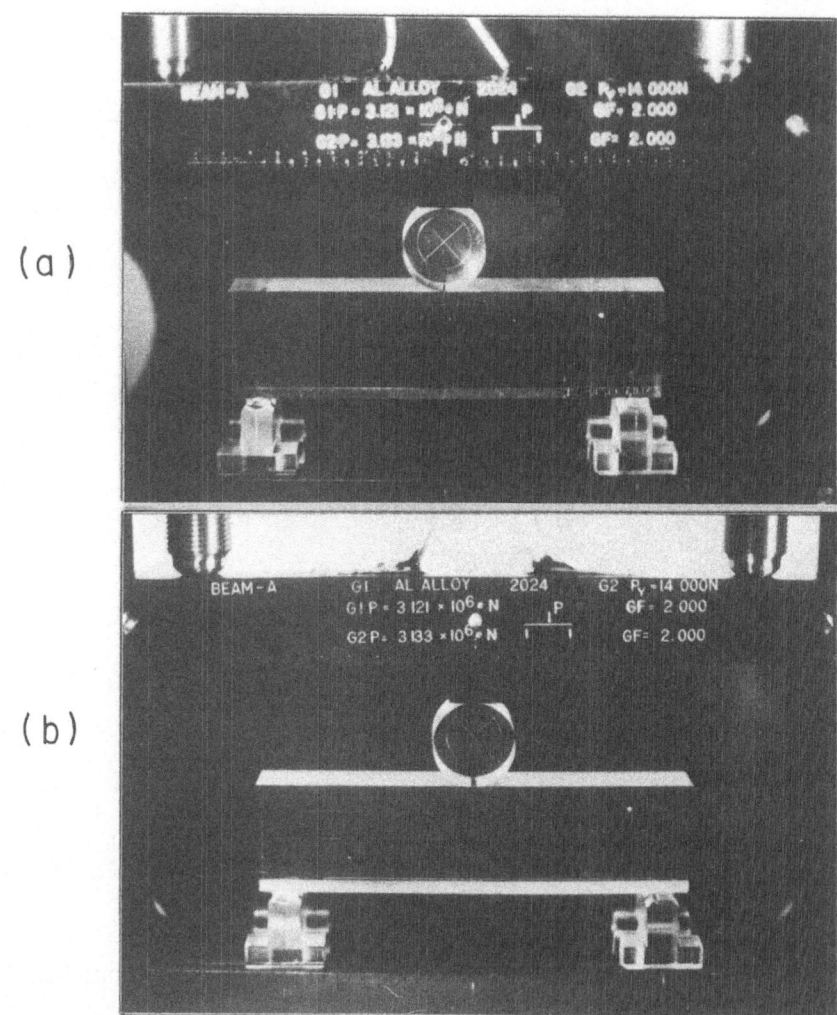

Fig. 7.19. Total reflection method of measuring contact areas.
(a) Load is low. (b) Load is 5 times higher.

7.3 References

[1] Chiang, F. P., "Differentiation of Moiré Patterns by Optical Spatial Filtering", *J. Appl. Mech.* **42**, Series E(1), 1975, pp. 25–28.
[2] Cloud, G. L. and Pindera, J. T., "Techniques in Infrared Photoelasticity", *Experimental Mechanics* **8**(5), 1968, pp. 193–201.
[3] Gilbert, J. A., "Differentiation of Holographic-Moiré Patterns", *Experimental Mechanics* **18**(11), 1978, pp. 436–440.
[4] Kobayashi, A. S. and Mall, S., "Dynamic Fracture Toughness of Homalite-100", *Experimental Mechanics* **18**(1), 1978, pp. 11–18.

[5] Parks, V. J. and Durelli, A. J., "Moiré Patterns of Partial Derivatives of Displacement Components", *J. Appl. Mech.* **33**, Series E(4), 1966, pp.901—906.

[6] Pindera, J. T., *Techniques of Photoelastic Investigations of Two-Dimensional Stress Problems* (in Polish), Engineering Transactions (Rozprawy Inzynierskie), Polish Acad. Sciences, **3**, (1) 1955, pp. 109—176.

[7] Pindera, J. T. and Kiesling, E. W., *On the Linear Range of Behaviour of Photoelastic and Model Materials*, VDI-Berichte No. 102, Experimentelle Spannungsanalyse, VDI-Verlag, Dusseldorf, 1966, pp. 89—94.

[8] Pindera, J. T. and Straka, P., "Response of the Integrated Polariscope", *J. of Strain Analysis* **8**(1), 1973, pp. 65—76.

[9] Pindera, J. T., Straka, P., and Krishnamurthy, A. R., "Rheological Responses of Materials Used in Model Mechanics", in: *Proc. of the Fifth International Conference on Experimental Stress Analysis*, CISM, Udine, 1974, pp. 2.85—2.98.

[10] Pindera, J. T. and Straka, P., "On Physical Measures of Rheological Responses of Some Materials in Wide Ranges of Temperature and Spectral Frequency", *Rheologica Acta* **13**(3), 1974, pp. 338—351.

[11] Pindera, J. T., Issa, S. S., and Krasnowski, B. R., "Isodyne Coatings in Strain Analysis", in: *Proc. of the 1981 Spring Meeting, SESA, Dearborn, Michigan, May 31—June 4, 1981*, Society for Experimental Stress Analysis, Brookfield Center, 1981, pp. 111—117.

[12] Pindera, J. T., Mazurkiewicz, S. B., and Krasnowski, B. R., "Determination of All Components of Plane Stress Fields Using Simple Technique of Differentiation of Photoelastic Isodynes", in: *Proc. of the 1981 Spring Meeting. SESA, Dearborn, Michigan, May 31—June 4, 1981*, Society for Experimental Stress Analysis, Brookfield Center, 1981, pp. 35—40.

[13] Pindera, J. T. and Mazurkiewicz, S. B., "Studies of Contact Problems Using Photoelastic Isodynes", *Experimental Mechanics* **21**(12), 1981, pp. 448—455.

[14] Pindera, J. T., "Foundations of Experimental Mechanics: Principles of Modelling, Observation and Experimentation", in: *New Physical Trends in Experimental Mechanics*, edited by J. T. Pindera, CISM Courses and Lectures No. 264, Springer-Verlag, 1981, pp. 199—327.

[15] Pindera, J. T., Krasnowski, B. R., and Pindera, M.-J., "Analysis of Models of Stress States in the Regions of Cracks Using Isodyne Photoelasticity", in: Jerzy T. Pindera (Ed.), *Modelling Problems in Crack Tip Mechanics*, Proc. of the 10th Canadian Fracture Conference, Martinus Nijhoff, The Hague, 1984.

[16] Pindera, J. T., Krasnowski, B. R., and Pindera, M.-J., "Theory of Elastic and Photoelastic Isodynes. Samples of Applications in Composite Structures", *Experimental Mechanics* **25**(3), September 1985, pp. 272—281.

[17] Pindera, J. T., "Device for Birefringence Measurements Using Three Selected Sheets of Scattered Light (Isodyne Selector, Isodyne Collector. Isodyne Collimator), United States Patent, No. 4,679,933 of July 14, 1987.

[18] Pindera, J. T., "Apparatus for Determination of Elastic Isodynes and of the General State of Birefringence Whole-Field-Wise, Using the Device for Birefringence Measurements in a Scanning Mode, United States Patent, No. 4,703,914 of Nov. 1987.

[19] Pindera, J. T. and Hecker, F. W., "Basic Theories and Experimental Methods of Strain Gradient Method", *Experimental Mechanics* **27**(3), 1987, pp. 314—327.

[20] Pirodda, L., "Optical Differentiation of Geometrical Patterns", *Experimental Mechanics* **17**(11), 1977, 427—432.

[21] Post, D., "Moiré Interferometry", in: Albert S. Kobayashi (ed.), *Handbook on Experimental Mechanics*, Prentice-Hall, Inc., Englewood Cliffs, New Jersey, 1981, pp. 314—387.

[22] Schwieger, H. and Haberland, G., "Application of Isodensitometry in Photoelasticity" (in German), Wiss. Z. Universität Halle-Wittenberg, *Math. Nat.* **4**, 1955, 853—858, and *Feingerätetechnik* **5**, 1956, 63—67, described in: "A New Application of Isodensitometry in the Moiré Technique", *Experimental Mechanics* **16**(7), 1976, 256—262.

[23] Sciammarella, C. A. and Chang, T. Y., "Optical Differentiation of the Displacement Pattern Using Shearing Interferometry by Wavefront Reconstruction", *Experimental Mechanics* **11**(3), 1971, 97—104.

[24] SGL Homalite, General Brochure, Wilmington, Delaware, 1976.

[25] Theocaris, P. S., *Moiré Fringes in Strain Analysis*, Pergamon Press, New York, 1969.
[26] Zandman, F., "Analysis des Contraintes par Vernis Photoélastique" (Stress Analysis with Photoelastic Coatings), *Groupement pour l'Avancement des Méthodes d'Analyse des Contraintes* **2**(6), 1956 (oral presentation 1955).
[27] Zandman, F., Redner, S. S., and Riegner, E. I., "Reinforcing Effect of Birefringent Coatings", *Experimental Mechanics* **2**(2), 1962.
[28] Zandman, F., Redner, S. S., and Dally, J. W., *Photoelastic Coatings*, The Iowa State University Press and Society for Experimental Stress Analysis, Westport, CT., 1977.

Perspectives

The concept of plane and differential analytical elastic isodynes can be generalized to three-dimensional stress states, and/or anelastic deformation states. From the point of view of stress states, it is convenient to distinguish two interrelated classes of generalized analytical elastic isodynes problems:

— general three-dimensional states;
— local three-dimensional states within two-dimensional stress fields.

From the point of view of deformation states, it is convenient to distinguish two other classes of generalized isodynes:

— analytical elastic isodynes related to elastic deformation states;
— analytical inelastic isodynes related, in general, to inelastic and anelastic deformation states.

8.1 Stress state approach

8.1.1 General three-dimensional problems

In such cases, six independent pieces of information are needed to determine the components of the elastic stress tensor without using any analytical relations. Making use of the equilibrium conditions, it is possible to reduce the number of independent pieces of experimentally obtained information to three.

Basically, no new theoretical problems exist in constructing more general mathematical models. However, the development of pertinent global relations may be time-consuming, and may lead to general expressions which violate the limitations imposed by the assumptions made in the course of developing the concept of plane elastic isodynes. Though, it follows from the theory of optical isodynes presented in Chapter 5 that local determination of stress components is feasible because optical isodynes can also supply information on the local principal directions.

8.1.2 Local three-dimensional states within plane stress fields

Such states called local effects, typically occur in plates in regions of notches and cracks, at the lamination planes in composite structures, in plies of composite structures when Poisson ratios are different, in regions of contact and of large deformations, and at the boundaries between the elastic and plastic regions. Observing the optical isodyne fields, it is easy to detect such regions and determine their boundaries.

Obviously, in such regions optical isodynes are not identical with the plane elastic isodynes — they are identical with the differential isodynes defined on the basis of relation (5.34). It is easy to see that in most practical cases some additional information can be obtained regarding principal direction(s), or the values of some components of stress tensor, etc. In such cases it is·sufficient to infer the meaning of optical isodynes on the basis of relation (5.34) when the basic assumptions regarding the state of birefringence are sufficiently closely approximated.

8.2 Deformation state approach

8.2.1 Linear elastic states

In such states the isothermal behaviour of a solid, homogeneous, and isotropic body is completely described by two elastic material parameters in the linear, time-invariant, relations between the components of stress and strain tensors. When the behavior is not isothermal due to the thermoelastic effect or to the external heat sources, and the heat convection is negligible, two more internal parameters influence the relations between stresses and deformations: the thermal expansion coefficient α_T which usually depends on the temperature, $\alpha_T = \alpha_T(T)$, and the specific heat coefficient at constant strain, c_ε, which also depends on the temperature. However, when a pertinent function of the heat conduction, radiation, and time is not negligible, several other material parameters also influence the linear elastic response of solid bodies to local or extended heat sources. Consequently, the similarity conditions for the actual thermal stress fields involve a greater number of physical parameters.

8.2.2 Anelastic states

It is common to use the term anelasticity to denote a particular group of inelastic states which exhibits the phenomenon of recovery. Some of the anelastic states can be analysed using the relations of linear viscoelasticity in the active domain (acting loads or imposed deformations). However, according to experience, the linear viscoelastic model is often inapplicable in the passive domain (recovery domain); in such domains the stress-strain relations are equivocal, even for the materials proven to be close to the linearly viscoelastic material in the active domain.

In general, relations between components of the strain and stress tensors cannot be reliably established in the anelastic state. Nevertheless, any information on the stress or strain components in the anelastic state of deformation supplied by optical isodynes may be very valuable for a theoretist or an engineer.

8.2.3 Plastic states

Thus far, no theoretically correct, sufficiently rigorous, and quantifiable definition of plastic deformation of real polymeric materials has been developed. Because of the lack of reliable theoretical-physical foundations, or of the sufficiently rigorous analytical-phenomenological relations, the various techniques of the so-called photo-plasticity and of model simulation of plastic deformations in engineering structures are closer to the category of technological tests rather than to empirical-analytical simulation. The needed theoretical foundation of the model experiments in the elastic-plastic region is still not developed.

Two-dimensional stress states

9.1 Introduction

The methods and techniques of elasticity based on the assumption of plane stress or strain continue to play a significant role in the solution of many problems of practical importance. The reason for this lies in the complexity of the three-dimensional boundary value problems of elasticity and the resulting lack of a general methodology of integrating the governing differential equations in three dimensions. On the other hand, the assumption of plane stress or strain greatly simplifies the governing differential equations which can be solved by a variety of techniques, including the general methodology based on the analytic function approach developed by Muskhelishvili [5]. However, it must be mentioned that the stress states commonly called two-dimensional or plane are, in fact, semi-plane stress states and should be treated as such as illustrated by one of the examples discussed in the sequel.

In view of the above, we begin the applications part of this monograph by comparing the predictions of the two-dimensional elasticity model with the results obtained using optical isodynes for two problems whose solutions are readily available. The first problem is the problem of a long beam loaded by diametrically opposite loads, either point loads or loads distributed over a small area. The solution to this problem has been obtained by Filon [1] in the case of a finite beam using the Fourier series expansion technique and later by Mesmer [4] who employed the infinite Fourier transform. The second problem is the problem of a circular disk loaded by diametrically opposite point loads. The solution to this problem is available in closed form in terms of the Airy's stress function or complex potentials [11, 12].

The above problems have been chosen for two reasons. First of all, since exact analytical solutions in the context of the plane elasticity model are available, the utility of the relations of plane isodynes can be readily established in regions where the two-dimensional model is applicable. As has been previously mentioned, these regions are characterized by relatively low stress gradients. Having established the efficacy of the experimental technique provides one with a useful tool to test the range of applicability of the plane analytical solutions in regions characterized by high inplane stress gradients. In the context of the chosen problems, these are the regions in the immediate vicinity of the concentrated loads where the state of stress becomes three-dimensional. As will be subsequently illustrated, the experimental data obtained on the basis of plane isodyne relations can be employed to clearly identify these regions upon comparison with the exact two-dimensional solution.

9.2 Beam loaded by concentrated forces

The problem of a long beam loaded by diametrically opposite concentrated loads was one of the first problems to be investigated on the basis of the relations of plane isodynes [8]. The objective was to compare the predictive power of the then newly developed method of plane isodyne stress analysis with the results of the exact two-dimensional solution in regions where such solution was valid, that is in the regions of relatively low stress gradients. It is useful to keep in mind that at the time of the above investigation the concept of differential isodynes was still not formulated. Thus no attempt was made here to investigate the stress field in the vicinity of the concentrated load where substantial experimental evidence gathered during the past fifty years indicates that the stress state is actually three dimensional. The results of such an investigation are presented in the second problem discussed in this chapter.

9.2.1 Experimental investigation

The beam under investigation was 200 mm long, 30 mm high and 10 mm thick. The load was applied through a circular disk 30 mm in diameter situated at the midspan of the beam directly in line with the bottom support. The thickness of the loading disk was the same as the thickness of the beam. Both the beam and the disk were fabricated using Palatal P6, a type of polyester resin. This material produces amplitude modulated fields of photoelastic isodynes. The elastic prarameters for this material are given in Table 9.1. The specimen geometry and the manner of loading are summarized in Figure 9.1.

The beam was loaded incrementally to the maximum load of 2668 N using increments of 667 N in order to establish the range of linearity of the stress

Table 9.1. Elastic parameters of the beam and loading disk.

Material type	E* Young's modulus (N/mm^2)	v* Poisson's ratio
Polyester resin Palatal P6	4462	0.38

* Test method — E is the instantaneous value obtained from creep tests. The value of v was supplied by the manufacturer.

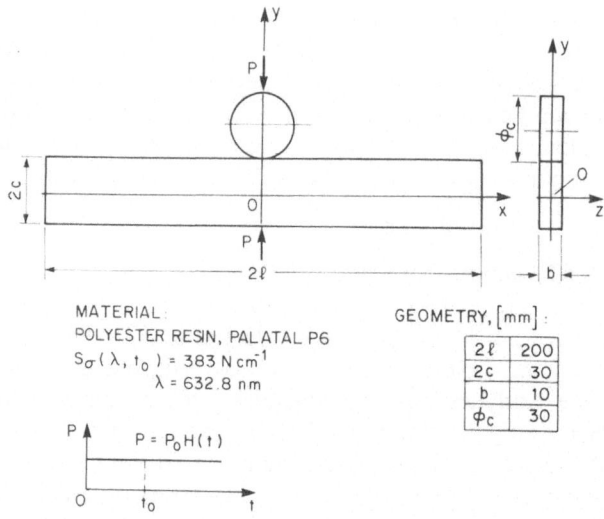

Fig. 9.1. Rectangular beam under point loads. Specimen geometry and loading conditions.

distribution with respect to the applied load. At each load increment isodyne recordings along the x-characteristic direction were taken in the $z = 0$ plane. The measurements were performed by taking into account the actual photovis-coelastic response of the tested material [10]. The time between loading and recording of isodyne fields at each load increment was fifteen minutes. This was sufficient to minimize the effect of creep during the recording period. Once the range of linearity of the stress distribution with respect to the applied load was established, the final recording of the x-isodynes was taken at the applied load of 2903 N in the $z = 0$ plane. The normal stress distribution $\sigma_{yy}(x, 0, 0)$ calculated from the x-isodynes is presented for this load level and subsequently compared with the predictions of Filon's solution. The actual value of the stress-optic coefficient $S_\sigma(\lambda, t)$ employed in the above calculation was deter-mined by using a creep compensator in the form of a circular disk [7]. Both the scattered light modulation and the transmitted light modulation was utilized for the above purpose. Both methods yielded 383 N/cm for the stress-optic coefficient at $t_0 = 15$ minutes and $\lambda = 632.8$ nm which was the wavelength of radiation employed in this investigation.

9.2.2 Experimental results

The experimentally obtained amplitude modulated x-isodyne fields in the $z = 0$ mm plane, and the corresponding transmission isochromatics, are presented in Figure 9.2 for the applied load of 2903 N. Graphical representation of the above fields obtained from the photographic recordings is illustrated in Figure 9.3.

In order to determine the in-plane stress component σ_{yy}, $y = y_0$ cross-

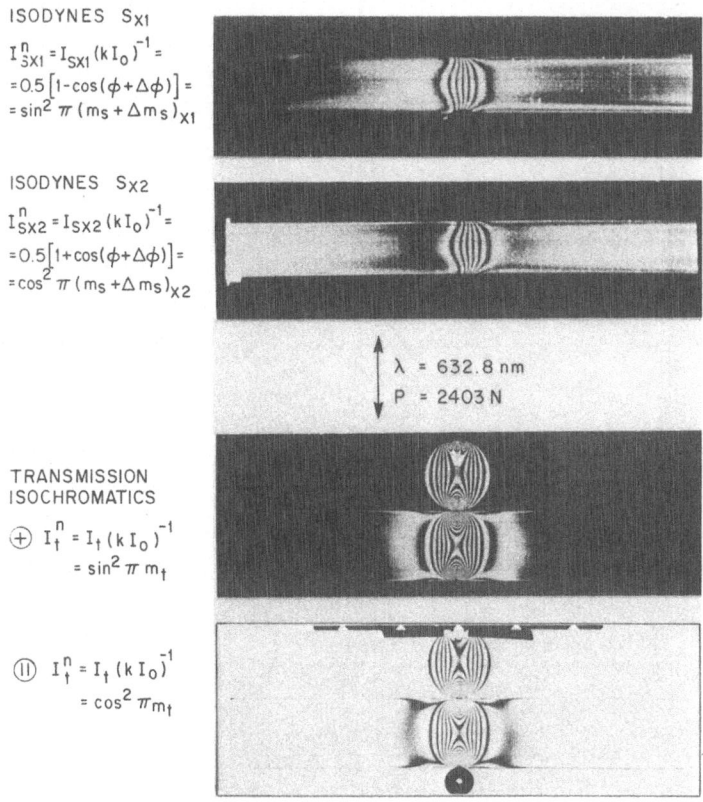

Fig. 9.2. Photographs of experimental amplitude modulated x-isodyne fields in the $z = 0.0$ mm plane at 2403 Newtons, and the corresponding transmission isochromatics.

sections of the x-isodyne fields are required for chosen z_0 distances. Differentiation of the particular $y = y_0$ cross-section with respect to the x-coordinate yields the normal stress distribution $\sigma_{yy}(x, y_0, z_0)$. Figure 9.4 illustrates the results of the above procedure for the $y = 0$ mm elevation. At this elevation the state of stress is two-dimensional and thus a direct comparison between the experimental results and the predicitions of the plane elasticity solution can be carried out in order to establish the efficacy of the optical isodynes method.

The $y = 0$ mm cross-sections of the x-isodyne fields determined from recordings obtained at four different load levels mentioned previously are presented in Figure 9.5. The results of differentiation of the four cross-sections at selected x_0 locations verify the linearity of the obtained normal stress distribution $\sigma_{yy}(x, 0, 0)$ at 2403 N. Further evidence of the linearity of the obtained stress distribution is provided by the locations of $\sigma_{yy}(x, 0, 0) = 0$ at the four load levels. Visual inspection of Figure 9.5 demonstrates that the location of the zero normal stress $\sigma_{yy}(x, 0, 0)$ does not change in the range of loads considered here. The above results establish the validity of the compar-

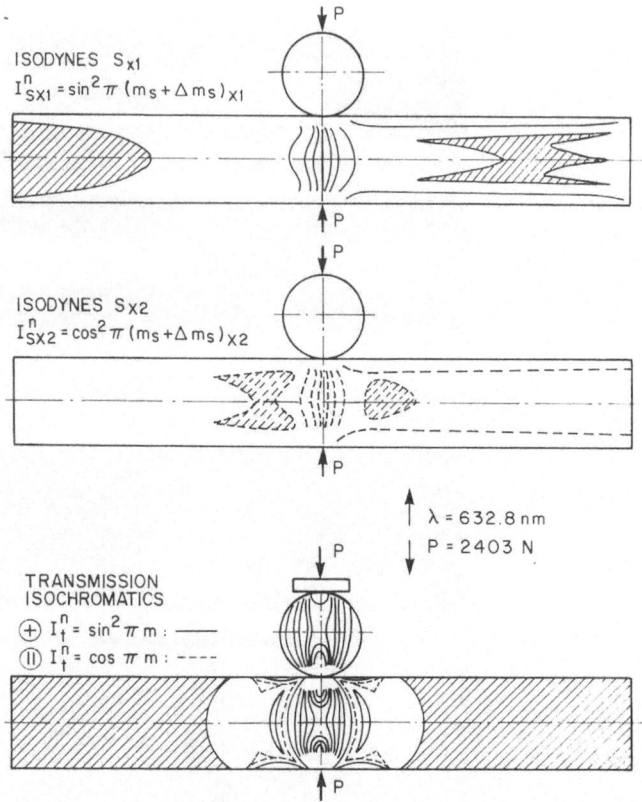

Fig. 9.3. Graphical representation of the x-isodyne fields and transmission isochromatics.

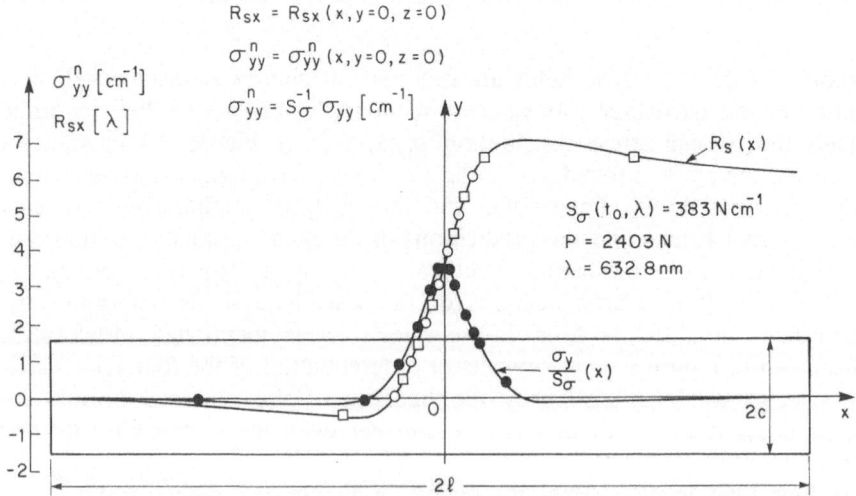

Fig. 9.4. Cross-section of x-isodynes at the $y = 0.0$ mm elevation in the $z = 0.0$ mm plane and the corresponding normalized distribution of the normal stress component $\sigma_{yy}(x, 0, 0)$.

ison of the experimentally obtained normal stress σ_{yy} with the predictions of the plane, linearly elastic solution to be presented in the following section.

9.2.3 Analytical model

The two-dimensional elasticity solution to the problem of a single infinite beam loaded by equal and opposite point loads has been obtained by Filon [1] who formulated the problem in terms of displacements and solved the resulting Navier equations using a Fourier series expansion technique. The solution was first developed for a finite-length beam of rectangular cross-section under any system of loads. Special consideration was given to concentrated or discontinuous loads.

For an infinite beam the solutions for the inplane displacements and stresses expressed in terms of Fourier series, where the coefficients are obtained directly from the boundary conditions, can be readily converted to infinite integrals. The expressions for the inplane stress components in this instance become:

$$
\begin{aligned}
\sigma_{xx} &= \frac{-2W}{\pi c} \int_0^\infty \left(\frac{\sinh u - u \cosh u}{\sinh 2u + 2u} \right) \cos \frac{ux}{c} \cosh \frac{uy}{c} \, du - \\
&\quad - \frac{2W}{\pi c} \int_0^\infty \frac{uy}{c} \left(\frac{\sinh u}{\sinh 2u + 2u} \right) \cos \frac{ux}{c} \sinh \frac{uy}{c} \, du \\
\sigma_{yy} &= \frac{-2W}{\pi c} \int_0^\infty \left(\frac{\sinh u + u \cosh u}{\sinh 2u + 2u} \right) \cos \frac{ux}{c} \cosh \frac{uy}{c} \, du + \\
&\quad + \frac{2W}{\pi c^2} \int_0^\infty \left(\frac{u \sinh u}{\sinh 2u + 2u} \right) \cos \frac{ux}{c} \sinh \frac{uy}{c} \, du \\
\sigma_{xy} &= \frac{2W}{\pi c} \int_0^\infty \left(\frac{u \cosh u}{\sinh 2u + 2u} \right) \sin \frac{ux}{c} \sinh \frac{uy}{c} \, du - \\
&\quad - \frac{Wy}{\pi c^2} \int_0^\infty \left(\frac{u \sinh u}{\sinh 2u + 2u} \right) \sin \frac{ux}{c} \cosh \frac{uy}{c} \, du
\end{aligned}
\tag{9.1}
$$

In the above equations, c is half the height of the beam and W is the applied load intensity (load per unit length), that is line load perpendicular to the plane of the strain.

The solution for the inplane stresses given by equation (9.1) is an approximate solution in the sense that the individual stress components are quantities that have been averaged through the beam's breadth, that is in the out-of-plane direction. This device allowed Filon to reduce the complicated three-dimensional problem of a beam subjected to an arbitrary system of loads to a two-dimensional one that could be managed analytically. In this way, it is not

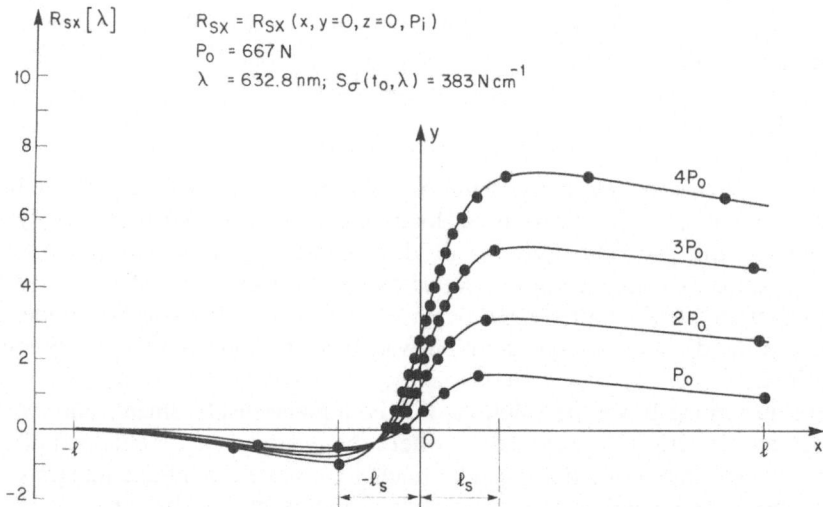

Fig. 9.5. Cross-section of x-isodynes at the $y = 0.0$ mm elevation for four different load levels.

necessary to satisfy the traction free lateral boundary conditions which make the problem a truly three-dimensional problem. The three-dimensional effects however, become dominant in the direct vicinity of the applied concentrated or discontinuous loads so that the solution obtained by Filon may be treated as a very good approximation to an actual distribution of stresses in a real rectangular beam sufficiently far away from the points of load applications. At this point, we defer the discussion of what we mean by sufficiently far till later.

9.2.4 Experimental/analytical correlation

The results of the experimentally obtained normal stress $\sigma_{yy}(x, 0, 0)$ are compared with the solution obtained by Filon in Figure 9.6. Both the experimental results presented in Figure 9.2 and the analytical solution indicate that for practical purposes the geometry of the employed beam in the x-direction can be taken as infinite for the given loading conditions. The results presented in Figure 9.6 have been normalized in the standard manner in order to eliminate the influence of the geometry and magnitude of the applied load from the data. The limited comparison between theoretical and experimental results outlined in this problem clearly illustrates the potential of isodyne stress analysis for the analysis of plane states of stress. The second problem outlined in this chapter presents more extensive evidence of the predictive capability of optical isodynes for both the analysis of two-dimensional states of stress and determination of the range of applicability of plane elasticity solutions.

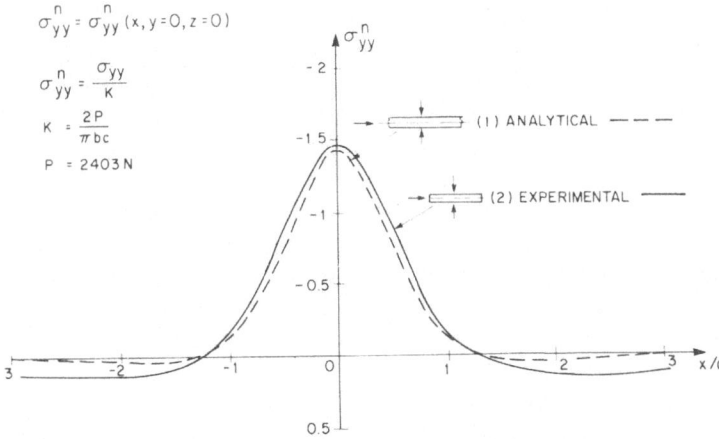

Fig. 9.6. Horizontal distribution of the normal stress σ_{yy} at the $y = 0.0$ mm elevation in the $z = 0.0$ mm plane. Comparison between analytical and experimental results.

9.3 Circular disk loaded by concentrated forces

The problem of a circular disk loaded by diametrically opposite concentrated forces belongs to a small group of problems in the theory of plane elasticity for which closed-form analytical solutions are available and for which the boundary conditions can be closely approximated experimentally. As a result, such disks have been used extensively to assess the reliability and accuracy of various photoelastic techniques [6]. Such disks have also been used as "creep compensators" or "creep calibrators" because they allow to account easily for optical and mechanical creep, and to evaluate reliably momentary values of the stress optic coefficient at the time of recording of data [7].

Since concentrated loads produce high stress and strain gradients in the regions of load applications, the question of reliability or range of applicability of plane analytic solutions within those regions becomes an important issue. Equally important is the influence of the deviation between the actual and hypothetical state of stress predicted by a two-dimensional solution in regions of high stress gradients on the far field stress state for a given geometry. The above problem is not a simple one to address since the reasons for discrepancy between the actual and a hypothetical plane elasticity solution in the vicinity of concentrated loads may be several-fold. First, solutions which utilize point loads for traction boundary conditions produce singular stresses at the points of load application which often must be excluded from global equilibrium considerations. In fact, inclusion of those points may violate global equilibrium. Secondly, singular stress fields violate the basic assumptions employed in deriving the governing differential equations of infinitesimal elasticity. An exact solution to a hypothetical boundary value problem that involves singular stress components is therefore outside of the theoretical framework of the linear model not only at the points of singularities, which must be excluded from consideration, but also in the direct vicinity of those singular points. Thirdly, local high stress gradients typically lead to a localized three-dimensional stress state for finite-thickness plane bodies loaded by tractions lying in their planes. Fourthly, the problem of applying a point load to a real object in the laboratory is an exceedingly difficult one. At best, the hypothetical point load can be approximated to a lesser or greater degree by a load distributed over a finite area whose dimensions depend on the geometry and material properties of the contacting bodies. These types of problems belong in the category of contact problems and require specialized techniques of analysis. Nevertheless, the treatment of contact problems on the basis of the two-dimensional model is still an approximate one which requires independent verification in view of the presence of high stress gradients in the vicinity of two contacting surfaces.

The example of a circular disk loaded by concentrated loads discussed in this section presents more extensive evidence of the efficacy of optical isodynes in the determination of planar states of stress. Here, normal stress components σ_{xx} and σ_{yy} along various cross-sections of the disk determined on the basis of the relations of plane isodynes are compared with the predictions of the exact solution of the hypothetical disk loaded by point loads. The comparison is carried out in regions removed from the locations of singular points where the state of stress is plane as well as in the immediate vicinity of the point loads. Clear evidence is presented that the optical isodyne method is an excellent tool in determining the range of applicability of plane solutions in regions of high stress gradients produced by concentrated loads.

In order to provide further evidence of the predictive power of optical isodynes, two additional independent experimental techniques were utilized in the analysis of the state of stress in the circular disk. Transmission photoelasticity was employed which provided the fields of isochromatics which were subsequently compared with the analytical isochromatics obtained from the exact two-dimensional elasticity solution. Strain-gages were also placed along the diameter of the disk coincident with the line of action of the applied loads, and along the diameter perpendicular to it. Normal stresses were subsequently evaluated from the strain data and compared with the isodyne and analytical results.

9.3.1 Experimental investigation

The circular disk under investigation was 100 mm in diameter and 9.75 mm thick [9]. For the optical isodynes and transmission photoelasticity studies the disk was fabricated using Palatal P6 polyester resin which is the same material employed in the investigation of the beam loaded by concentrated forces outlined in the preceding section. For tensometric studies an aluminum alloy (AL 2024 T3) specimen 120 mm in diameter and 15 mm thick was instrumented with longitudinal and transverse strain gages along the $x = 0$ and $y = 0$ axes of the disk. The specimen geometry, manner of loading and strain gage locations for the investigated disks are summarized in Figure 9.7. The elastic parameters for Palatal P6 have been given in Table 9.1, those for the aluminum alloy are given in Table 9.2.

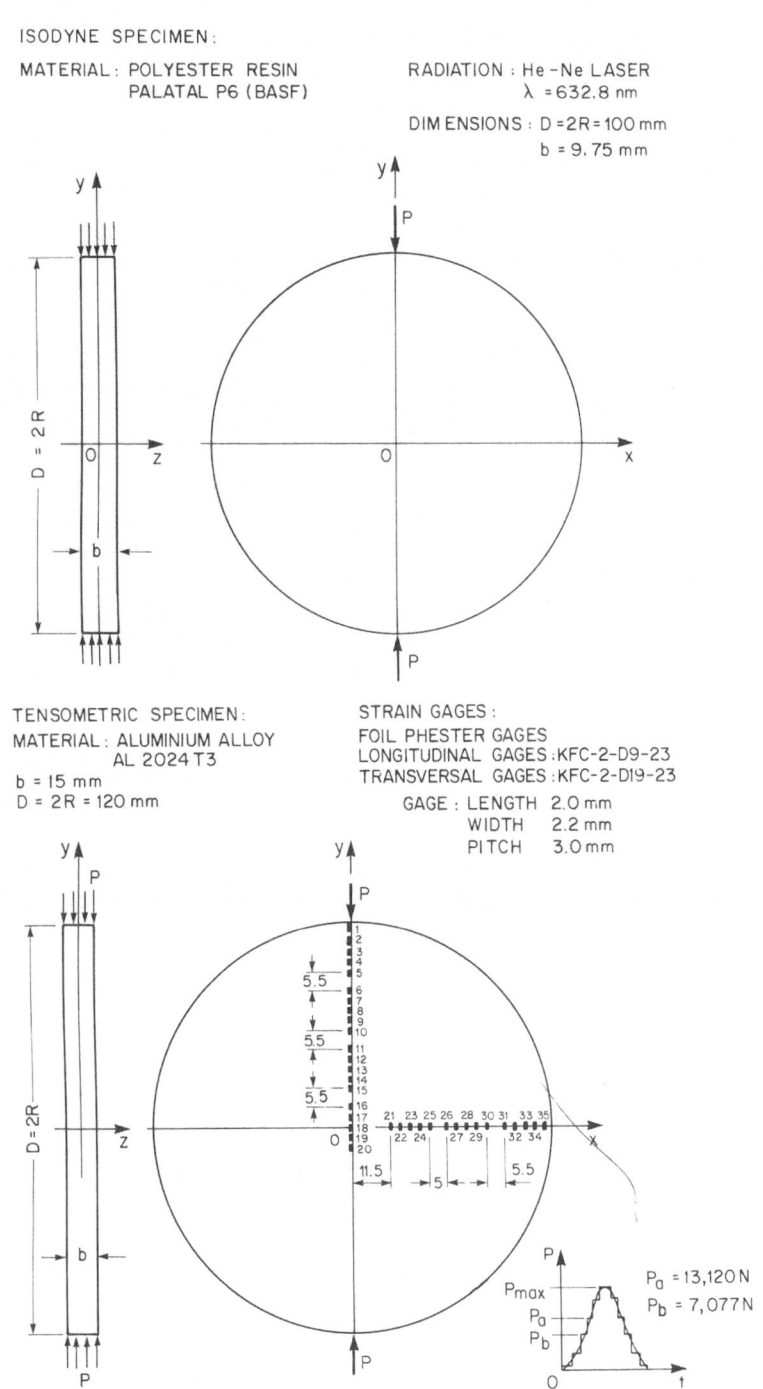

Fig. 9.7. Circular disk under point loads. Specimen geometry, loading conditions and strain gage locations. (a) — isodyne specimen, (b) — tensometric specimen.

Table 9.2. Elastic parameters of the tensometric disk.

Material type	E* Young's modulus (N/mm^2)	ν^* Poisson's ratio
Aluminum alloy AL 2024 T3	74,700	0.32

* Test method — Manufacturer's data.

In loading the specimens, care was taken to ensure that the measurements were performed in the linear range. The specimens were examined after unloading and no permanent effects were observed in the plane of the specimens as well as out of plane. The loading and recording of data in the case of the polyester specimen followed the procedure outlined in the preceding section. To minimize the effects of mechanical and optical creep, birefringence measurements were performed at practically constant deformation, in accordance with the procedure outlined by Pindera and Straka [10]. In order to eliminate the influence of the deformation of disk surfaces, the recording of isodyne fields was performed with the loaded disk submerged in an immersion tank. Both the x- and y-isodyne fields were recorded in the midplane of the disk specified by $z = 0$ mm (see Figure 9.7). The medium employed for the recording of isochromatics was air.

The aluminum specimen was loaded to the maximum load of approximately 17,000 N and subsequently unloaded using six load increments on the way up and down. Recording of strains was performed at each load increment. The strain values at all load levels were linearly related. The strain and stress data obtained from the strain gages is presented for the load levels 7,020 N and 13,140 N.

9.3.2 Experimental results

The experimentally obtained x- and y-isodyne fields in the $z = 0$ mm plane, and the corresponding transmission isochromatics, are presented in Figure 9.8. Graphical representation of the isodynes fields is provided in Figure 9.9 along with the orders of the individual isodynes. Slopes of the constant $x = x_0$ and $y = y_0$ cross-sections of the y- and x-isodynes, respectively, yield the corresponding distributions of the normal stress components $\sigma_{xx}(x_0, y, 0)$ and $\sigma_{yy}(x, y_0, 0)$. These distributions will be compared with the predictions of the analytical model, stresses calculated from the strain-gage data and the results obtained from the isochromatics fields along the major diameters parallel and normal to the applied load in Section 9.3.4.

The normal strain components along the major diameters parallel and normal to the applied load and the corresponding normal stresses calculated from the strain-gage data are given in Figure 9.10. Since the state of stress on lateral surface of the disk is plane and since aluminum is an isotropic material,

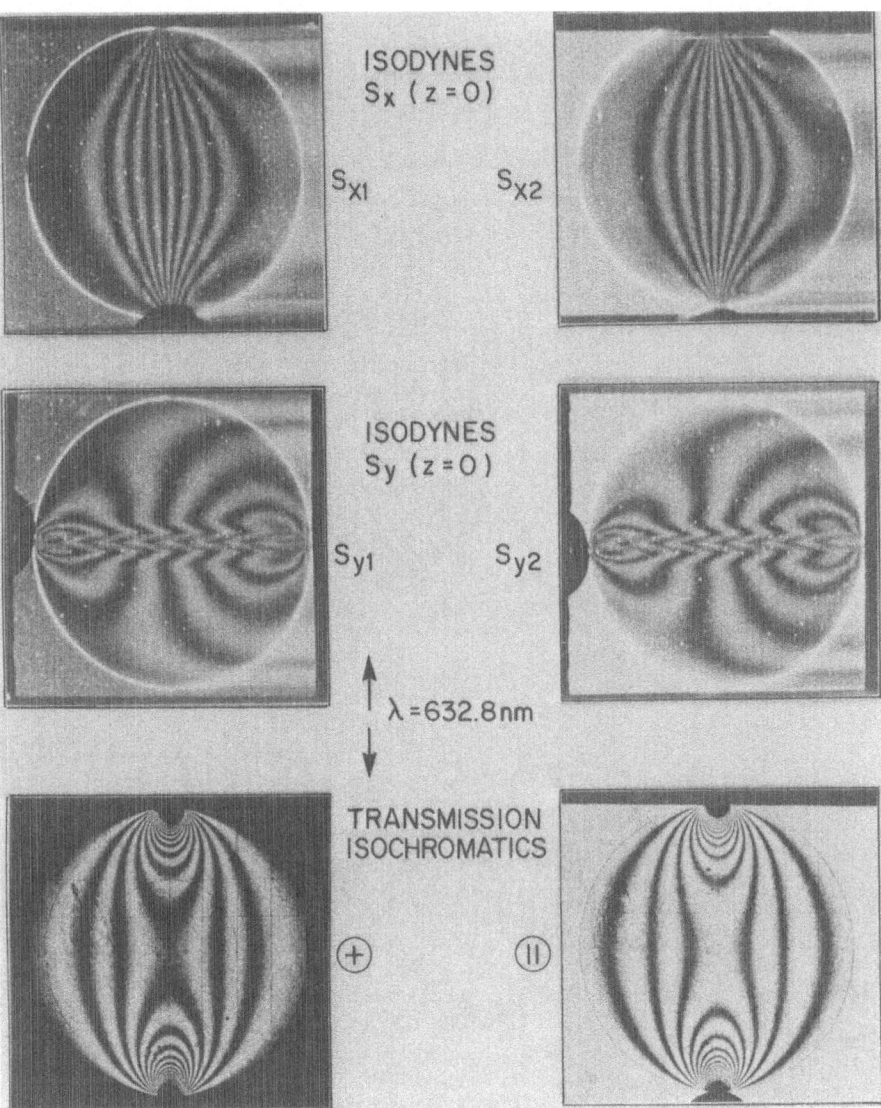

Fig. 9.8. Photographs of experimental amplitude modulated *x*- and *y*-isodyne fields in the $z = 0.0$ mm plane, and the corresponding transmission isochromatics.

the knowledge of ε_{xx} and ε_{yy}, along with the material parameters given in Table 9.2, is sufficient to calculate the normal inplane stresses. These normal stress components are also principal stresses because shear strains vanish along the two major diameters due to the symmetry of the specimen and the applied load. This makes it possible to compare the analytical predictions of the principal stress differences along the two major diameters with the experimental results obtained with the three independent techniques employed in this investigation.

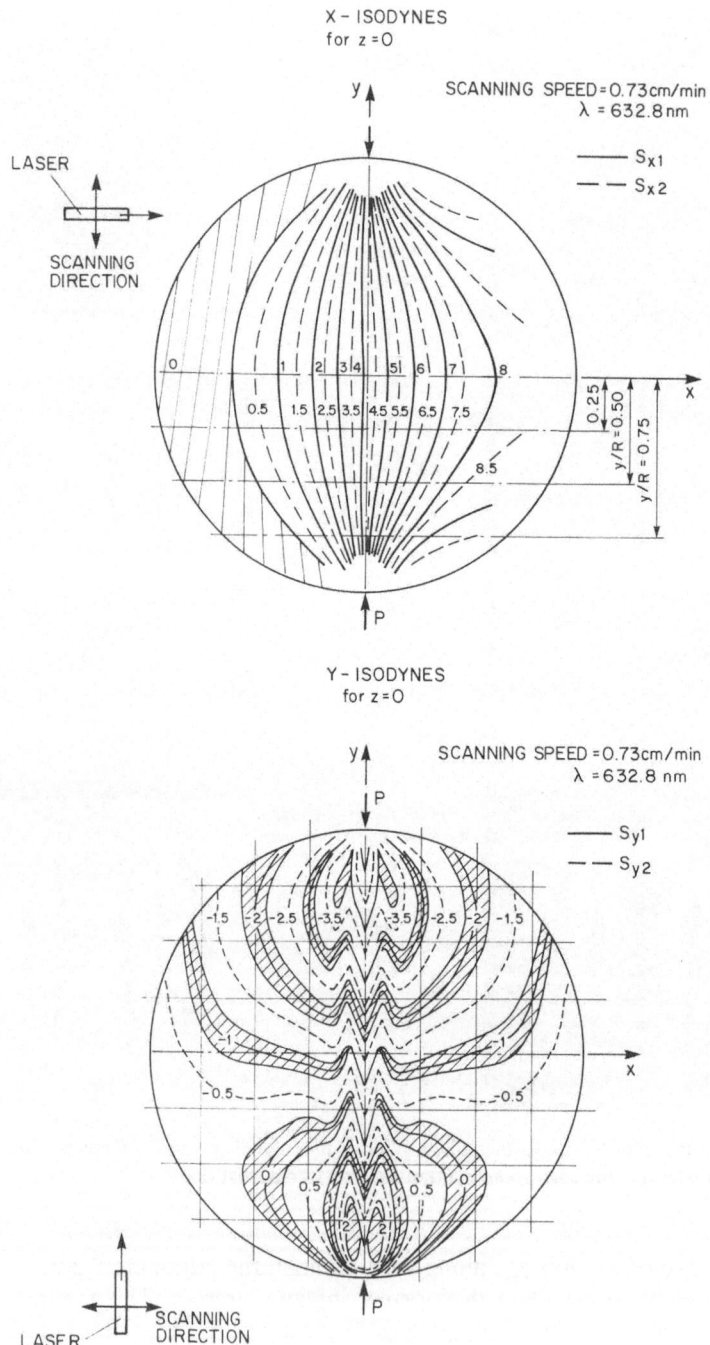

Fig. 9.9. Graphical representaion of the *x*- and *y*-isodyne fields.
(a) *x*-isodynes (b) *y*-isodynes.

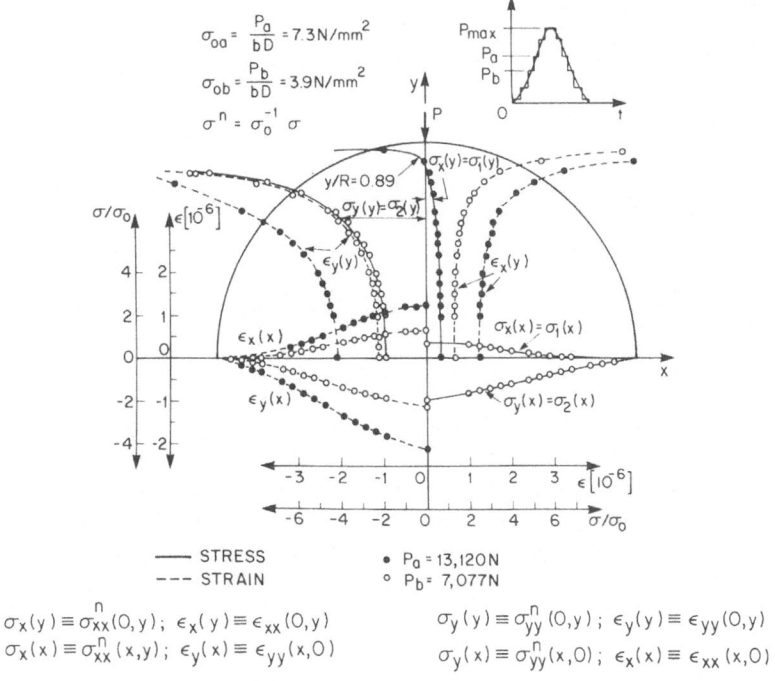

Fig. 9.10. Normal strain and stress components along major diameters parallel and perpendicular to the load axis determined from strain gage data.

9.3.3 Analytical model

There is no analytical solution available for the problem of a locally three-dimensional stress state in a circular disk of a finite thickness, loaded diametrically by opposite concentrated forces. Two convenient analytical solutions to the problem of a plane stress state in such a disk are available: the Hondros solution [3] in which the concentrated loads are modeled by radial tractions distributed over a finite arc length, and the classical solution [11, 12] in which the applied load is taken in the form of point load. Since the second solution is extensively used in the various investigations discussed previously, it will be employed in the present analysis for the sake of illustrating the limitations of this type of approach of solving plane elasticity problems.

The two-dimensional elasticity solution to the problem of a circular disk loaded by diametrically opposite point forces is given in terms of the Airy's stress function given below,

$$\psi(x, y) = \frac{P}{\pi bD} r^2 + \frac{P}{\pi b} (r_1 \phi_1 \cos \phi_1 + r_2 \phi_2 \cos \phi_2) \tag{9.2}$$

where the variables r, r_1, r_2, ϕ_1 and ϕ_2 have been introduced for the convenience of notation and are defined in Figure 9.11. In terms of the Cartesian co-ordinates these variables are:

$$r = \sqrt{x^2 + y^2}$$

$$r_1 = \sqrt{x^2 + (R - y)^2} \qquad r_2 = \sqrt{x^2 + (R + y)^2} \qquad (9.3)$$

$$\phi_1 = \tan^{-1}\left(\frac{R - y}{x}\right) \qquad \phi_2 = \tan^{-1}\left(\frac{R + y}{x}\right)$$

The Airy stress function defined by equation (9.2) satisfies the biharmonic compatibility equation and the associated traction boundary conditions and thus provides exact solution to the hypothetical plane boundary value problem almost everywhere within the region bounded by $r \leqslant R$. The points that must be excluded from consideration are the singular points $(0, \pm R)$ where the normal stress component $\sigma_{yy}(0, R)$ becomes undefined as seen in the expressions for the stress components derived from the above stress function given below.

$$\sigma_{xx} = \frac{\partial^2 \psi}{\partial y^2} = \frac{2P}{\pi b D}\left[1 - Dx^2\left(\frac{R - y}{r_1^4} + \frac{R + y}{r_2^4}\right)\right]$$

$$\sigma_{yy} = \frac{\partial^2 \psi}{\partial x^2} = \frac{2P}{\pi b D}\left[1 - D\frac{(R - y)^3}{r_1^4} - D\frac{(R + y)^3}{r_2^4}\right] \qquad (9.4)$$

$$\sigma_{xy} = -\frac{\partial^2 \psi}{\partial x \partial y} = -\frac{2Px}{\pi b}\left[\frac{(R - y)^2}{r_1^4} - \frac{(R + y)^2}{r_2^4}\right]$$

In fact, inclusion of the singular point in the equilibrium check for the x direction along the $x = 0$ line leads to violation of the equilibrium requirement. That is, setting $x = 0$ in equation (9.4) produces $\sigma_{xx}(0, y) = 2P/\pi b D$ whose integral along the diameter of the disk is not zero as required by global equilibrium. This type of problem is common with the so-called exact solutions that involve singularities produced by point loads. For instance, similar phenomenon is encountered in the plane solution of the Flamant–Boussinesq problem [11].

Since the solution to the circular disk problem given by equation (9.2) is not valid at the points $(0, \pm R)$, and by extension in the immediate vicinity of these points, it is certainly valid to inquire how closely the singular points can be approached with the given analytical solution. This question will be answered in the following section.

Equation (9.4) can be employed directly in comparing the experimental

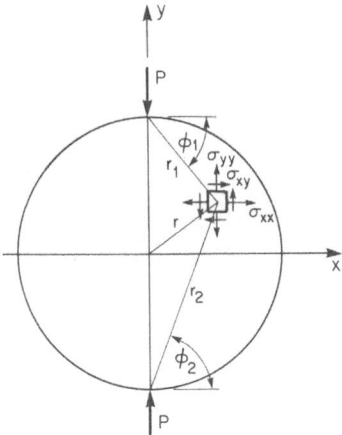

Fig. 9.11. Geometrical variables of the Airy's stress function given by equation (9.2).

results for the normal stresses along various $x = x_0$ and $y = y_0$ cross-sections obtained using optical isodynes with the analytical model. Also, since the shear stresses vanish along the major diameters defined by $x = 0$ and $y = 0$, the principal stresses and their differences can be evaluated along those lines using the results of the three independent experimental techniques and subsequently compared with the analytical predictions. However, in order to compare the isochromatics with the analytical solution it is necessary to evaluate the lines of constant difference of the principal stresses from equation (9.4). Using the definition of isochromatics given by $\sigma_1 - \sigma_2 = mS_\sigma/b$, where m is the order of isochromatics and S_σ is the stress-optic coefficient, leads to the equation for the family of isocromatics with each curve of the family defined by a particular value of m:

$$m^2(S_\sigma/b)^2 = (\sigma_{xx} - \sigma_{yy})^2 + 4\sigma_{xy}^2 \qquad (9.5)$$

Substituting for the inplane stress components given by equation (9.4) in the above equation yields:

$$\frac{m}{A} = \frac{R^2 - (x^2 + y^2)}{(R^2 + x^2 + y^2)^2 - 4R^2y^2} \qquad (9.6)$$

where

$$A = -\frac{2DP}{\pi S_\sigma}$$

The above equation can be simplified by assuming that the zero order isochromatics m_0 passes through the center of the disk. In this case $A = R^2 m_0$, and the coordinates of isochromatics of arbitrary orders can be calculated from the following relations:

$$\frac{x}{R} = \pm\left(\sqrt{\frac{1}{4}\left(\frac{m_0}{m}\right)^2 + 2\frac{m_0}{m} + 4\left(\frac{y}{R}\right)^2} - \left(1 + \frac{1}{2}\frac{m_0}{m} + \left(\frac{y}{R}\right)^2\right)\right)^{1/2}$$

or

$$\frac{y}{R} = \pm\left(\left(1 - \frac{1}{2}\frac{m_0}{m}\right) - \left(\frac{x}{R}\right)^2 \pm \sqrt{\frac{1}{4}\left(\frac{m_0}{m}\right)^2 - 4\left(\frac{x}{R}\right)^2}\right)^{1/2} \qquad (9.7)$$

for

$$-1 \leqslant \frac{x}{R} \leqslant +1 \text{ and } -1 \leqslant \frac{y}{R} \leqslant +1$$

9.3.4 Experimental/analytical correlation

We begin the discussion of the correlation between experimental and analytical results by comparing the fields of isochromatics obtained from equation (9.7) with the corresponding quantities obtained on the basis of transmission photo-elasticity. The comparison is presented in Figure 9.12. The apparent excellent correlation between the predictions of analytical model and experiment in the vicinity of the applied load could be misleading because in those regions the photoelastic isochromatics do not reliably represent analytical isochromatics. However, the excellent correlation between analytical and experimental results in regions remote from local effects establishes the reliability of the employed experimental and analytical methodologies and the ensuing results, and sets the

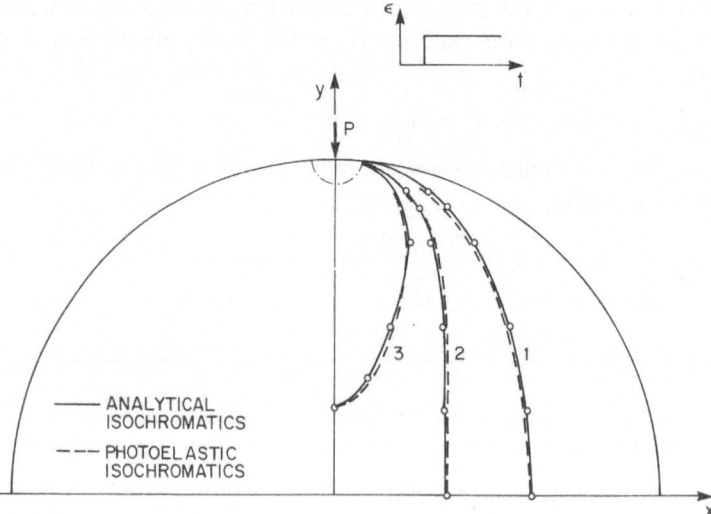

Fig. 9.12. Analytical and photoelastic isochromatics fields. Comparison between analytical and experimental results.

stage for subsequent correlation of the individual stress components along various cross-sections obtained with the different techniques.

The comparison between the analytical predictions of the normal stress components $\sigma_{xx}(x_0, y)$ and $\sigma_{yy}(x, y_0)$ and the corresponding values calculated from the respective isodyne cross-sections is presented in Figure 9.13. The distribution of the normal stress $\sigma_{xx}(x_0, y)$ was calculated for $x_0/R = 0, 0.05,$ 0.25, 0.50 and 0.75, Figure 9.13a, while the distribution of the normal stress $\sigma_{yy}(x, y_0)$ was obtained for $y_0/R = 0.0, 0.25, 0.50$ and 0.75, Figure 9.13b. Generally, the agreement between analytical model and experiment is very good in regions sufficiently removed from the applied concentrated load. As the point of load application is approached, the deviation between analytical and isodyne results increases as clearly observed in the two figures. The most striking discrepancy between analytical model and experiment occurs in the case of the normal stress $\sigma_{xx}(0, y, 0)$ along the major diameter coincident with the direction of applied load, Figure 9.13a. The experimental result calculated from the isodyne fields indicates that $\sigma_{xx}(0, y, 0)$ changes sign at the point $y/R = 0.90$ whereas the analytical model predicts a constant stress which violates the global equilibrium of the disk along the $x = 0$ cross-section discussed previously. It is not difficult to see, in fact, that the stress component normal to the load axis must change sign in the vicinity of the applied load if equilibrium of the disk in the x direction is to be preserved for the $x = 0$ cross-section. The experimental results obtained with the isodyne technique predict this change correctly.

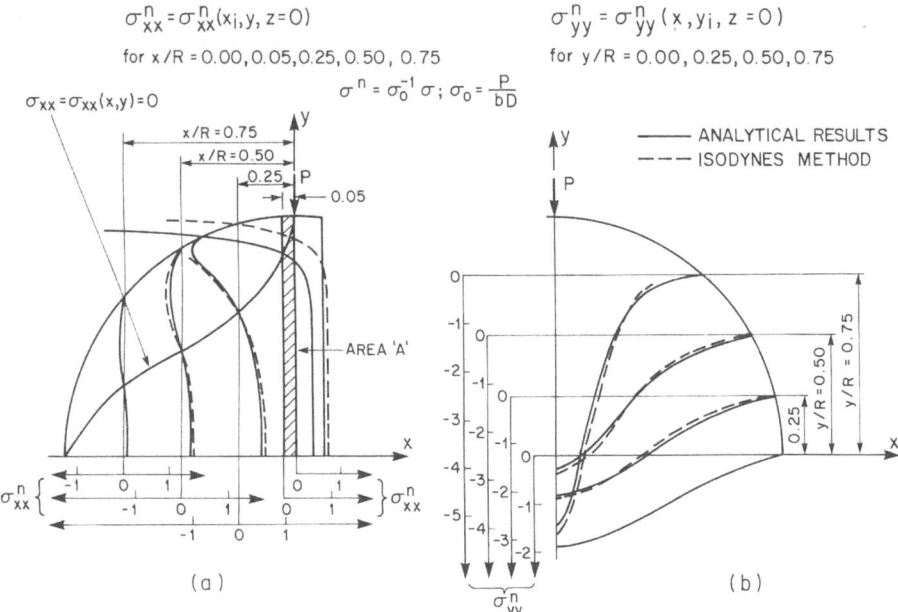

Fig. 9.13. Normal stress distributions along selected $x = $ const and $y = $ const cross-sections. Comparison between analytical predictions and optical isodynes data.

Figure 9.14 presents the two normal stress components σ_{xx} and σ_{yy} along the major diameters normal and perpendicular to the load axis obtained from the analytical solution, equation (9.4), isodyne fields and strain gage data. The correlation between the stress values obtained with the three techniques is satisfactory with the exception of the immediate vicinity of the point of load application approximately defined by the distance $0.2R$. It should be noted that the strain gage measurements yield lower values of the normal stress σ_{xx} than the isodyne calculations in regions where the stress gradient in the direction normal to the line load is high. Two possibilities exist that may be employed to explain this difference. First, the width of the strain gages may be a contributing factor and second; the state of stress on the lateral surface may not be identically the same as the state of stress in the $z = 0$ mm plane where the isodyne data was recorded. However, it is significant that the location of the point where the normal stress $\sigma_{xx}(0, y, 0)$ changes sign determined from the strain gage data is $0.89R$. This compares favorably with $0.90R$ obtained from the isodyne fields.

The final comparison between the analytical model and the experimental results obtained with the three techniques is presented in Figure 9.15 for the principal stress difference $\sigma_1 - \sigma_2$ along the two major diameters. Virtually no discrepancy is observed in the principal stress difference calculated on the basis of the four independent techniques along the diameter perpendicular to the load

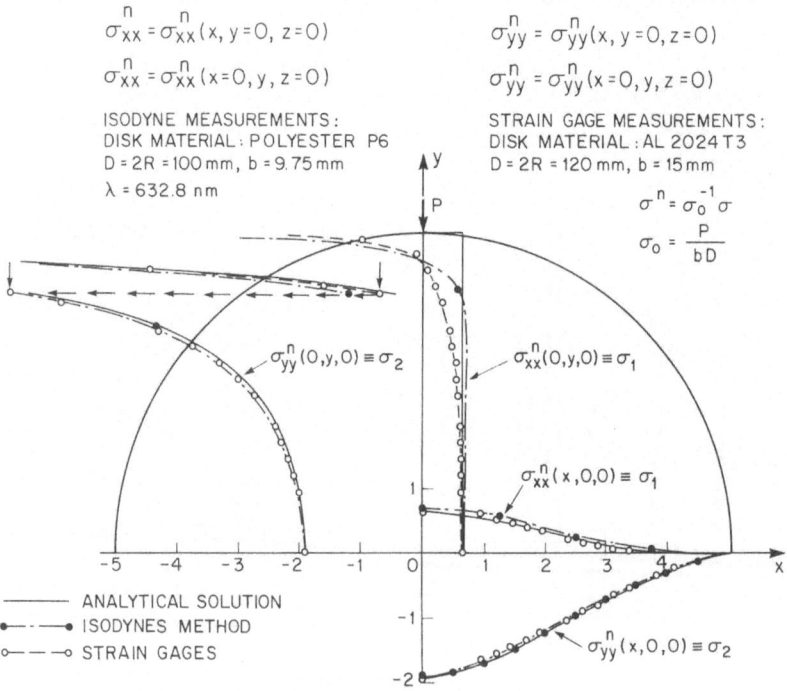

Fig. 9.14. Normal stress distributions along the major diameters parallel and perpendicular to the load axis. Comparison between analytical predictions, optical isodynes data and strain gage data.

$$(\sigma_1 - \sigma_2)^n = \Delta\sigma^n(x) = \Delta\overset{n}{\sigma}(x, y=0, z=0)$$

$$(\sigma_1 - \sigma_2)^n = \Delta\overset{n}{\sigma}(y) = \Delta\overset{n}{\sigma}(x=0, y, z=0)$$

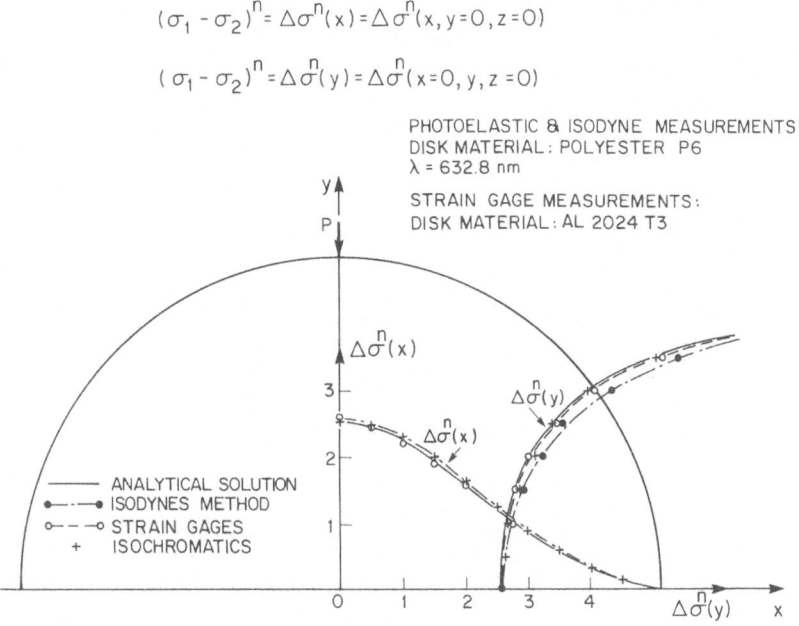

Fig. 9.15. Distributions of principal stress difference along the major diameters parallel and perpendicular to the load axis. Comparison between analytical predictions, optical isodynes, transmission photoelasticity and strain gage data.

line. However, deviations between results of the four methods become apparent along the major diameter parallel to the load line with decreasing distance from the concentrated load.

9.4 Closing comments

The examples presented in this chapter provide ample evidence of the applicability of the relations and techniques of optical isodynes in the analysis of planar states of stress. The accuracy of this method in evaluating inplane stress components in regions where the state of stress is clearly two-dimensional has been demonstrated by comparing the results obtained with this technique with the predictions of exact plane elasticity solutions as well as two other independent experimental techniques. Further, the results of the outlined investigations establish the feasibility of employing optical isodynes in testing the range of applicability of plane elasticity solutions in regions where there is sufficient reason to believe that a particular solution may not be valid. As already stated, such regions are characterized by high stress gradients produced by concentrated or discontinuous loads, cracks or material discontinuities and so on. In fact, comparison of the experimental results with the analytical solution for the second problem discussed in this chapter clearly demonstrated the problem that one encounters when dealing with singular solutions. Since in practical applications the regions where high stress gradients are present are of immediate interest, it is imperative that the range of applicability of singular solutions be quantified. The optical isodyne method is certainly a useful tool in accomplishing this.

9.5 References

[1] Filon, L. N. G., "On an Approximate Solution for the Bending of a Beam of Rectangular Cross-Section under any System of Load, with Special Reference to Points of Concentrated or Discontinuous Loading", *Royal Society of London, Phil. Trans.* **201**, 63—155 (1903).

[2] Frocht, M. M., *Photoelasticity*, Vol. II, John Wiley, New York (1948).

[3] Hondros, G., "The Evaluation of Poisson's Ratio and the Modulus of Materials of a Low Tensile Resistance by the Brazilian (Indirect Tensile) Test with Particular Reference to Concrete", *Australian J. Appl. Sci.* **10**(2), 243—268 (1959).

[4] Mesmer, G., "Vergleichende spannungsoptische Untersuchungen und Fliessversuche unter konzentriertem Druck", Ph.D. Dissertation, University of Gottingen (1929).

[5] Muskhelishvili, N. I., *Some Basic Problems of the Mathematical Theory of Elasticity*, P. Noordhoff, Groningen, The Netherlands (1953).

[6] Pindera, J. T., *Outline of Photoelasticity*, P. W. T., Warsaw (1953, in Polish).

[7] Pindera, J. T., "Technique of Photoelastic Investigations of Plane Stress Fields", *Engineering Transactions*, (*Rozprawy Inzynierskie*), Polish Acad. Sciences, Vol. 3, No. 1, 109—176 (1955, in Polish).

[8] Pindera, J. T. and Mazurkiewicz, S. B., "Integrated Plane Photoelastic Methods — Applications of Scattered Photoelastic Isodynes", Proc. 1977 SESA Spring Meeting, Paper No. D-16 Dallas, Texas, May 15—20, 1977.

[9] Pindera, J. T., Mazurkiewicz, S. B. and Khattab, M. A., "Stress Field in Circular Disk Loaded Along Diameter: Discrepancies Between Analytical and Experimental Results", Proc. 1978 SESA Spring Meeting, Paper No. CR-10, Wichita, Kansas, May 14—19, 1978.

[10] Pindera, J. T. and Straka, P., "On Physical Measures of Rheological Responses of Some Materials in Wide Ranges of Temperature and Spectral Frequency", *Rheologica Acta* **13**, 846—859 (1974).

[11] Sokolnikoff, I. S., *Mathematical Theory of Elasticity*, McGraw-Hill, Toronto (1956).

[12] Timoshenko, S. P. and Goodier, J. N., *Theory of Elasticity*, McGraw-Hill, Toronto (1970).

10

Contact problems

10.1 Introduction

The correlation between the experimental results obtained using the techniques of isodynes and the two-dimensional elasticity solutions to selected problems presented in the preceding chapter establishes isodyne stress analysis as a quantitative tool in the analysis of plane states of stress. In view of the previous discussion regarding the range of applicability of a certain class of plane solutions, such an independent experimental technique is indeed very useful. There are, of course, other classes of plane elasticity problems whose solutions require independent verification or for which solutions are not readily available. One such class of problems which is the subject of the present chapter comprises a widely varied group of problems known as contact problems. These problems deal with localized stress distributions in the immediate vicinity of two contacting bodies whose deformation along the line of contact is taken into account in the course of generating a solution.

One may argue, in fact, that the majority if not all of stress analysis problems involving mechanical loads encountered in practice are indeed contact problems since mechanical loads are transmitted by direct contact to the medium of interest. For instance, the two examples discussed in the preceding chapter fall into the category of contact problems regarding the actual distribution of stresses in the immediate vicinity of points of load application as mentioned previously. Often, it is the actual distribution of local stresses in the immediate vicinity of contacting surfaces that is of immediate interest if, for instance, those areas are nucleation sites of failure. In other situations, use is made of Saint Venant's Principle and the actual distribution of tractions is replaced by a system of statically equivalent tractions. Problems involving point loads or loads distributed uniformly or parabolically over a small interval fall into that category.

As mentioned previously, an exact solution to a contact problem may not always be available due to complications arising from the geometry of the contacting surfaces, finiteness of the regions, geometry-induced nonlinearities and so on. Even when a solution is available, often it needs to be verified experimentally in order to confirm not only the correctness of the employed assumptions but also the accuracy of the technique employed to generate numerical values. The latter precaution arises because a significant number of contact problems are solved by reducing the governing differential equations to a system of integral equations involving the unknown surface tractions and displacements which in turn are solved numerically using devices such as the series expansion technique for instance [1]. The former precaution is necessary because it is often not clear what constraints to impose on the variables of interest in the contact region. Solutions to contact problems are therefore, by necessity, solutions to mathematical models that need to be verified by independent means in each particular situation.

In this chapter, two contact problems for which solutions are not readily available are analyzed using isodyne stress analysis [2]. These two problems are natural extensions of the example problem presented in the preceding chapter that dealt with a single beam subjected to diametrically opposite, concentrated loads. Here, we consider two and three unbonded beams stacked on top of each other subjected to the same loading as in the above problem. Initially, the beams are in contact with each other along the entire interface when no load is applied. Under the action of applied load the interfaces separate partially as predicted by Filon's solution for a single homogeneous beam (see [1], Chapter 9) and the load is transmitted from one beam to another through the remaining contact region. The separation of the beams is caused by the presence of tensile normal stresses along sections of the common interfaces that would arise had the interfaces been bonded together.

In the present problem, the quantities of interest are the distribution of the normal stresses in the contact area and the effect of the interfacial separation on the redistribution of stresses in the individual beams for the given external loading. To investigate those quantities, the experimental results for the normal stress distribution $\sigma_{yy}(x, y_0, z_0)$ along various elevations obtained using isodyne stress analysis will be compared with the results of Filon's solution for a single beam of identical dimensions as the dimensions of the two and three unbonded beams considered in this investigation.

10.2 Experimental investigation

The specimen geometry and loading conditions of the two and three unbonded beams stacked on top of each other and subjected to diametrically opposite concentrated loads are summarized in Figure 10.1. Each beam was 200 mm long, 30 mm high and 10 mm thick, which are the dimensions of the single beam subjected to concentrated loads treated in Chapter 9. The same experimental procedure was employed to record the fields of isodynes and isochromatics in the present examples as that in the case of the single beam. The data was recorded at the load of 2403 N and the results are presented for this load level for direct comparison with the example of Chapter 9.

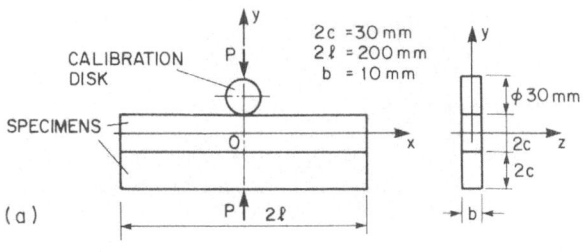

MATERIAL: POLYESTER RESIN
PALATAL P6

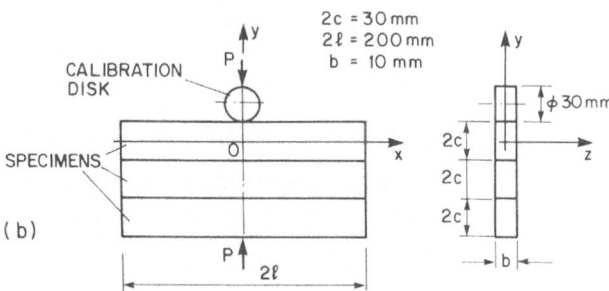

Fig. 10.1. Two and three unbonded beams in contact under point loads. Specimen geometry and loading conditions.

10.3 Two beams in contact

10.3.1 Experimental results

Figure 10.2 presents the recordings of amplitude-modulated x-isodyne fields in the $z = 0.0$ mm plane at the applied load of 2403 Newtons for the two-beam contact problem. The transmission isochromatics, which clearly illustrate the relative separation of the two beams along the initially common interface, are also included in the figure. The graphical representation of the x-isodyne and isochromatics fields is given in Figure 10.3 from which the various cross-sections of x-isodynes, called x-isodyne functions, at fixed elevations can be evaluated. An example of such procedure is given in Figure 10.4 for the $y = 0.0$ mm elevation. Differentiation of the x-isodyne function $m_{sx}(x, y_0, 0)$ with respect to the x coordinate yields the normal stress component $\sigma_{yy}(x, y_0, 0)$ which is included in Figure 10.4.

10.3.2 Experimental/analytical correlation

The normal stress distributions $\sigma_{yy}(x, y_0, 0)$ have been determined from the isodyne functions $m_{sx}(x, y_0, 0)$ at the elevations $y_0 = 0.0$ and -15.0 mm, that is,

ISODYNES S_{X1}

$I_{SX1}^{n} = I_{SX1}(kI_0)^{-1} =$
$= 0.5\left[1 - \cos(\phi + \Delta\phi)\right] =$
$= \sin^2 \pi (m_s + \Delta m_s)_{X1}$

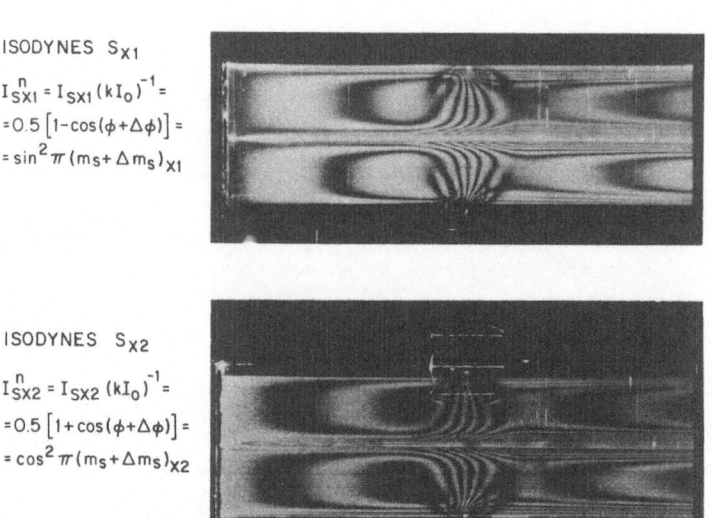

ISODYNES S_{X2}

$I_{SX2}^{n} = I_{SX2}(kI_0)^{-1} =$
$= 0.5\left[1 + \cos(\phi + \Delta\phi)\right] =$
$= \cos^2 \pi (m_s + \Delta m_s)_{X2}$

TRANSMISSION
ISOCHROMATICS

$\oplus \ I_t^{n} = I_t(kI_0)^{-1} =$
$= \sin^2 \pi m_t$

Fig. 10.2. Recordings of amplitude modulated x-isodyne fields in the $z = 0.0$ mm plane at 2403 Newtons, and the corresponding transmission isochromatics. Two unbonded beams in contact under point loads.

half-way across the top beam and along the common interface, respectively. The distributions are presented in Figure 10.5a, b. Included in the figure are the corresponding predictions of Filon's solution for a single beam of equivalent dimensions subjected to the same loading conditions (cf. Eq. 9.1, Chapter 9). The experimental and analytical results have been normalized by using the nondimensional quantities x/h for the abscissa and $\sigma_{yy}(x, y_0, 0)/(2P/\pi bh)$ for the ordinate (where h is one-half of the beam's height) in order to eliminate the effect of geometric and load variables on the ensuing stress fields.

Figure 10.5a indicates that the separation of the interface between the top and bottom beam results in the redistribution of the normal stress $\sigma_{yy}(x, y_0, 0)$ half-way across the top beam in the following manner. The maximum stress directly under the point load is slightly smaller than the corresponding stress in the fully bonded beams and the distribution profile is fuller. The fuller distribution profile results in the normal stress becoming tensile at the distance of approximately $x/h = 0.84$ in order to maintain the upper beam in vertical equilibrium. On the other hand, it is clear from Figure 10.5b that the vertical

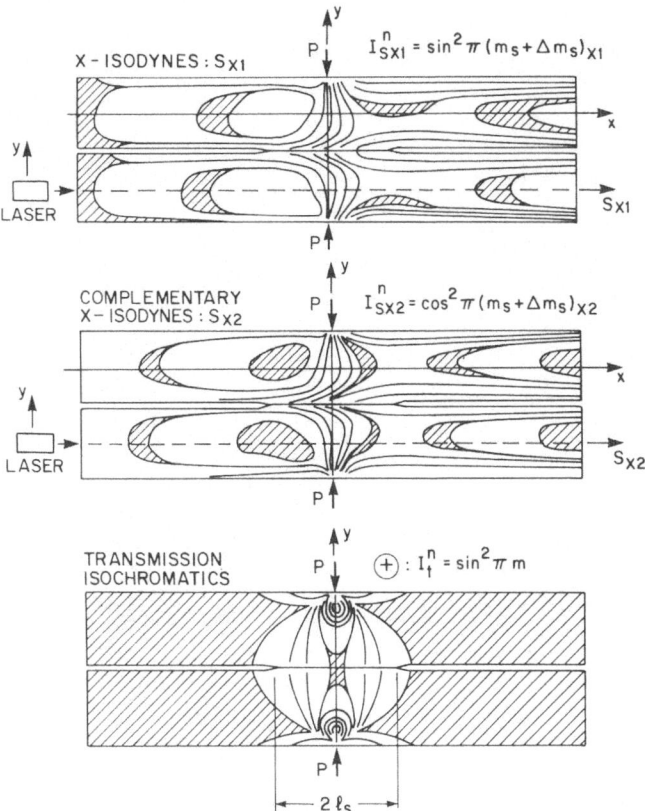

Fig. 10.3. Graphical representation of the x-isodyne fields and transmission isochromatics.

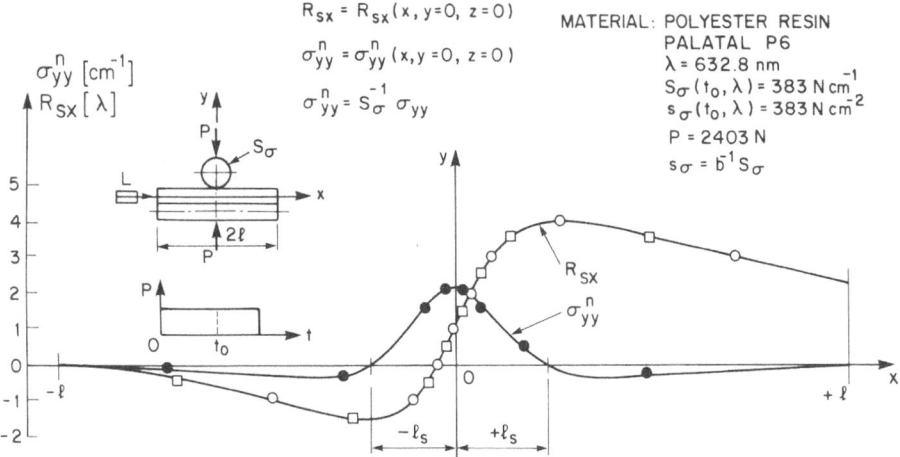

Fig. 10.4. Cross-section of x-isodynes at the $y = 0.0$ mm elevation in the $z = 0.0$ mm plane and the corresponding normalized distribution of the normal stress component $\sigma_{yy}(x, y_0, 0)$.

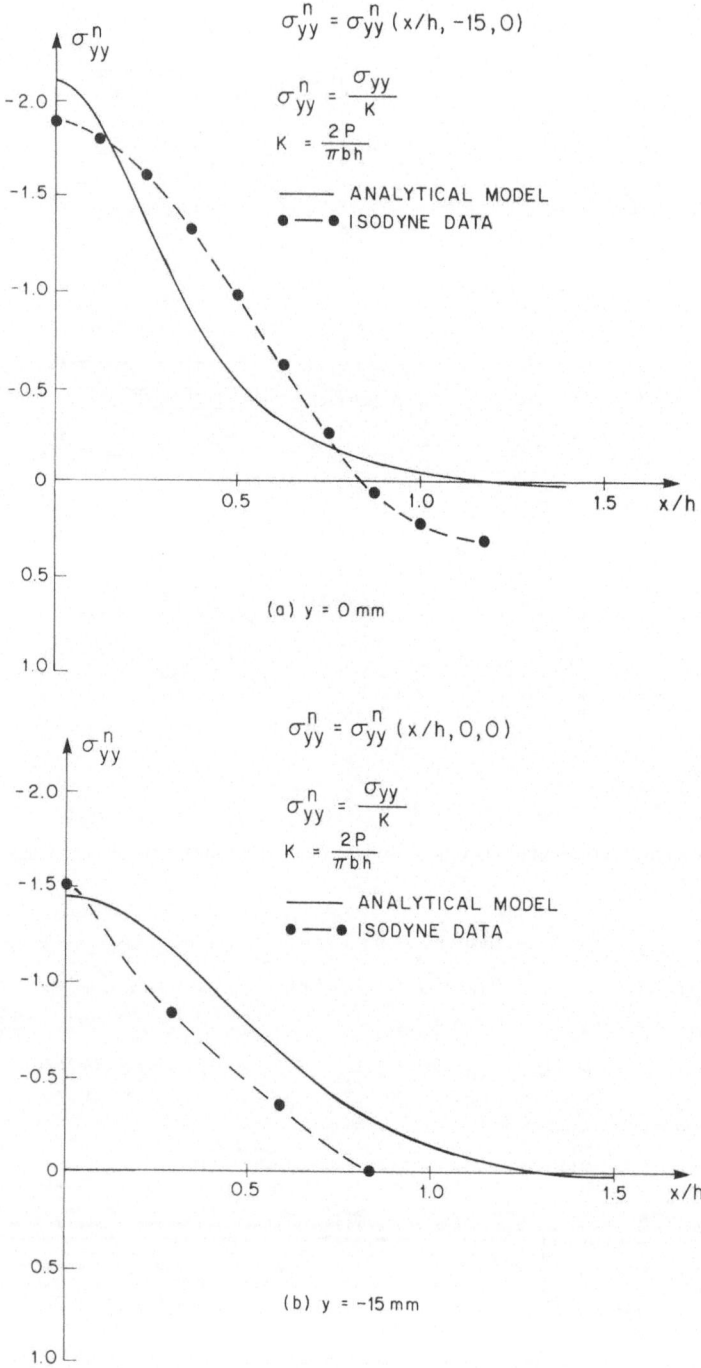

Fig. 10.5. Horizontal distributions of the normal stress $\sigma_{yy}(x, y_0, 0)$ at two elevations. Comparison between analytical predictions and experiment.

equilibrium along the common interface of the two beams is not satisfied if it is assumed that a plane state of stress exists through the thickness of the contact region and that the experimentally obtained isodyne fields are plane. It appears that the isodyne functions in the contact region must be interpreted as differential isodynes rather than plane isodynes. In such case, the experimental results presented in Figure 10.5b pertain to the difference in the normal stress distribution, $\sigma_{yy}(x, y_0, 0) - \sigma_{zz}(x, y_0, 0)$, rather than the normal stress distribution $\sigma_{yy}(x, y_0, 0)$.

10.4 Three beams in contact

10.4.1 Experimental results

The recordings of amplitude-modulated x-isodyne fields in the $z = 0.0$ mm plane at the applied load of 2403 Newtons for the three-beam contact problem are presented in Figure 10.6, along with the corresponding transmission

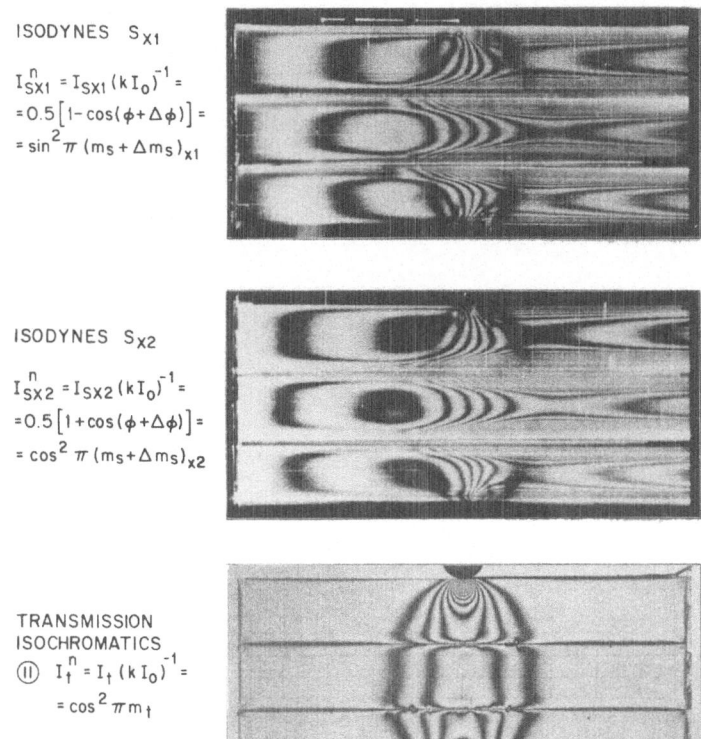

ISODYNES S_{X1}

$$I_{SX1}^{n} = I_{SX1}(kI_0)^{-1} =$$
$$= 0.5[1 - \cos(\phi + \Delta\phi)] =$$
$$= \sin^2 \pi (m_s + \Delta m_s)_{x1}$$

ISODYNES S_{X2}

$$I_{SX2}^{n} = I_{SX2}(kI_0)^{-1} =$$
$$= 0.5[1 + \cos(\phi + \Delta\phi)] =$$
$$= \cos^2 \pi (m_s + \Delta m_s)_{x2}$$

TRANSMISSION
ISOCHROMATICS
$$\text{(II)} \quad I_t^{n} = I_t(kI_0)^{-1} =$$
$$= \cos^2 \pi m_t$$

Fig. 10.6. Recordings of amplitude modulated x-isodyne fields in the $z = 0.0$ mm plane at 2403 Newtons, and the corresponding transmission isochromatics. Three unbonded beams in contact under point loads.

isochromatics. Again, the relative spearation of the common interfaces initially in contact is clearly visible. Two cross-sections of the *x*-isodynes half-way across the height of the top and middle beams are given in Figure 10.7 along with the corresponding normal stress distributions $\sigma_{yy}(x, y_0, 0)$ obtained by differentiating the above isodyne functions with respect to the axial corrdinate.

10.4.2 Experimental/analytical correlation

The experimentally obtained normal stress distribution $\sigma_{yy}(x, y_0, 0)$ at the elevation $y_0 = -30.0$ mm, that is, half-way across the middle beam, is compared with the prediction of Filon's solution for a single beam of equivalent dimensions and loading conditions in Figure 10.8. In this case, the effect of the interfacial separation on the normal stress distribution $\sigma_{yy}(x, y_0, 0)$ half-way across the middle beam results in an increase in the magnitude of the maximum stress directly under the applied load which is compensated by a decrease in the magnitudes of the normal stress in the interval $0.28 \leqslant x/h \leqslant 0.78$. Beyond $x/h = 0.78$, the experimentally determined normal stress becomes tensile while the analytical model for the fully bonded beams indicates significant compressive stresses. The redistribution of the normal stress caused by the interfacial separation yields an apparent total vertical force that is smaller than the corresponding vertical force resultant in the fully bonded case. It appears therefore that the vertical equilibrium is not satisfied if the experimental data is interpreted on the basis of the plane isodyne model. Apparently, the influence of three-dimensional contact stresses at the top and bottom interfaces of the middle beam extends deeply into the interior of the middle beam.

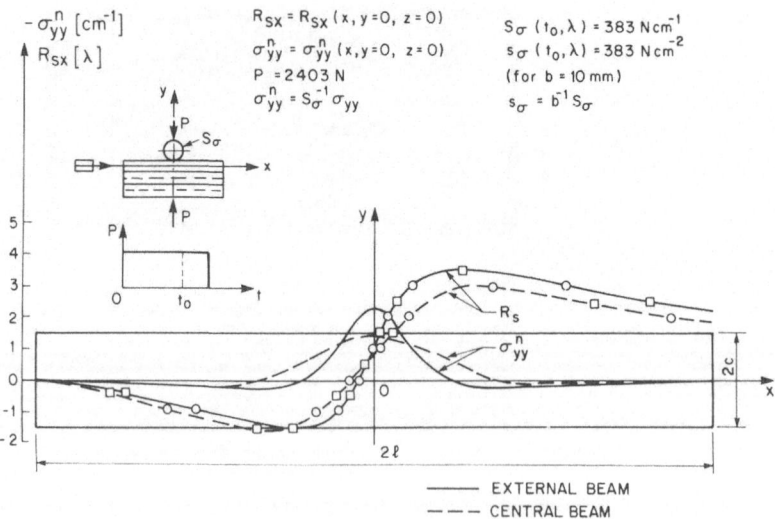

Fig. 10.7. Cross-sections of *x*-isodynes at two elevations in the $z = 0.0$ mm plane and the corresponding normalized distributions of the normal stress components $\sigma_{yy}(x, y_0, 0)$.

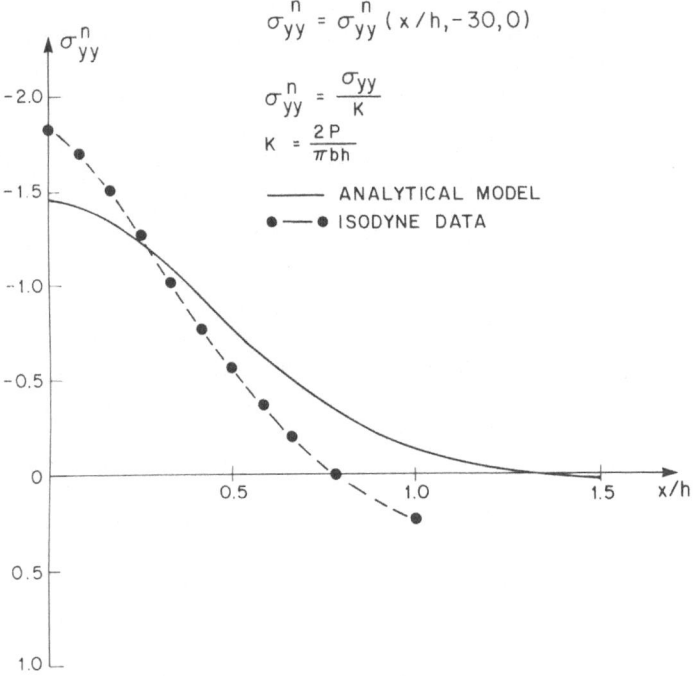

Fig. 10.8. Horizontal distribution of the normal stress $\sigma_{yy}(x, y_0, 0)$ half-way across the middle beam. Comparison between analytical prediction and experiment.

10.5 Closing comments

The two examples presented in this Chapter were one of the earliest cases to be investigated using isodyne stress analysis. At that time, the concept of differential isodynes was not formulated and thus no attempt was made to investigate the three-dimensional aspects of the stress fields in the regions of contacting bodies, together with the influence of contact stresses on the through-the-thickness stress distribution in the interior. The outlined results suggest that the actual stress distributions in the type of contact problems considered here are three-dimensional; these can readily be characterized by performing isodyne data collection in the different through-the-thickness planes. The results presented in Chapter 11 clearly demonstrate through-the-thickness variations of stress fields in beams and plates in the regions of high stress gradients caused by notches and concentrated loads.

10.6 References

[1] Gladwell, G. M. L., *Contact Problems in the Classical Theory of Elasticity*, Sijthoff & Noordhoff, Alphen aan den Rijn, The Netherlands (1980).

[2] Pindera, J. T. and Mazurkiewicz, S. B., "Integrated Plane Photoelastic Methods — Applications of Scattered Photoelastic Isodynes", Proc. 1977 SESA Spring Meeting, Paper No. D-16, Dallas, Texas, May 15—20, 1977.

Three-dimensional local effects

11.1 Introduction

The three-dimensional effects in beams or plates in the regions of high stress gradients caused by notches or concentrated loads are not easily quantifiable analytically since a three-dimensional solution to the apparently simple problem of a beam or plate loaded nonuniformly along its top or bottom faces is not available at present. When dealing with problems of this type, recourse is typically made to treating the problem as a generalized plane stress or plane deformation problem, or assuming some specific through-the-thickness stress distribution that satisfies certain of the field equations of elasticity [2, 7]. Thus far, relatively little experimental work has been carried out to verify the validity of the various plane or approximate three-dimensional analytical solutions in regions where three-dimensional effects may be present. A factor which certainly has contributed to the slow progress in the experimental investigation of three-dimensional stress fields in engineering structures is the lack of simple, reliable and nondestructive experimental methodologies, based on noncontradictory and self-contained assumptions, which are suitable for three-dimensional analysis.

The results of experimental investigations using the techniques of optical isodynes presented thus far have been limited to analysis of stress components lying in a single plane. This plane was typically chosen to be the plane of symmetry of an object of constant thickness subjected to inplane loading. Although experimental evidence was presented which indicated that in regions of high stress gradients the actual state of stress deviated noticeably from the solution of the corresponding two-dimensional elasticity model, no attempt was made to experimentally investigate the three-dimensional effects in those regions. As already mentioned, regions of high stress gradients in plates or beams subjected to inplane loading are characterized by localized three-dimensional stress fields which, in turn, lead to variations in the inplane stress components through the investigated object's thickness. The optical isodyne method is ideally suited for investigating such effects.

Two most common sources of high stress gradients in plates or beams are cracks and concentrated loads. The present chapter outlines the results of experimental investigations involving the above two sources of localized three-dimensional stress fields. The first example deals with a notched, rectangular beam subjected to four-point bending and the problems associated with the determination of Mode I stress intensity factor using the two-dimensional elasticity approach. It is illustrated that a pronounced through-the-thickness variation in the axial stress component exists at the tip of the notch which is accompained by an out-of-plane normal stress component. The pronounced through-the-thickness variation in σ_{xx}, along with the presence of the out-of-place stress σ_{zz}, introduces an ambiguity in the determination of the stress intensity factor based on the common two-dimensional approach. In the second example, the inplane stress distributions in the vicinity of a concentrated load are investigated in different through-the-thickness planes for a rectangular beam subjected to three-point bending. Variations in the inplane stress components are observed in the different through-the-thickness planes which increase with decreasing distance from the point of load application. Greatest variations are obtained for the inplane shear stress component evaluated on the basis of the relations of plane isodynes. The experimental results are subsequently compared with the predictions of the two-dimensional elasticity solution and it is illustrated that the differences between theory and experiment increase with decreasing distance from the location of applied load.

11.2 Four-point bending of a notched beam

The present example deals with the application of optical isodynes to determination of Mode I stress intensity factor for a notched beam subjected to four-point loading. For a rectangular beam with a surface notch perpendicular to the boundary subjected to the above loading, the stress field in the immediate vicinity of the notch is inversely proportional to the square root of the distance from the crack tip according to the plane elasticity model. The stress intensity factor, which gives the strength of the stress singularity, can therefore be represented in the form

$$K_I = \lim_{r \to 0} \sqrt{2\pi r} \; \sigma_{xx}(r, \theta = 0) \tag{11.1}$$

where r and θ are polar coordinates in the $x - y$ plane of the beam with the origin at the crack tip, and y-axis is collinear with the crack direction. Reliable and accurate determination of the stress intensity factor from experimental data depends therefore on accurate and reliable determination of the functional relationship between the normal stress component σ_{xx} and the distance from the crack tip along the line of symmetry $\theta = 0$.

If the characteristic directions are aligned with the x and y-axis of the beam, the normal stress component $\sigma_{xx}(r, 0)$ can be directly obtained by differentiating the $x = 0$ cross-section of the y-isodynes with respect to the y corrdinate provided that the state of stress is truly plane. Multiplying the thus determined normal stress by the factor $\sqrt{2\pi r}$ yields an experimental function which in the immediate vicinity of the crack tip should attain a constant value if the two-dimensional elasticity model is valid. On the other hand, if three-dimensional effects are present in the vicinity of the crack tip, differentiation of the *y*-isodynes in the manner outlined above yields the normal stress difference $\sigma_{xx}(r, 0, 0) - \sigma_{zz}(r, 0, 0)$ in the $z = 0.0$ mm plane. Multiplication of this difference by the factor $\sqrt{2\pi r}$ may or may not produce a region where the product of the two functions is constant, depending on the functional form of the out-of-plane normal stress component $\sigma_{zz}(r, \theta, z)$ as well as the inplane normal stress component $\sigma_{xx}(r, \theta, z)$ in the vicinity of the crack tip. The presence of three-dimensional effects can be easily ascertained by recording the *x*- and *y*-isodyne fields in different planes situated at different through-the-thickness locations with respect to the beam's or plate's plane of symmetry, or by evaluating the out-of-plane normal stress component $\sigma_{zz}(r, 0, z)$ from the *x*-isodyne fields at the tip of the notch for any given cross-section.

In the course of determining the stress intensity factors using experimental data obtained from tests conducted on real materials, one additional aspect needs to be taken into account. High stresses in the vicinity of crack tips result in material and geometric nonlinearities which distort the elastic stress fields. The state of stress in the immediate vicinity of the crack tip is thus a very complex phenomenon involving material and geometric factors which provide additional constraints with regard to the local three-dimensional effects. Since optical isodynes can produce directly the normal stress component associated with a chosen characteristic direction provided that the state of stress is two-dimensional, the region in the vicinity of the crack tip where the stress field is elastic and plane can be readily identified for the purpose of determining the stress intensity factor.

In the outlined investigation, the results of the determination of Mode I stress intensity factor in a notched beam under four-point loading on the basis of optical isodynes are compared with the asymptotic plane solution of Benthem and Koiter [1]. In order to explain the discrepancy between analytical predictions and experimental data taken in the plane of symmetry of the beam, additional recordings were taken in four different through-the-thickness planes. The data obtained from these recordings indicates the presence of the out-of-plane normal stress component σ_{zz} along with a pronounced through-the-thickness variation of the axial stress component σ_{xx} at the tip of the notch. The above findings point to conceptual problems that arise if the common experimental methods based on the assumption of plane stress are employed in the course of determining stress intensity factors.

11.2.1 Experimental investigation

The experiments performed in this investigation were conducted on two different beam configurations. The external inplane dimensions of the two configurations and the loading conditions were the same. The beam configurations differed in the thickness and the length of the symmetrically located edge notch. The overall length of the beams was 160 mm and the height 30 mm. The distance between inner supports was 110 mm while between the outer supports the distance was 140 mm. Type 1 beam was 9.5 mm thick and had a 3 mm notch located on the compressive side of the beam. The radius of the notch tip was 0.1 mm which resulted in the notch width of 0.2 mm. These dimensions were common for the two tested configurations. Type 2 beam was 20 mm thick and had a 6 mm notch located on the compressive side of the beam. The geometry and loading conditions for the two beam configurations are summarized in Figure 11.1.

Type 1 beam was fabricated using Homalite H100 while Palatal P6 was used to fabricate Type 2. The stress-optic coefficient for Palatal P6 was determined to be 36.8 N/mm while for Homalite H100 it was 58.6 N/mm. The wavelength of the radiation employed to record the isodyne fields was 632.8 nm. In order to eliminate the influence of the time-dependent response of Palatal and Homalite during recording of the isodyne fields, the load was applied in the constant overall deformation mode in which the time-alteration of the strain-optic coefficient is negligible. In addition, the load relaxation was monitored and measurements were taken when the relaxation process became negligible. The recording of the isodyne fields was carried out at the applied load P of 1,725 N, Figure 11.1. In the case of the thin beam, Type 1, isodyne fields were recorded in the plane of symmetry of the beam, that is $z = 0.0$ mm plane. In the case of the thick beam, Type 2, the recording of isodyne fields was carried out in four different through-the-thickness planes defined by the locations $z = 0.0, 4.0, 6.0$ and 9.0 mm.

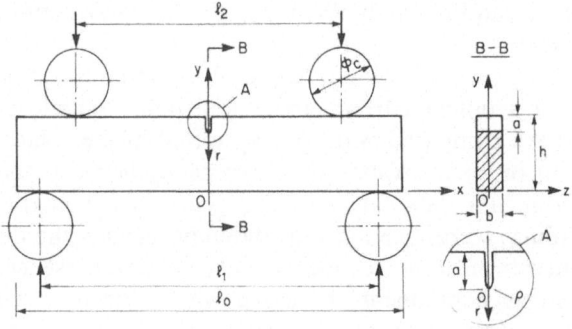

SPECIMEN TYPE	MATERIAL POLYESTER RESINS:	GEOMETRY								PHYSICAL DATA	
		ℓ_0	ℓ_1	ℓ_2	h	b	a	ρ	ϕ_c	$(S_\sigma)_\epsilon$	λ
		mm								$N\,mm^{-1}$	nm
TYPE 1	HOMALITE 100	160	140	110	30	9.5	3	0.1	30	58.6	632.8
TYPE 2	PALATAL P6	160	140	110	30	20	6	0.1	30	36.8	632.8

Fig. 11.1. Four-point bending of a notched beam. Specimen configurations and loading conditions.

11.2.2 Experimental results

Recordings of the x- and y-isodyne the fields in the $z = 0.0$ mm plane for the Type 1 beam are presented in Figure 11.2. Transmission isochromatics are included in the figure for comparison. The axial stress distribution $\sigma_{xx}(r, 0, 0)$ along the vertical axis of symmetry of the beam has been evaluated from the y-isodyne fields by differentiating the $x = 0.0$ mm cross-section of the y-isodynes with respect to the y coordinate. It should be noted that in the immediate vicinity of the notch tip the differentiation of the y-isodyne function yields the normal stress difference $\sigma_{xx}(r, 0, 0) - \sigma_{zz}(r, 0, 0)$. The maximum value of $\sigma_{zz}(r, 0, 0)$ occurs at the notch tip. This was evaluated from the x-isodyne fields since at the notch tip the normal stress component σ_{yy} vanishes. In order to separate the σ_{xx} and σ_{zz} stresses in the vicinity of the notch tip, an assumption was made that the variation of the out-of-plane stress with the radial distance from the notch tip is parabolic and that $\sigma_{zz}(r, 0, 0)$ vanishes at the

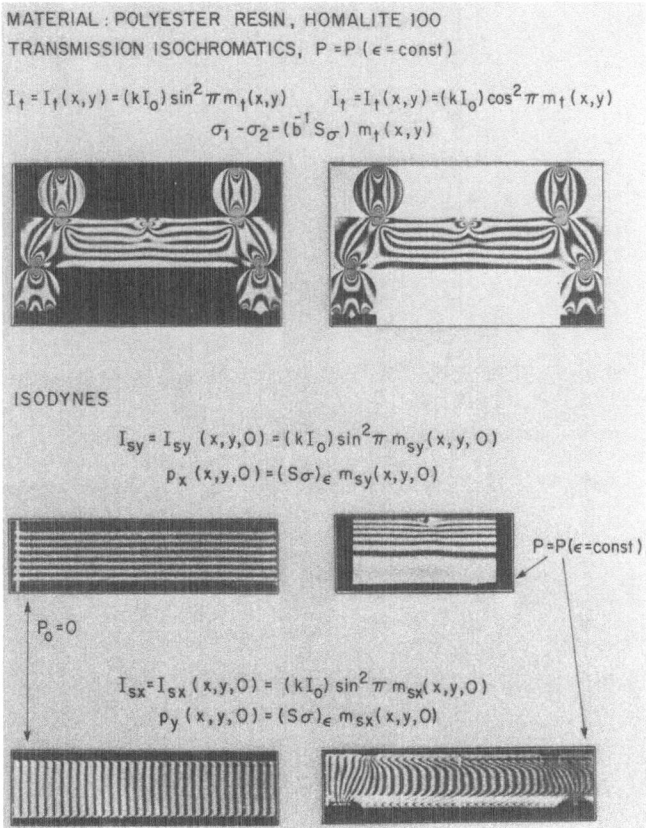

Fig. 11.2. Recordings of spatial frequency modulated x- and y-isodyne fields in the $z = 0.0$ mm plane at 1,725 Newtons for Type 1 beam.

distance of one and one-half beam thickness. Based on the above approach the results for the axial stress distribution $\sigma_{xx}(r, 0, 0)$ for the Type 1 beam are presented in Figure 11.3.

In order to determine Mode I stress intensity factor from the experimental data presented in Figure 11.3, the axial stress $\sigma_{xx}(r, 0, 0)$ in the immediate vicinity of the notch tip was multiplied by the factor $\sqrt{2\pi r}$. The resulting experimentally obtained function $K_I(r)$, denoted here stress intensity function,

$$K_I(r) = \sqrt{2\pi r}\ \sigma_{xx}(r, 0, 0) \tag{11.2}$$

is presented as a function of the radial distance r from the notch tip in Figure 11.4. The limit of this function as the radial distance tends to zero is defined in

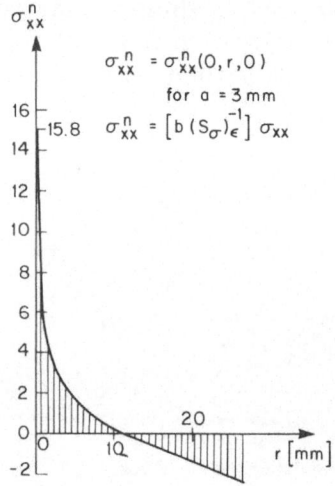

Fig. 11.3. Axial stress distribution $\sigma_{xx}(r, 0, 0)$ in the $z = 0.0$ mm plane along the axis colinear with the edge notch for Type 1 beam.

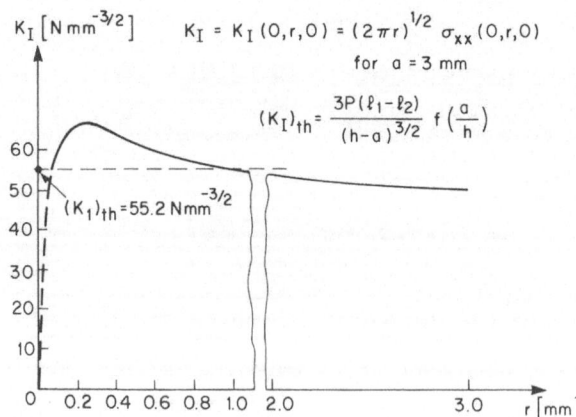

Figure 11.4. The function $K_I(r)$ in the vicinity of the notch tip for Type 1 beam.

classical fracture mechanics based on the linear two-dimensional elasticity model as the stress intensity factor, equation (11.1). In practice however, the presence of highly localized nonlinear effects and subsequent blunting of the crack tip together with three-dimensional stress fields excludes the immediate vicinity of the crack tip from consideration. Thus the determination of the stress intensity factor has to be carried in the region where the two-dimensional elastic model is applicable if indeed such a region exists.

The results presented in Figure 11.4 illustrate the problems that may be encountered in identifying the regions where the concepts of two-dimensional linear elastic fracture mechanics are applicable. The variation of the function $K_I(r)$ in the vicinity of the notch tip suggests that the actual normal stress distribution $\sigma_{xx}(r, 0, 0)$ departs from the functional form of the singular stress field predicted by the plane elasticity model. It ought to be mentioned that the character of the function $K_I(r)$ would not change significantly had a different variation of $\sigma_{zz}(r, 0, 0)$ commensurate with the physics of the problem been assumed. In fact, the presence of the out-of-plane stress component appears to reduce the variation of the function $K_I(r)$ with regard to the magnitude of the local maximum seen in Figure 11.4.

The presence of three-dimensional stress fields in the vicinity of the notch tip is illustrated in Figure 11.5 where the recordings of y-isodyne fields are presented in four different through-the-thickness planes for the Type 2 beam. The normal stresses σ_{xx} and σ_{zz} at the notch tip have been evaluated in the different planes from the x- and y-isodyne fields and are presented in Figure 11.6 as a function of the out-of-plane coordinate z. In this case, the normal stress component σ_{zz} can be determined directly at the notch tip from the x-isodyne fields since the normal stress σ_{yy} vanishes at this location. Consequently, the normal stress σ_{xx} at the notch tip can be separated from the normal stress difference $\sigma_{xx} - \sigma_{zz}$ obtained from the differential y-isodynes at the notch tip.

The pronounced variation of σ_{xx} through the thickness of the specimen along with the not insignificant variation of σ_{zz} provides conclusive evidence that the

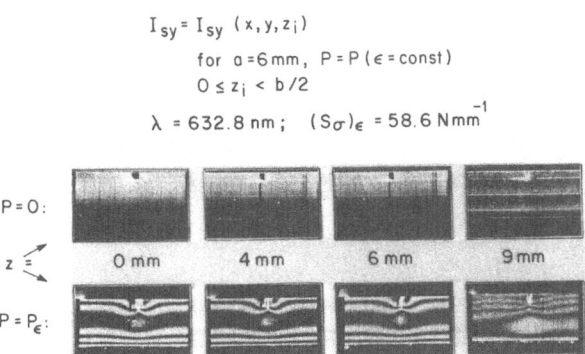

Fig. 11.5. Recordings of amplitude modulated y-isodyne fields in four through-the-thickness planes for Type 2 beam.

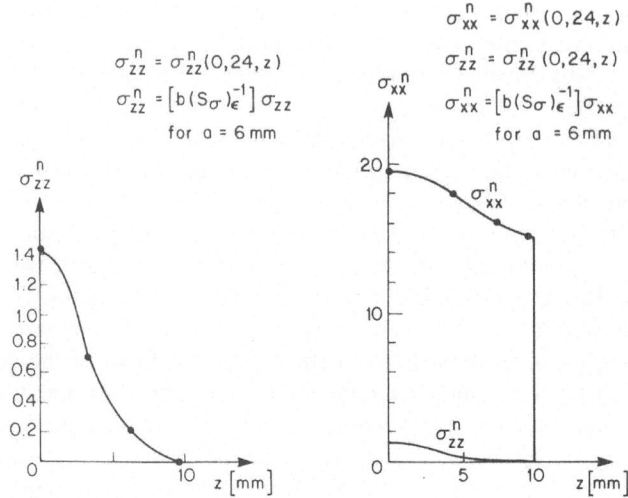

Fig. 11.6. Normal stress distributions σ_{xx} and σ_{zz} at the tip of the notch across Type 2 beam's thickness.

state of stress in the vicinity of the notch tip is three-dimensional. This raises a number of questions regarding the accuracy and reliability of common experimental methods based on the assumption of plane elasticity which are currently being employed to determine stress intensity factors in plates and beams. The strong variation in σ_{xx} across the thickness of the beam illustrated in Figure 11.6 along with the absence of easily identifiable region where the function $K_I(r)$ attains a constant value suggests that the values of K_I obtained on the basis of the typical two-dimensional experimental techniques such as the transmission photoelasticity ought to be interpreted with caution. This is discussed further in the following section.

11.2.3 Experimental/analytical correlation and discussion

The experimental results obtained in the preceding section for the function $K_I(r)$ are compared with the predictions of the asymptotic plane elasticity solution for the singular stress field in the vicinity of an edge crack in an infinite strip subjected to pure bending obtained by Benthem and Koiter [1]. The asymptotic form for the singular stress σ_{xx} for $a/h \to 0$ is given by the above authors by the following expression:

$$\sigma_{xx} \to 1.122 \frac{6M}{h^2} \left(1 - 1.217 \frac{a}{h}\right) \frac{\sqrt{a}}{\sqrt{2r}} \left[1 + 0\left(\frac{a^2}{h^2}\right)\right] \tag{11.3}$$

where M is the applied bending moment, h is the height of the beam and a is the depth of the edge crack. The theoretical stress intensity factor K_I^{th} is obtained by multiplying the above expression by $\sqrt{2\pi r}$ according to equation

(11.1). This yields:

$$K_I^{th} = 1.122 \sqrt{\pi} \frac{6M}{h^2} \left(1 - 1.217 \frac{a}{h} \right) \sqrt{a} \tag{11.4}$$

neglecting terms of the order (a^2/h^2).

For the present loading and geometry, the value of K_I^{th} for the Type 1 beam evaluated from the above equation is 55.2 N/mm$^{3/2}$. This value is indicated by the dashed line in Figure 11.4. The ambiguity that may result with regard to the value of the stress intensity factor obtained on the basis of plane elasticity hypothesis is clearly evident in Figure 11.4.

One may argue, on the basis of the functional form of $K_I(r)$ presented in Figure 11.4, that it is indeed possible to employ the two-dimensional approach for the determination of K_I in the region $r \geqslant 1.0$ mm provided that non-singular terms are employed in the expansion for the stress field around the crack tip as suggested by Sanford [5]. Since the contribution of the nonsingular terms in this region may be as important as the singular term, the variation of $K_I(r)$ in the region $r \geqslant 1.0$ mm may be pronounced and expected. Further, it may be argued that the variation of $K_I(r)$ in the region $0 < r \leqslant 1.0$ mm is due to notch tip bluntness and nonlinear effects as suggested by Smith and Kobayashi [6]. Here, we choose a different line of reasoning. The fact that the function $K_I(r)$ exhibits a strong, local maxima indicates that the functional form of σ_{xx} in the vicinity of the crack tip departs significantly from the two-dimensional solution. This is due to the presence of three-dimensional stress fields caused by high stress gradients and possibly nonlinear effects at the notch tip. Since the variation of $K_I(r)$ is very pronounced, the two-dimensional approach employed in the determination of the stress intensity factors may not be very meaningful. This is further supported by the data presented in Figure 11.6 which suggests that the function $K_I(r)$ defined by equation (11.2) is not unique in the different $z = z_0$ planes. Thus the question arises how to understand the values of the stress intensity factors obtained on the basis of the various experimental techniques which employ the assumptions of plane elasticity. In addition, we note that if the presence of plasticity and geometric effects was the only factor influencing the functional form of $K_I(r)$ in the immediate vicinity of the notch tip within the two-dimensional framework, one would not expect $K_I(r)$ to exhibit a local maxima, but to decrease. In other words, the results presented in Figure 11.4 suggest that the common technique of accounting for the existence of geometric and nonlinear effects at the crack tip by defining an effective crack length which includes the size of the plastic zone may not be very meaningful.

11.3 Three-point bending of an unnotched beam

The various cases treated in the isodyne investigations outlined in the preceding chapters and the preceding section illustrated the versatility of the technique for two-dimensional and three-dimensional stress analysis of selected problems.

However, relatively little systematic comparison between experimental and analytical results was provided. Furthermore, attention was directed towards determination of only the normal stress components.

The investigation outlined in this section was undertaken in order to provide a systematic comparison between the experimental results obtained with isodyne methods and the predictions of the widely employed two-dimensional elasticity model [4]. The problem under consideration is the fundamental problem of a simply-supported homogenous beam subjected to a centrally-applied concentrated load. The objective was to determine the distributions of all the inplane stress components in the different through-the-thickness planes of the beam and subsequently to compare the experimental data with the results of the two-dimensional analytical solution.

11.3.1 Experimental investigation

The beam under investigation was 220 mm long, 30 mm high and 9.4 mm thick. It was placed on supports 180 mm apart situated symmetrically with respect to the midspan of the beam. The load was applied through a circular disk 30 mm in diameter at the midpoint of the beam. The thickness of the disk was the same as the thickness of the beam. Both the beam and the loading disk were fabricated using Homalite 100. The elastic parameters for this material are given in Table 11.1. The specimen geometry and the loading conditions employed in this study are summarized in Figure 11.7.

Table 11.1. Elastic parameters of the beam and loading disk

Material type	E* Young's modulus (N/mm^2)	v* Poisson's ratio
Homalite 100	2413	0.38

* ASTM D638 Tensile test method.

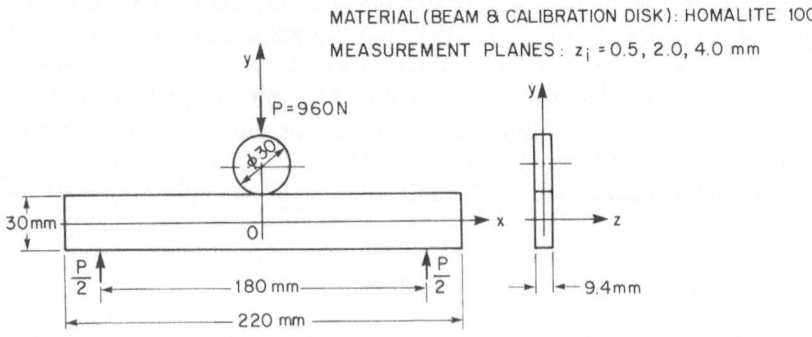

Fig. 11.7. Three-point bending of an unnotched beam. Specimen geometry and loading conditions.

The specimen was loaded incrementally up to the maximum load of 1226 Newtons. The presented results pertain to the load of 960 Newtons. At each load increment, isodyne recordings along the x- and y-characteristic directions were taken in three planes situated at different thought-the-thickness distances from the beam's vertical midplane. The through-the-thickness coordinates of these planes were: $z = 0.5$ mm, 2.0 mm and 4.0 mm. The above procedure was carried out in order to check the linearity of the stress distributions with respect to the applied load. In order to eliminate the influence of the time-dependent response of Homalite 100 during recording of the isodyne fields, the load was applied in the constant overall deformation mode, and additionally the load relaxation was monitored at each loading stage. Measurements in the three planes were taken when the relaxation of the applied load became negligible. Figure 11.8 illustrates the loading and recording history employed in the present investigation. The apparent relaxation processes depicted in Figure 11.8 are actually mixed processes involving relaxation and creep. The related optical processes are much weaker than the mechanical processes because the optical relaxation modulus of polymers is practically constant within the visible band of electromagnetic radiation [3].

The choice of the material used in the present study was dictated by the desire to obtain high-resolution recordings in the vicinity of the applied load. It will be recalled that two basic isodyne measurement techniques have been developed thus far: the light-intensity modulation technique and the modulation of the spatial frequency of residual isodynes technique. The spatial frequency modulation technique was chosen since it gives resolution which is about one order of magnitude higher than the resolution of the amplitude modulation technique. Homalite 100 is a particularly useful material for the frequency

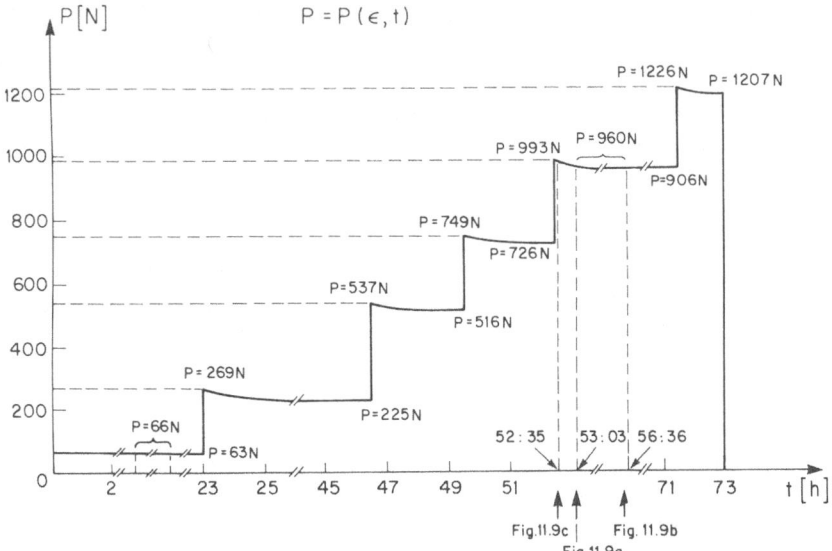

Fig. 11.8. Loading and recording history.

modulation technique because it possesses a highly ordered field of residual isodynes in the unloaded state. The radiation used in the present study was produced by helium laser. The wavelength of the radiation was 632.8 nm while the diameter of the information collecting light beam was 0.3 mm.

11.3.2 Experimental results

The experimentally obtained spatial frequency modulated x- and y-isodyne fields in the $z = 0.5$ mm plane at the applied load of 960 Newtons, and the corresponding transmission isochromatics, are presented in Figure 11.9. Similar recordings have been obtained in the remaining two planes. It will be recalled that the x-isodynes are produced by projecting the light beam in the x-characteristic direction while scanning in the y-direction. Similarly, the y-isodynes are produced by projecting the light beam in the y-characteristic direction while scanning in the x direction.

In order to obtain the various stress components in the planes coplanar with the beam's midplane, cross-sections of the x- and y-isodyne fields at a number of $x = x_0$ and $y = y_0$ locations have been determined from the experimentally obtained spatial frequency modulated isodyne fields. Selected $x = x_0$ cross-sections obtained from the y-isodyne fields are presented in Figure 11.10 as a function of the y-coordinate for the three planes $z = 0.5$, 2.0 and 4.0 mm investigated. The σ_{xx} stress component, in the plane stress situation, or the difference $\sigma_{xx} - \sigma_{zz}$ in the three-dimensional stress situation, is obtained from the cross-sections given in Figure 11.10 upon differentiating the curves with respect to the y-coordinate. These will be presented and compared with the results of the two-dimensional elasticity solution in the following sections. At this point we note that some differences exist in the presented $x = x_0$ cross-sections of the y-isodyne fields in the different $z = z_0$ planes.

Selected $y = y_0$ cross-sections of the y-isodyne fields are presented in Figure 11.11 as a function of the x-coordinate for the investigated planes at different through-the-thickness distances from the beam's midplane. These cross-sections have been determined by evaluating the values of the corresponding $x = x_0$ cross-sections of the y-isodynes at various $y = y_0$ locations. In order to obtain very accurate results close to the centrally-applied load, readings have been taken every half millimeter in the interval $0 \leqslant x \leqslant 1$ mm, every millimeter in the interval $1 \leqslant x \leqslant 12$ mm and every two millimeters in the interval $12 \leqslant x \leqslant 24$ mm. It will be recalled that in the case of plane stress, the derivative of the above cross-sections with respect to the x-coordinate yields the inplane shear stress component σ_{xy}. In the case of three-dimensional fields in beams or plates, the present theory may not yield the inplane shear stress component when the $y = y_0$ cross-sections of the y-isodynes are differentiated with respect to the x-coordinate as mentioned previously. The evaluated derivatives of the above cross-sections will be presented in the following sections along with the results of the plane elasticity solution. At this point, we make an observation that in the vicinity of the applied load, large differences exist between the $y = y_0$ cross-sections of the y-isodynes in the different planes. These differences are

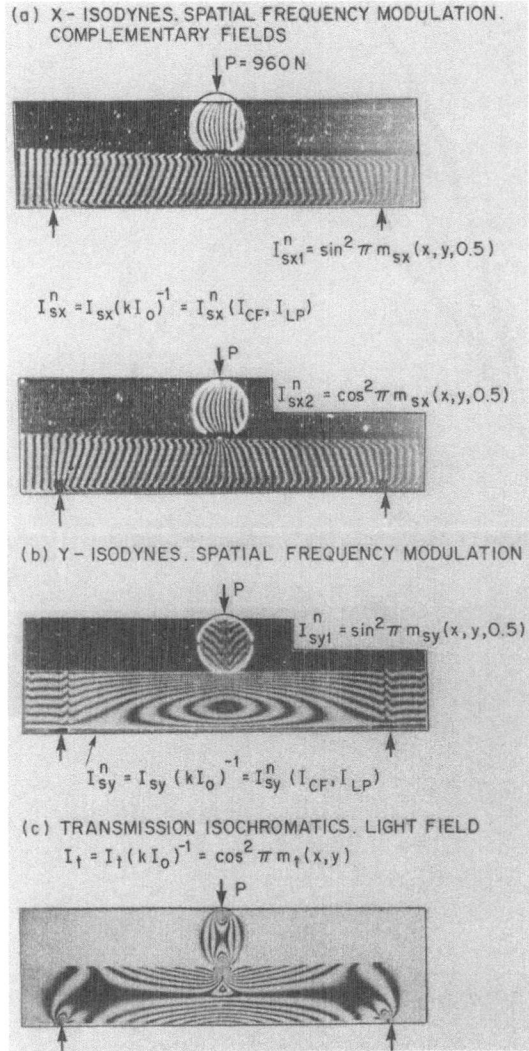

Fig. 11.9. Recording of spatial frequency modulated *x*- and *y*-isodyne fields in the $z = 0.5$ mm plane at 960 Newtons, and the corresponding transmission isochromatics. (a) Complementary fields of spatial frequency modulated *x*-isodynes. (b) Spatial frequency modulated *y*-isodynes. (c) Light field transmission isochromatics.

much more significant than the differences observed between the $x = x_0$ cross-sections of the *y*-isodynes in the three planes $z = 0.5$, 2.0, and 4.0 mm.

The $y = y_0$ cross-sections of the *x*-isodynes have been determined in a

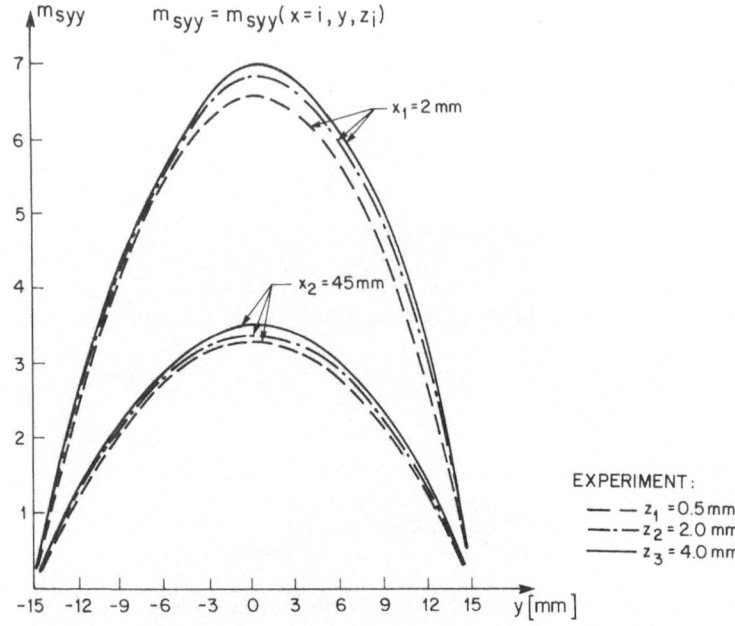

Fig. 11.10. Cross-sections of y-isodyne field at various $x = x_0$ and $z = z_0$ locations.

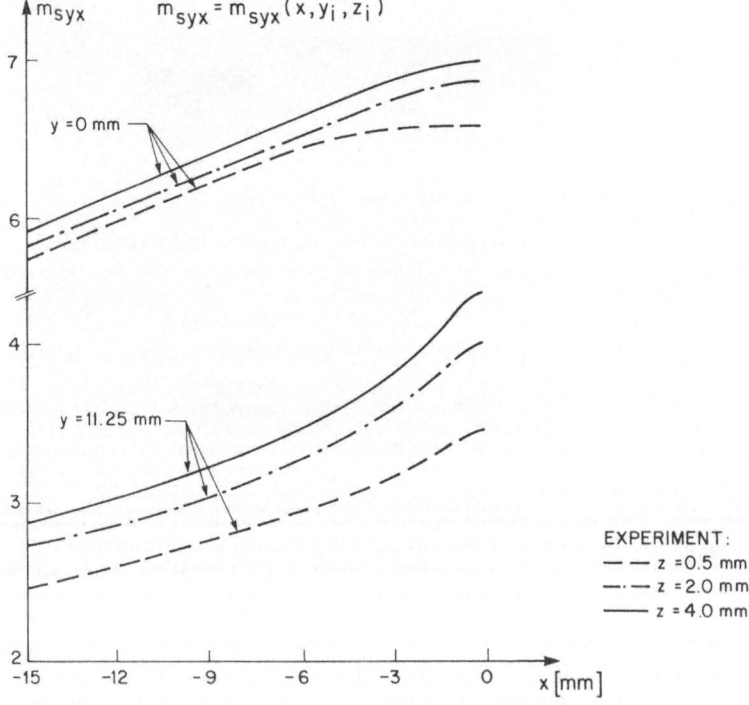

Fig. 11.11. Cross-sections of y-isodyne field at various $y = y_0$ and $z = z_0$ locations.

similar fashion as the $x = x_0$ cross-sections of the y-isodynes. The σ_{yy} stress component in the case of plane stress, and the difference $\sigma_{yy} - \sigma_{xx}$ in the case of three-dimensional state of stress, have been determined in the different planes from the x-isodynes by differentiating the $y = y_0$ cross-sections with respect to the x-coordinate. Results and comparison with the plane elasticity solution will be presented in the following sections.

11.3.3 Analytical model

The analytical model employed in the present investigation is the plane elasticity model for which the governing differential equations for the inplane stress distributions are the familiar equilibrium equations given below

$$\frac{\partial \sigma_{xx}}{\partial x} + \frac{\partial \sigma_{xy}}{\partial y} = 0$$

$$\frac{\partial \sigma_{xy}}{\partial x} + \frac{\partial \sigma_{yy}}{\partial y} = 0$$

(11.5)

The boundary conditions for the present loading are specified in terms of the applied tractions in the following manner:

$$\sigma_{yy}\left(x, +\frac{h}{2}\right) = \frac{P}{\delta} \quad \text{for} \quad |x| < \frac{\delta}{2}$$

$$= 0 \quad \text{otherwise}$$

$$\sigma_{yy}\left(x, -\frac{h}{2}\right) = \frac{P}{2\delta} \quad \text{for} \quad |\pm x - 90| < \frac{\delta}{2}$$

(11.6)

$$= 0 \quad \text{otherwise}$$

$$\sigma_{xy}\left(x, \pm\frac{h}{2}\right) = 0$$

These boundary conditions are not the actual boundary conditions in the experimental model since the distribution of stresses in the contact area between the loading disk and the beam is governed by the local deformation of the two bodies. These types of problems belong in the category of contact problems of elasticity and are outside the scope of the outlined investigation. Naturally, the stress distribution in the immediate vicinity of the contact area will be strongly influenced by the actual distribution of tractions in the contact area. Consequently, comparison between predictions of the analytical model and experiment must be carried out for regions which are sufficiently removed from the contact area in order to be able to draw correct conclusions regarding the degree of correspondence between the analytical model and experimental results. Discussion of this issue is deferred to the subsequent sections.

A number of techniques is available for the solution of the equilibrium equations, equation (11.5), subject to the specified boundary conditions. For this problem, the displacement formulation was chosen in conjunction with the infinite Fourier transform technique. That is, the equilibrium equations were expressed in terms of the appropriate displacement gradients using the stress-strain and strain-displacement relations, and the Fourier transform was applied to the resulting two-dimensional Navier equations in order to eliminate the x-dependence in the inplane displacements. This effectively reduces the equilibrium equations to two coupled ordinary differential equations in the transform domain s-y, where s is the Fourier transform variable. The solution of the problem follows readily by solving the two equations in the transform domain s-y, subject to the transformed boundary conditions, and subsequently inverting back to the x-y domain.

In order to ensure convergence and accuracy of the solution for the inplane stresses, 50 integration intervals, with each interval having 64 integration points, were employed in the Gaussian integration scheme in the process of inverting the various Fourier-transformed quantities. The integration intervals were chosen on the basis of the functional dependence of the Fourier-transformed inplane stresses on the transform variable s. As a final check of the accuracy of the analytical solution, equilibrium of the beam along various sections was verified to within a few tenths of one percent.

The solution of the problem outlined above is strictly applicable when the length of the beam is infinite or at least very large. In order to establish the extent of geometric similitude between the analytical and experimental models, the distribution of stresses in regions outside the bottom supports was calculated at various elevations as a function of the axial coordinate x. It was found that the inplane stress components vanished approximately 10 mm away from the bottom supports. Since in the experimental model the length of the overhang is 20 mm, high degree of geometric similitude between the experimental and analyical model is expected, particularly in the vicinity of the applied load which is the region of primary interest in the present investigation.

11.3.4 Experimental/analytical comparison

For the comparison of the predictions of the analytical model with the experimental results of optical isodynes, it was necessary to estimate the length δ of the contact area between the loading disk and the beam employed in the boundary conditions given by equation (11.6). In view of the fact that it was not possible to determine this value with great precision, a set of values for δ ranging from 0.5 mm to 2.5 mm was employed in the analytical model to generate the inplane stress distributions in the cross-sections for which the comparison between theory and experiment was carried out. Virtually no difference in the predicted stress distributions was observed in those cross-sections. The maximum value for delta of 2.5 mm was estimated to be an upper bound on the basis of the photograph of the deformed beam subjected to the load of 960 Newtons. The comparison between theory and experiment is carried out for this load level. As mentioned previously, linearity of the stresses with respect to the applied load was verified up to this load level by recording the isodyne fields at several intermediate load levels and comparing the resulting values at selected locations.

The comparison between the predictions of the plane elasticity model and the results obtained using optical isodynes is given in Figures 11.12—11.14. Figure 11.12 presents the results of the vertical distribution of the normal stress σ_{xx} in the four cross-section $x = 45, 10, 6$ and 2 mm. It is observed that far away from the central load, Figure 11.12a, very little difference exists between the experimentally obtained normal stress in the three different through-the-thickness planes. The correlation between predictions of the analytical model and experiment is excellent. The differences in the experimentally determined normal stress in the three $z = z_0$ planes become somewhat more apparent closer to the $x = 0$ axis, Figures 11.12b—d. Generally, these differences are more pronounced in the vicinity of the top and bottom surfaces of the beam, with greater differences observed near the top surface. In the immediate vicinity of the top and bottom surfaces, the normal stress in the outer planes is higher than in the plane close to the vertical midplane of the beam (i.e. $z = 0.5$ mm plane). Comparing the experimental results with the predictions of the analytical model as the $x = 0$ axis is approached, it is seen that the correlation is very good with some differences observed close to the top and bottom surfaces. Near the bottom surface, the experimentally obtained normal stress in the two outer planes is closer to the plane elasticity solution in the cross-sections $x = 10, 6$ and 2 mm. Near the top surface however, this trend appears to be reversed. The results presented in Figure 11.12 indicate that the functional form of the normal stress σ_{xx} distribution obtained using the method of optical isodynes has a slightly different form near the top and bottom surfaces of the beam than the predictions of the analytical model.

Figure 11.13 presents the vertical distribution of the shear stress σ_{xy} in the same $x = x_0$ cross-sections as mentioned in the preceding paragraph. Far away from the applied load, Figure 11.13a, very little difference is observed in the experimentally obtained shear stress in the three through-the-thickness planes. Again, the correlation between the experimental results and the predictions of the two-dimensional elasticity model is very good. As the centrally-applied load is approached, Figures 11.13b—d, considerable differences in the experimental values of the shear stress in the three through-the-thickness planes are observed. These differences are much more pronounced above the "neutral axis" than below it and increase with decreasing distance from the central load. While no clear trend exists in the experimentally obtained shear stress in the three $z = z_0$ planes below the neutral axis, above it the values of the shear stress increase with increasing z-coordinate. This trend is well-defined for the investigated $x = x_0$ cross-sections. Comparing the experimental results with the analytical predictions, it is seen that generally good correlation is obtained below the $y = 0$ axis. Above this axis, large differences exist depending on the particular $y = y_0$ location. No apparent trend between the experimental results obtained for any given $z = z_0$ plane at a particular y-coordinate and the analytical model

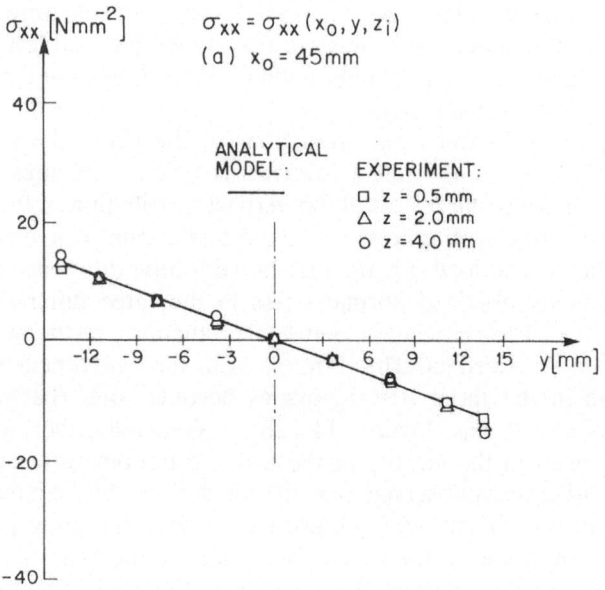

11.12a

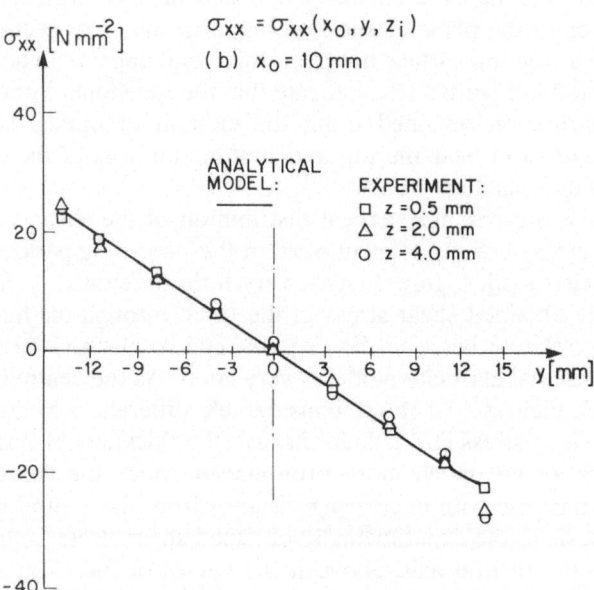

11.12 b

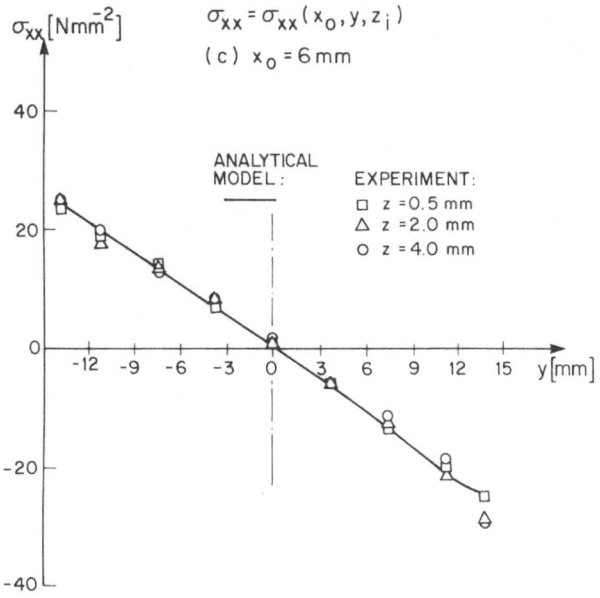

11.12 c

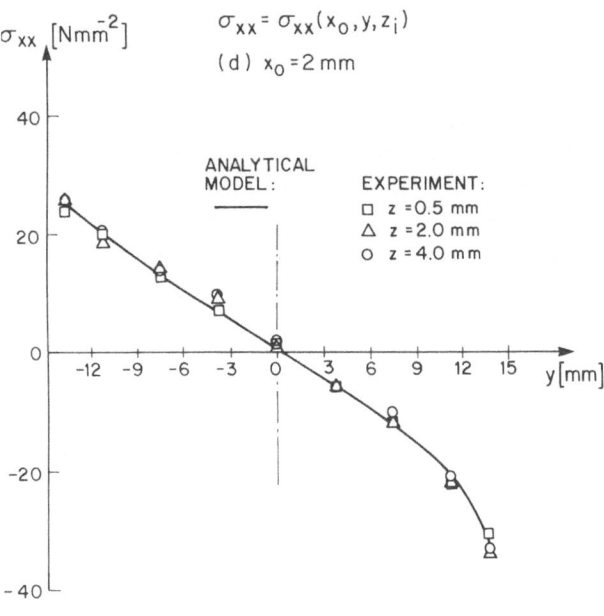

11. 12 d

Fig. 11.12. Vertical distribution of the normal stress σ_{xx} at various $x = x_0$ and $z = z_0$ locations. Comparison between predictions of an analytical model and experiment.
(a) $x = 45.0$ mm. (b) $x = 10.0$ mm. (c) $x = 6.0$ mm. (d) $x = 2.0$ mm.

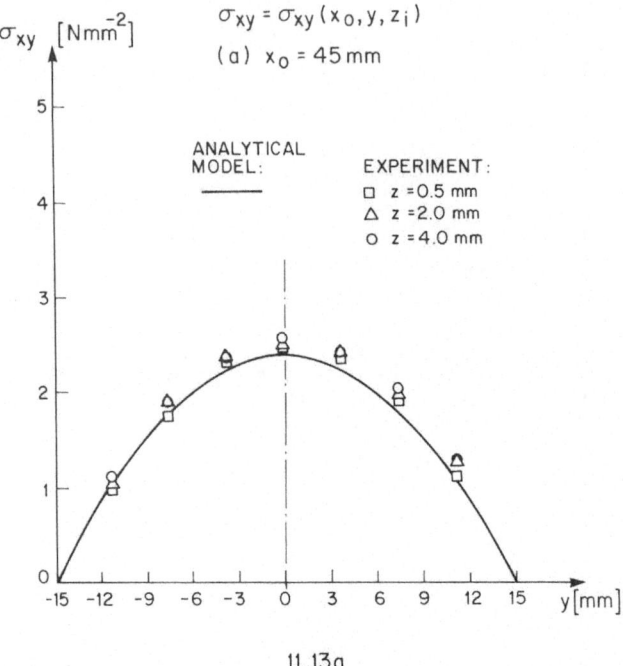

11.13a

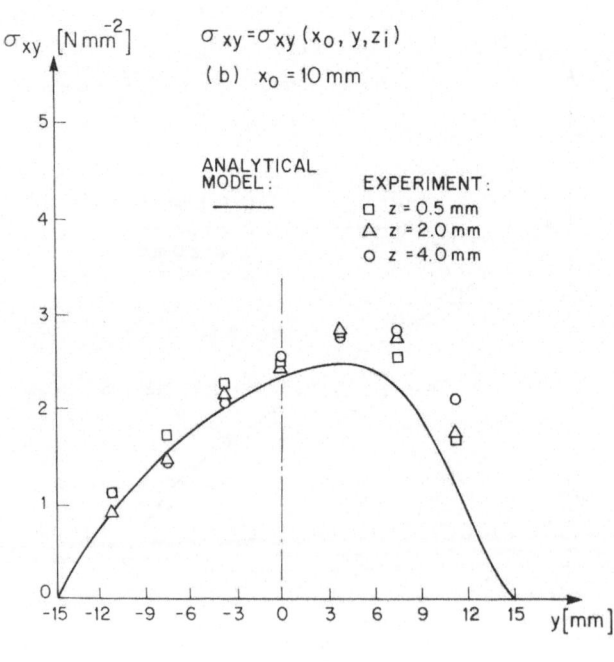

11.13b

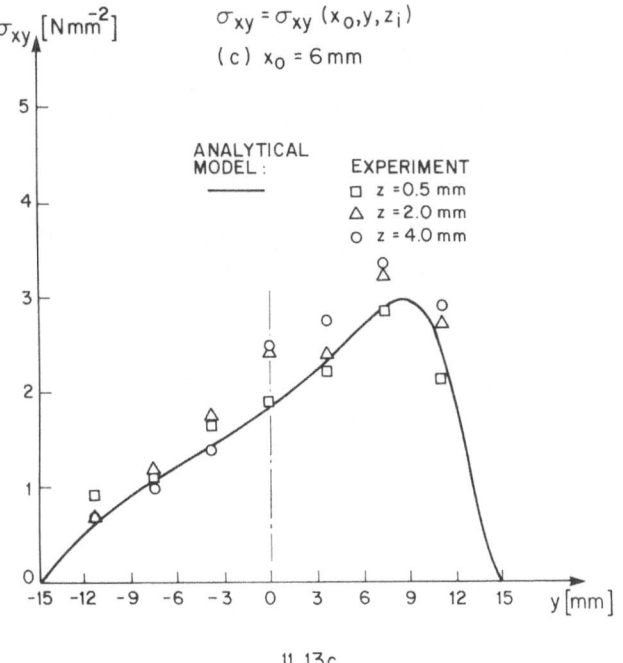

11. 13c

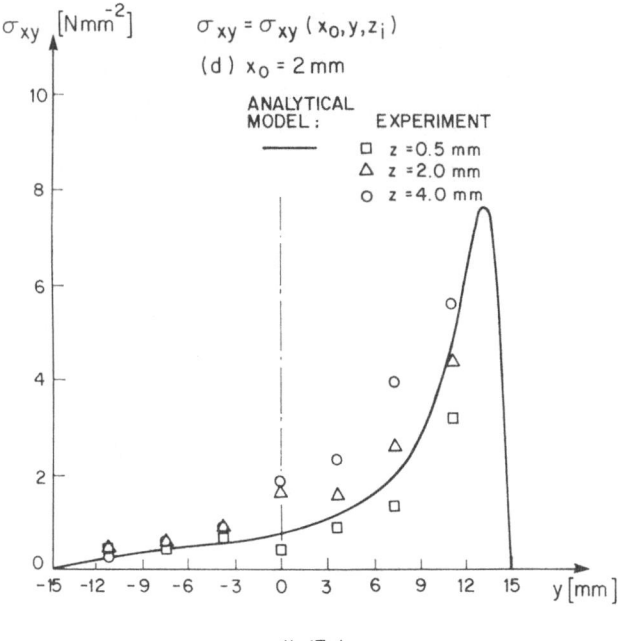

11. 13 d

Fig. 11.13. Vertical distribution of the shear stress σ_{xy} at various $x = x_0$ and $z = z_0$ locations. Comparison between predictions of an analytical model and experiment.
(a) $x = 45.0$ mm. (b) $x = 10.0$ mm. (c) $x = 6.0$ mm. (d) $x = 2.0$ mm.

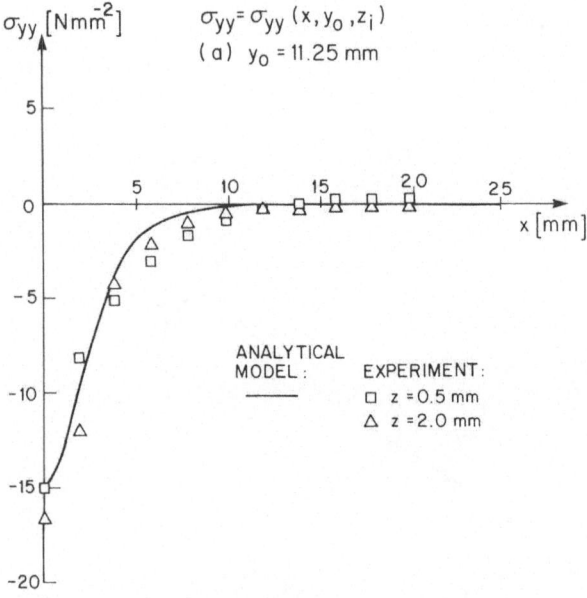

11.14 a

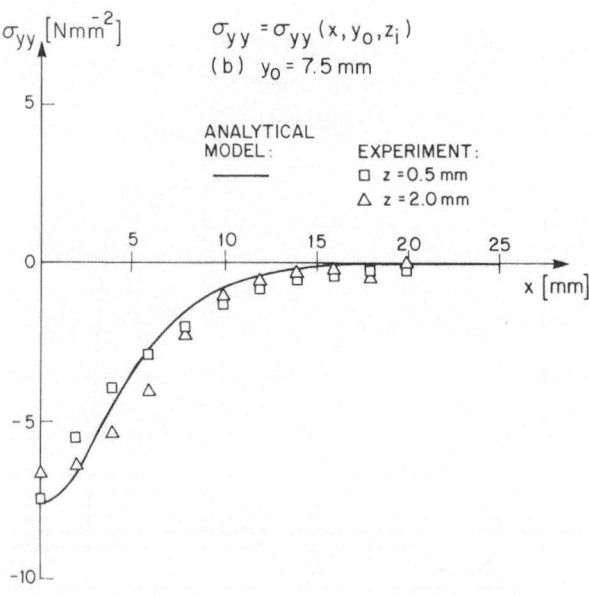

11.14 b

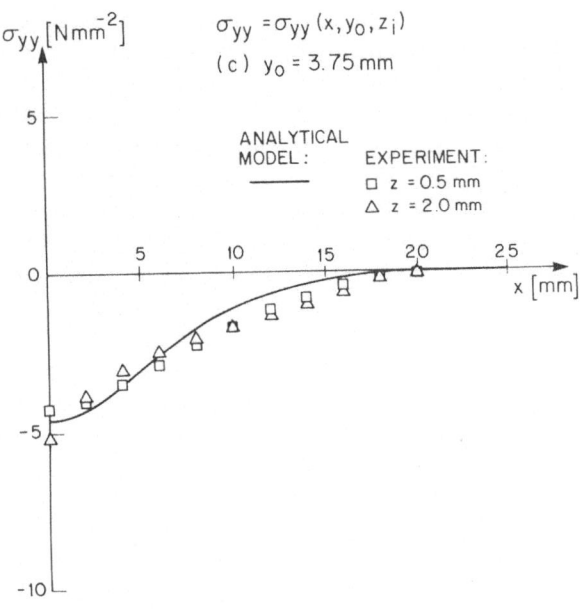

11.14 c

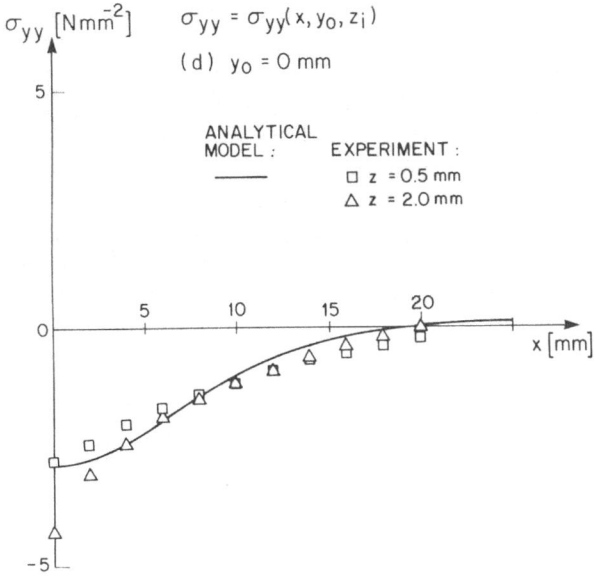

11.14 d

Fig. 11.14. Horizontal distribution of the normal stress σ_{yy} at various $y = y_0$ and $z = z_0$ locations. Comparison between predictions of an analytical model and experiment.
(a) $y = 11.25$ mm. (b) $y = 7.5$ mm. (c) $y = 3.75$ mm. (d) $y = 0.0$ mm.

is observed. For the $x = 10$ mm cross-section the plane elasticity model predicts lower shear stresses above the $y = 0$ axis than the shear stresses obtained experimentally in the three through-the-thickness planes. In the case of the $x = 6$ and 2 mm cross-sections on the other hand, the analytical results fall between the results obtained experimentally, with no apparent trend between analysis and experiment. In other words, it cannot be concluded that the experimentally obtained shear stress distribution in any given plane is systematically closer to the plane elasticity solution than the stress distribution in the other planes. It appears that the functional form of the experimentally obtained shear stress above the $y = 0$ axis in the $z = z_0$ planes close to the applied load is significantly different than the results of the two-dimensional elasticity solution.

Figure 11.14 presents the horizontal distribution of the normal stress σ_{yy} in the $y = 11.25$, 7.5, 3.75 and 0.0 mm cross-sections. At the present, the experimental results are available in the $z = 0.5$ and 2.0 mm planes. At points sufficiently away from the applied load, the differences in the experimentally obtained values of the normal stress do not differ significantly in the two through-the-thickness planes. As the $x = 0$ axis is approached, differences in the two planes become relatively significant along the various $y = y_0$ elevations. For the $y = 3.75$ and 0.0 mm cross-sections, Figures 11.14c and d, these differences are greatest directly under the applied load, with the outer plane yielding higher values of the normal stress. For the $y = 11.25$ and 7.5 mm cross-sections, Figures 11.14a and b, the above differences are greatest at some distance away from the $x = 0$ axis. Interestingly, the normal stress directly under the applied load for the $y = 7.5$ elevation is greater in the inner plane than the outer plane, which is not the case for the remaining cross-sections. Comparing the experimental results with the analytical solution, it is observed that generally good correlation exists along the various elevations. The observed differences between analysis and experiment do not appear to follow a general trend which would indicate that the experimental results in a given plane are systematically closer to the two-dimensional elasticity solution than in the remaining plane. In summary, we note that the relative differences between analysis and experiment in this case are generally smaller than the corresponding differences observed in the shear stress distribution discussed previously.

11.3.5 Discussion

The comparison between experimentally determined inplane stress components in the three different through-the-thickness planes and the two-dimensional elasticity solution indicates that large differences between analysis and experiment exist in regions close to the applied load for certain stress components. Furthermore, there is a significant variation in the individual, experimentally obtained, stress components in the different planes. The relative magnitude of the experimentally determined differences is greatest for the inplane shear stress component σ_{xy}, followed by the normal stress component σ_{yy} and finally by σ_{xx}. Far away from the centrally-applied load however, no significant variation in the individual stress components as a function of the z-coordinate is observed.

In those regions, the agreement between the experimental results and the predictions of the plane elasticity solution is very good.

The observed differences cannot be explained by scatter in the experimental data or insufficiently accurate knowledge of the actual traction conditions under the applied load. In the first instance, great care has been taken to ensure that a sufficient number of data points in the vicinity of the applied load has been employed so that sufficiently smooth isodyne cross-sections could be obtained, Figures 11.10 and 11.11. This is very important since the isodyne cross-sections have to be differentiated in order to obtain the various stress components. In the second instance, the comparison between analysis and experiment was carried out in regions unaffected by different boundary conditions for the normal traction component, as predicted by the analytical model. No attempt was made to account for the shear tractions that may exist in the contact area between the loading disk and the top surface of the investigated beam. However, it must be noted that even if the existence of shear tractions *did* influence the various stress distributions in the considered cross-sections for which comparison between theory and experiment was carried out, it would still not explain the strong though-the-thickness variation of the experimentally determined stresses.

The differences between the experimentally obtained inplane stress components in the different planes, as well as the corresponding differences between analysis and experiment, can be explained by the presence of the out-of-plane normal stress σ_{zz}. In fact, differentiation of the various isodyne cross-sections with respect to the corresponding geometric variable in the case of the normal stress components yields the differences $\sigma_{xx} - \sigma_{zz}$ and $\sigma_{yy} - \sigma_{zz}$ in regions where a three-dimensional state of stress exists. The experimental data strongly suggests that in the vicinity of the applied load a three-dimensional state of stress exists and thus the isodyne functions have to be interpreted as differential isodynes rather than plane isodynes. This is particularly evident in the case of the inplane shear stress σ_{xy}. In this case, using the differentiation procedure valid for plane isodynes leads to significant variation in the experimentally obtained distribution of σ_{xy} in the different through-the-thickness planes, and thus significant variation between analysis and experiment. As, at present, no relationship is available between the differential isodynes and the inplane shear stress σ_{xy}, no conclusion can be reached regarding the meaning of the thus obtained "shear stress". However, the through-the-thickness variation in the slopes of the isodyne cross-sections $m_{sx}(x_0, y)$ and $m_{sy}(x, y_0)$, and the resulting differences between theory and experiment, provide one with a very sensitive technique of determining the regions where the state of stress is clearly three-dimensional.

Two hypotheses may be postulated regarding the presence of three-dimensional state of stress in the vicinity of the applied load. It is possible that the loading disk tends to contract differently than the beam in the contact area, thus providing a geometric constraint. In the present investigation, the loading disk and the beam had the same thickness and were fabricated from the same material. The tendency towards differential contraction would thus have to result from different stress distribution in the contact area (excluding the various traction components). On the other hand, a hypothesis can be advanced according to which the three-dimensional state of stress results from high inplane stress gradients caused by the concentrated load. In this scenario, the presence of high inplane stress gradients results in the tendency of the material to contract differently at different distances from the point of load application, thus also providing a kind of geometric constraint. Consequently, three-dimensional state of stress arises in order to satisfy the three-dimensional compatibility conditions. It would appear that the second scenario may be more plausible based on the experimental data and discussion presented here, as well as the results of previous investigations obtained with the method of isodynes. Further investigation of the stress components in the contact area is obviously required to determine the exact cause of the three-dimensional state of stress in the investigated regions.

11.4 References

[1] Bentham, J. P. and Koiter, W. T., "Asymptotic Approximations to Crack Problems", in *Mechanics of Fracture*, Vol. 1, G. C. Sih (ed.), Noordhoff International Publishing (1973).

[2] Clark, R. A., "Three-Dimensional Corrections for a Plane Stress Problem", *Int. J. Solids Structures* **21**(1), 3—10 (1985).

[3] Pindera, J. T. and Straka, P., "On Physical Measures of Rheological Responses of Some Materials in Wide Ranges of Temperature and Spectral Frequency", *Rheologica Acta* **13**(3), 338—351 (1974).

[4] Pindera, M.-J., Pindera, J. T. and Ji, Xinhua, "Three-Dimensional Effects in Beams: Isodyne Assessment of a Plane Solution", *Experimental Mechanics* **29**(1), 23—31 (1989).

[5] Sanford, R. J., "The Influence of Nonsingular Stresses on Experimental Measurements of the Stress Intensity Factor", in *Modelling Problems in Crack Tip Mechanics*, J. T. Pindera (ed.), Martinus Nijhoff Publishers (1984).

[6] Smith, C. W. and Kobayashi, A. S., "Experimental Fracture Mechanics", in *Handbook on Experimental Mechanics*, A. S. Kobayashi (ed.), Society for Experimental Mechanics ed., Prentice-Hall (1987).

[7] Sokolnikoff, I. S., *Mathematical Theory of Elasticity*, McGraw-Hill (1956).

Stresses in composite structures

12.1 Introduction

A number of interesting problems in the mechanics of composite structures requires the determination of three-dimensional stress states in the course of analyzing failure characteristics of these advanced materials. The departure from plane strain or plane stress fields occurs naturally, for example, in the so-called boundary layer close to the free edge due to the mismatch in properties between adjacent laminae, or in the vicinity of defects that that may have the form of transverse cracks created by locally failed plies or local delaminations along material interfaces. The above regions exhibit significant stress gradients, suggesting singular-like behavior outside of the locally nonlinear or plastic zone.

Two model problems in the mechanics of composite structures have been chosen to illustrate the efficacy of the isodyne technique in analyzing stress fields in different planes within a given laminated structure. The first problem involves a symmetric three-ply construction with the middle ply containing through-the-thickness transverse crack normal to the external plies and extending across the width of the structure. The laminated plate is subjected to inplane tension directed along the longitudinal axis. The second problem deals with a bi-material beam subjected to three-point bending and containing two interlaminar disbonds located symmetrically with respect to the central load.

The above two problems have been chosen for several reasons. First of all, they represent models of two different but complementary problems of technological importance encountered in the field of composites. Interlaminar and transverse cracks in composite materials and structures are often difficult to avoid and thus accurate characterization of stress fields in the vicinity of such defects is indispensable from the point of view of defining flaw criticality of advanced composites under various loading conditions. Secondly, the functional form of stress distributions in the vicinity of crack tips for the above configurations as predicted by quasi-state plane elasticity solutions is markedly different, thus presumably leading to different failure modes. As at present there are no generally accepted theories to predict the direction of crack growth and the critical loads for the above cases, the experimental results are of theoretical interest to fracture mechanicians. Thirdly, the possibility of a three-dimensional state of stress in the vicinity of the simulated defects, as has already been demonstrated in the case of homogeneous beams containing surface notches, raises the issue of the range of applicability of the plane elasticity solutions in problems of this type. Lastly, experimental models utilized in this study, being inherently three-dimensional, can be taken as iconic models of real structures. Consequently, extrapolation of the experimentally obtained stress distributions along various planes within the model to the actual engineering structures is a direct process, a feature of optical isodynes which illustrates both the power and versality of the developed technique.

12.2 Three-ply structure with a transverse crack

The present investigation was undertaken to study the redistribution of stresses in the outer ply of a symmetric three-ply structure subjected to an inplane axial load caused by a through-the-thickness transverse crack extending across the entire width of the middle ply, Figure 12.1. This is a model problem with applications to technologically important problems involving bonded joints, cracking in laminated plates and fracture of single fibers. In the absence of the crack, or far away from it, the state of stress in the individual plies is plane and is charac-

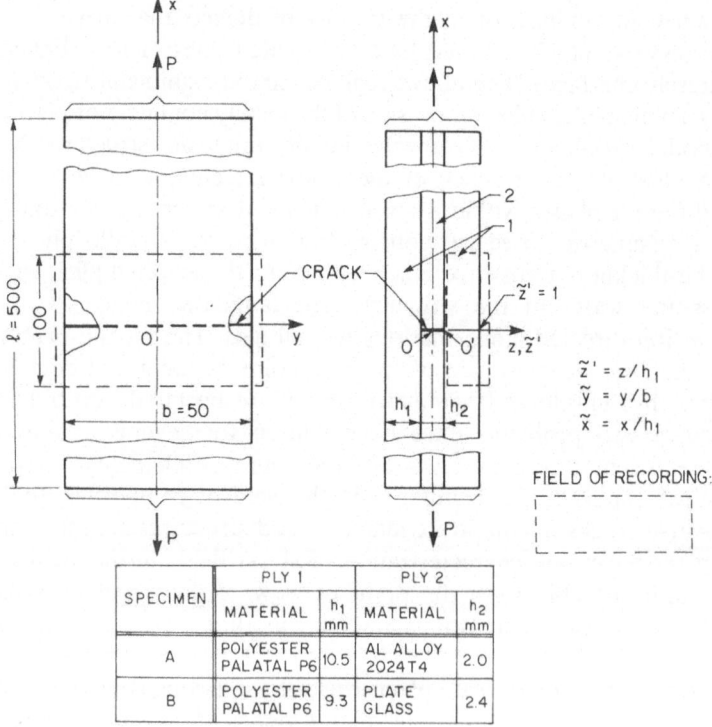

SPECIMEN	PLY 1		PLY 2	
	MATERIAL	h_1 mm	MATERIAL	h_2 mm
A	POLYESTER PALATAL P6	10.5	AL ALLOY 2024T4	2.0
B	POLYESTER PALATAL P6	9.3	PLATE GLASS	2.4

Fig. 12.1. Three-ply composite structure with a transverse crack. Specimen geometry and loading conditions.

terized by the normal components σ_{xx} and σ_{yy} away from the free edges. The presence of the crack disturbs the plane state of stress in the immediate vicinity of the cracked ply and gives rise to out-of-plane normal and shear stresses σ_{zz} and σ_{xz}. The state of stress in the outer ply directly above the crack front is characterized by large stress gradients which must be evaluated with sufficient accuracy in order to determine the load-bearing capacity of the structure.

The stress distribution in the outer ply of a three-ply structure with a cracked middle ply was investigated for two material configurations with different Poisson's ratios for the inner and outer plies. The first configuration was a polyester-glass-polyester construction and the second a polyester-aluminum-polyester construction. Since the Young's moduli of glass and aluminum are nearly the same, Poisson's ratio of the cracked ply was the only material parameter that was significantly different in the two configurations. The experimentally obtained stress distributions in the different planes of the outer ply above the crack front are subsequently compared with the predictions of two plane elasticity models developed to analyze different response characteristics of the considered structure.

12.2.1 Experimental investigation and results

The length and width of both the polyester-glass-polyester and polyester-aluminum-polyester configurations was 500 mm and 50 mm, respectively. In the polyester-glass-polyester lay-up the thickness of the outer polyester plies was 9.3 mm while the thickness of the inner glass ply was 2.4 mm. In the polyester-aluminum-polyester lay-up the thickness of the outer polyester plies was 10.5 mm and the thickness of the inner aluminum ply was 2.0 mm. The geometry of the two configurations is summarized in Figure 12.1. Figure 12.2 presents the elastic parameters of the polyester, glass and aluminum layers employed in the two configurations.

Recordings of the y-isodyne fields have been taken in several $z = z_0$ planes of the outer ply directly above the crack front, with the field of recording extending 50 mm along the axial direction on either side of the crack, Figure 12.1. Figure 12.2 presents the recorded y-isodyne fields at four different elevations above the crack front for the two material configurations. The significant influence of the Poisson's ratio on the internal stress distribution in the vicinity of the crack is clearly evident in the recordings.

Differentiation of the $x = x_0$ cross-sections of the y-isodyne fields with respect to the y-coordinate yields the difference of the secondary principal stresses $\sigma_1(x_0, y, z_0) - \sigma_3(x_0, y, z_0)$ in the $x - z$ plane, normal to the direction of the primary beam. This difference is equal to twice the maximum shear stress acting in the $x - z$ plane. Carrying out the differentiation procedure on different $x = x_0$ cross-sections in the different $z = z_0$ planes yields the maximum shear stress in the $x - z$ plane at any point in the outer ply. As an illustration, Figure 12.3 presents the distributions of the maximum shear stress $\sigma_{13}^{\max}$ in the $y = 0$ mm plane of the outer ply as a function of the axial coordinate x at different elevations above the crack front evaluated from the recordings

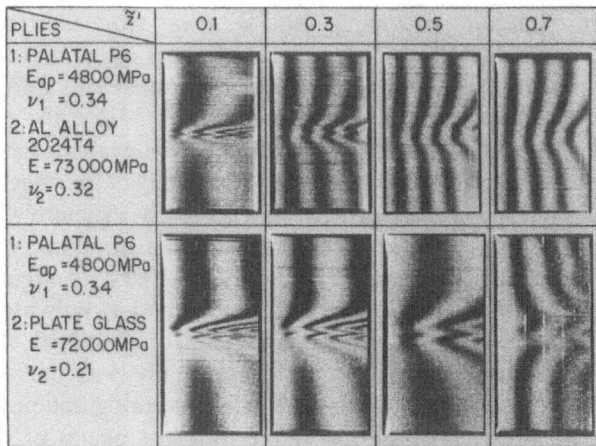

Fig. 12.2. Recordings of y-isodyne fields in different planes of the outer ply above the transverse crack of the inner ply.

presented in Figure 12.2. The experimental results have been normalized with respect to the far-field axial stress σ_{xx}^o in the outer ply. It is interesting to note that the loci of the maximum values of the shear stress σ_{13}^{max} occur at different axial distances from the crack front in the two material configurations. It appears that this is a direct consequence of the differences in the Poisson's ratio of the middle ply in the two lay-ups.

12.2.2 Analytical models

The two analytical models employed for comparison with the experimental data are the approximate model proposed by Vasilev *et al.* [8] and the local plane

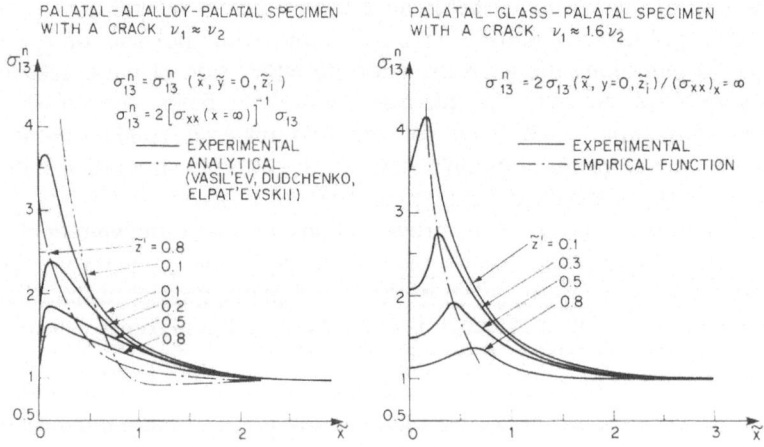

Fig. 12.3. Distributions of the maximum shear stress in different planes of the outer ply. Comparison between predictions of an analytical model and experiment.

elasticity solution for the stress distribution around a crack perpendicular to a bi-material interface developed by Zak and Williams [10]. The model proposed by Valilev *et al.* was developed to predict the average response of symmetric cross-ply laminates under inplane axial loading in the presence of transverse cracks in the middle plies. Most recently, the same approach was taken by Hashin [5] to estimate the changes in the laminate elastic parameters as a function of the crack density. The central assumption employed by the above authors is that there is no variation of the normal axial stress σ_{xx} across the ply thickness in the vicinity of the crack front in the inner and outer plies, i.e. $\sigma_{xx} = \sigma_{xx}(x)$ only due to the appearance of a crack in the middle ply. The above assumption allows determination of the functional form of the remaining inplane stress components with the aid of the stress equilibrium equations in the $x - z$ plane in terms of the single unknown function $\sigma_{xx}(x)$. Subsequent minimization of the strain energy function for the cracked laminate yields a fourth order ordinary differential equation for the functional form of the disturbed axial stress $\sigma_{xx}(x)$ due to the presence of the crack. Solution of the above governing differential equation yields the following expressions for the inplane stress components σ_{xx}, σ_{zz} and σ_{xz} in the outer ply when $a > b$, which is the case for the lay-ups in the present investigation:

$$\sigma_{xx}(x) = \sigma_{xx}^o \left\{ 1 + \left(\frac{1}{\gamma} \right) \left(\frac{h_2}{h_1} \right) \left[\left(\frac{k_{22}}{k_{22} - k_{11}} \right) e^{-k_{11}\lambda} - \left(\frac{k_{11}}{k_{22} - k_{11}} \right) e^{-k_{22}x} \right] \right\}$$

$$\sigma_{zz}(x, z) = \sigma_{xx}^o \frac{1}{2} \left(\frac{1}{\gamma} \right) \left(\frac{k_{11} k_{22}}{k_{22} - k_{11}} \right) (k_{11} e^{-k_{11}x} - k_{22} e^{-k_{22}x}) [z - (h_1 + h_2)]^2 \qquad (12.1)$$

$$\sigma_{xz}(x, z) = -\sigma_{xx}^o \left(\frac{1}{\gamma} \right) \left(\frac{h_2}{h_1} \right) \left(\frac{k_{11} k_{22}}{k_{22} - k_{11}} \right) (e^{-k_{22}\lambda} - e^{-k_{11}x}) [z - (h_1 + h_2)]$$

where

$$k_{11} = \sqrt{a^2 - \sqrt{a^4 - b^4}} \qquad k_{22} = \sqrt{a^2 + \sqrt{a^4 - b^4}}$$

$$\gamma = \left(\frac{E_1}{E_2} \right) \left\{ \frac{1 + \left(\frac{E_1}{E_2} \right) \left[\left(\frac{h_2}{h_1} \right) (1 - v_2^2) - v_2^2 \right]}{1 + \left(\frac{E_1}{E_2} \right) \left(\frac{h_2}{h_1} \right) - v_1^2 \left(1 + \frac{h_2}{h_1} \right)} \right\}$$

$$a^2 = \frac{2 \left\{ \frac{E_2}{3G_2} \left(\frac{h_2}{h_1} \right)^3 + \frac{E_2}{3G_1} \left(\frac{h_2}{h_1} \right)^2 - v_1 \left[\frac{2}{3} \left(\frac{h_2}{h_1} \right)^3 + \left(\frac{h_2}{h_1} \right)^2 \right] + \frac{1}{6} \left(\frac{E_2}{E_1} v_1 + v_2 \right) \left(\frac{h_2}{h_1} \right)^2 \right\}}{h_1^2 \left\{ \frac{1}{5} \left(\frac{h_2}{h_1} \right)^5 - \frac{2}{3} \left(\frac{h_2}{h_1} \right)^4 \left(1 + \frac{h_2}{h_1} \right) + \left(\frac{h_2}{h_1} \right)^3 \left(1 + \frac{h_2}{h_1} \right)^2 + \frac{1}{5} \left(\frac{h_2}{h_1} \right)^2 \right\}}$$

$$b^4 = \frac{4 \left(\frac{h_2}{h_1} \right) \left\{ 1 + \left(\frac{E_2}{E_1} \right) \left(\frac{h_2}{h_1} \right) \right\}}{h_1^4 \left\{ \frac{1}{5} \left(\frac{h_2}{h_1} \right)^5 - \frac{2}{3} \left(\frac{h_2}{h_1} \right)^4 \left(1 + \frac{h_2}{h_1} \right) + \left(\frac{h_2}{h_1} \right)^3 \left(1 + \frac{h_2}{h_1} \right)^2 + \frac{1}{5} \left(\frac{h_2}{h_1} \right)^2 \right\}}$$

In the above relations, the subscripts 1 and 2 refer to the outer and inner plies, respectively, and σ_{xx}^o is the undisturbed far-field axial stress in the outer ply. Using equation (12.1), the maximum shear stress σ_{13}^{max} is obtained as a function of the axial coordinate x for different $z = z_0$ elevations from the relation:

$$\sigma_1 - \sigma_3 = 2\sigma_{13}^{max} = \sqrt{(\sigma_{xx} - \sigma_{zz})^2 + 4\sigma_{xz}^2} \tag{12.2}$$

which can readily be normalized with respect to the far-field stress σ_{xx}^o for comparison with the experimental data presented in Figure 12.3.

The solution obtained by Zak and Williams gives the local stress field in the immediate vicinity of the crack tip for a crack perpendicular to a bi-material interface separating two semi-infinite half-spaces with different elastic constitutive parameters. The tip of the crack is situated directly on the interfacial line separating the two materials and the crack is loaded in a symmetric fashion. The local solution satisfies the biharmonic equation, continuity of tractions and displacements across the bi-material interface and traction-free conditions on the crack faces. Consequently, it is an exact solution to the idealized mathematical model of plane elasticity. In terms of a local polar coordinate system (r, θ) situated at the crack tip, with the angle θ measured in a counterclockwise fashion from the reference line colinear with the crack, the local inplane stresses are given by the following expressions:

$$\sigma_{\theta\theta} = r^{\lambda-1}\lambda(\lambda+1)F_2(\theta)$$
$$\sigma_{rr} = r^{\lambda-1}[F_2''(\theta) + (\lambda+1)F_2(\theta)] \tag{12.3}$$
$$\sigma_{r\theta} = -r^{\lambda-1}\lambda F_2'(\theta)$$

where

$$F_2(\theta) = d_2[\gamma \cos(1+\lambda)\theta + \cos(1-\lambda)\theta]$$

In the above, d_2 is a constant which depends on the type of the external loading and λ is a material parameter which depends on the properties of the adjacent media. The parameter λ is obtained from the characteristic equation given below which is derived from homogeneous boundary conditions ensuring continuity of interfacial tractions and displacements and traction-free crack faces:

$$2\alpha \cos \pi\lambda - (\beta\lambda^2 + \gamma) = 0 \tag{12.4}$$

where

$$\alpha = (m + k_2)(1 + mk_1)$$
$$\beta = -4(m + k_2)(1 - m)$$
$$\gamma = (1 - m)(m + k_2) + (1 + mk_1)(m + k_2) - m(1 + k_1)(1 + mk_1)$$
$$m = G_2/G_1$$
$$k_{1,2} = 3 - 4\nu_{1,2} \text{ plane strain}$$
$$k_{1,2} = (3 - \nu_{1,2})/(1 + \nu_{1,2}) \text{ plane stress}$$

In the present context, λ is a real number which lies between 0 and 1. For the polyester-aluminum-polyester laminate under plane stress conditions in the $x - z$ plane for example, the value of λ calculated from equation (12.4) is 0.2061.

In order to determine the maximum shear stress σ_{13}^{max} in the $x - z$ plane in the immediate vicinity of the crack tip, the following relation can be used:

$$\sigma_1 - \sigma_3 = 2\sigma_{13}^{max} = \sqrt{(\sigma_{rr} - \sigma_{\theta\theta})^2 + 4\sigma_{r\theta}^2} \tag{12.5}$$

Using the relation $\tan\theta = x/z_0$ in equation (12.3), the values of the maximum shear stress σ_{13}^{max} can be generated from equation (12.5) at fixed $z = z_0$ locations above the crack for different values of the axial coordinate x.

12.2.3 Experimental/analytical correlation and discussion

The predictions of the approximate analytical model developed by Vasil'ev *et al.* are included with the experimental results presented in Figure 12.3. As the variation of the normal axial stress across the ply thickness has been neglected in the above approximate model, the correlation between the analytical solution and isodyne data is mainly of a qualitative nature at points sufficiently remote from the crack. In the region close to the crack tip however, the empirical and analytical results differ radically. Whereas the approximate analytical model yields continuously increasing values for σ_{13}^{max}, the experimental data indicate local maxima as already mentioned. On the other hand, the local solution for the stress distribution around a crack perpendicular to a bi-material interface developed by Zak and Williams does indeed exhibit such local maxima of σ_{13}^{max} near the crack tip, Figure 12.4. However, the influence of the Poisson's ratio on the localized crack tip stresses predicted by the above plane elasticity solution does not appear to be as significant as indicated by the experimental results.

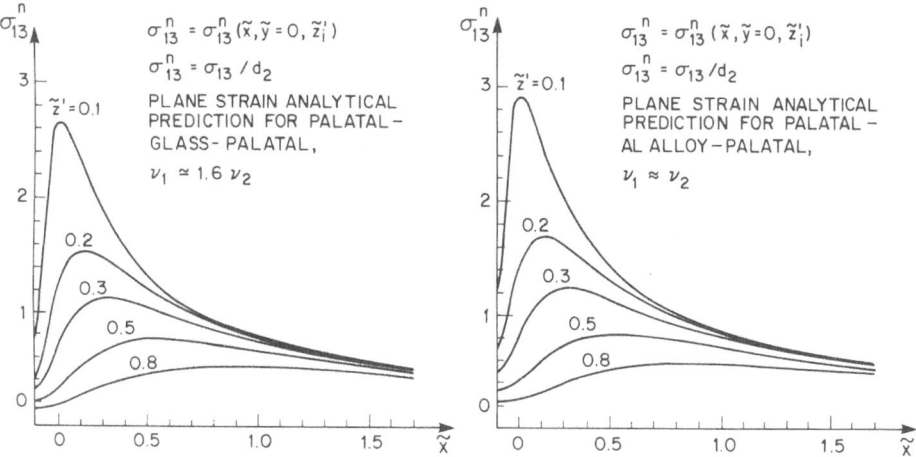

Fig. 12.4. Distributions of the maximum shear stress at different elevations from the crack tip according to Zak and Williams [10].

The fracture surface of the polyester-glass-polyester configuration is shown in Figure 12.5. As seen in the figure, there is no evidence of debonding along the bi-material interface and the fracture surface has a very smooth appearance away from the crack tip. Evidence of intense stress concentrations in the immediate vicinity of the crack tip is indicated by the local undulations in the fracture plane. These undulations may be caused by the presence of a highly inhomogeneous three-dimensional state of stress with a pronounced through-the-width variation in the vicinity of the crack tip. The extent of variation of the maximum shear stress σ_{13}^{max} across the width of the laminate above the crack front can be readily determined using the results obtained with the isodyne technique.

12.3 Laminated beam with interlaminar disbonds

In this section we discuss the application of the isodyne technique to the technologically important problem of a crack along an interface separating two dissimilar materials. This problem is characteristically different from the classical fracture mechanics problems in that coupling exists between normal and shear stresses in the vicinity of the crack tip along the bi-material interface under the action of Mode I or II far-field loading. Consequently, the classical crack extension criteria based on the concept of self-similar crack growth cannot be applied in a straightforward fashion. This, in turn, has lead to the introduction of new concepts for mixed-mode crack growth [1, 2, 7] which must be tested experimentally. The first step in this direction is the verification of the interfacial stress fields in the vicinity of the crack tip obtained from plane elasticity solutions using independent experimental techniques which do not assume a priori functional forms of existing stresses. This investigation was undertaken to illustrate the efficacy of the isodyne technique in the determina-

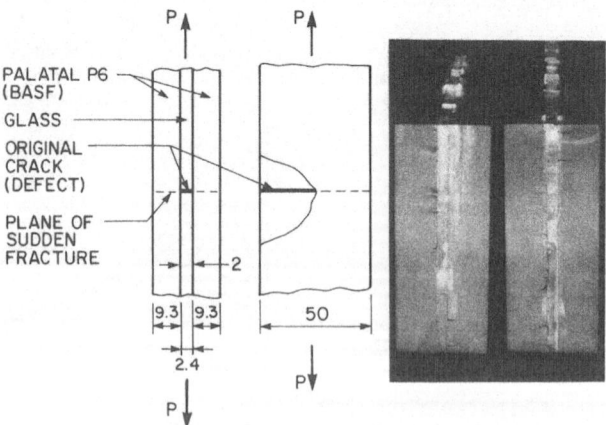

Fig. 12.5. Fracture surface of a three-ply composite structure with a transverse crack.
Left: sketch of specimen. Right: picture of a catastrophic fracture in the plane of the initial crack.

tion of interfacial stresses in a laminated beam with interlaminar disbonds under three-point bending, Figure 12.6. The experimentally obtained interfacial tractions are subsequently compared with a plane elasticity solution which incorporates the singular features of the stress field in the vicinity of the crack tip.

12.3.1 Experimental investigation and results

The overall length of the laminated beam was 220 mm, the height was 34 mm and the width was 10 mm. The beam was laminated by bonding an aluminum strip (2024 alloy) 6 mm in height to a 28 mm polyester (Palatal P6) beam. The bonding agent was an epoxy resin (Araldite 6010) and the bonding procedure resulted in a bond thickness that was on the order of a hundreth of a millimeter. Interlaminar disbonds 10 mm long and centered symmetrically 45 mm from the midpoint of the composite beam were introduced with the aid of thin teflon strips which were subsequently removed after the lamination procedure was completed.

The laminated beam was placed symmetrically on supports situated 180 mm apart and the load was applied through a 30 mm loading disk placed in contact with the upper aluminum layer. The geometry of the beam and loading conditions are summarized in Figure 12.6. The recording of data was carried out in an immersion tank filled with Dow Corning vacuum pump fluid no. 704 in order to minimize the reflection problems due to the laminated construction of the specimen. The recordings of isodyne fields and the experimental results which will be subsequently presented have been obtained at the total applied load of 2233 N.

Figure 12.7 presents the recordings of the x- and y-isodyne fields obtained in the $z = 0.0$ mm plane, along with the recordings of the dark and light isochromatics fields included for comparison. The disturbances in the stress

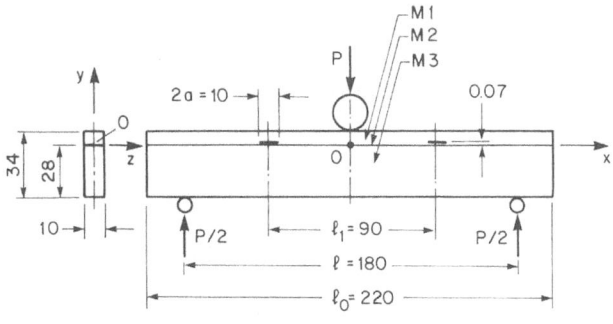

M1 : AL ALLOY 2024
M2 : EPOXY ARALDITE 6010 (CIBA)
M3 : POLYESTER PALATAL P6 (BASF)

Fig. 12.6. Laminated beam with interlaminar disbonds. Specimen geometry and loading conditions.

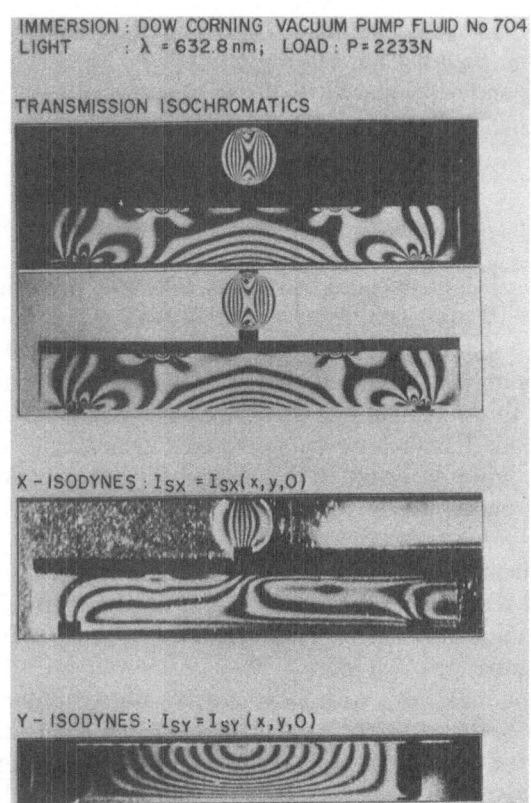

Fig. 12.7. Recordings of x- and y-isodyne fields in the $z = 0.0$ mm plane and the corresponding transmission isochromatics.

fields caused by the presence of the two interlaminar disbonds are clearly visible in the two sets of recordings, which also illustrate the near-perfect symmetry of the stress distributions with respect to the centrally applied load. The inter-laminar tractions evaluated from the isodyne data along the interface separating the upper and lower layers of the composite beam are presented in Figure 12.8.

12.3.2 Analytical model

The analytical model chosen for comparison with the experimental results is based on the two-dimensional elasticity formulation which has been proposed by Erdogan and Gupta [4], among others. It employs an infinite Fourier transform technique and thus is strictly applicable to finite-height, laminated plane geometries extending to $\pm \infty$ in the x-direction. Consequently, the experimental-analytical correlation is not expected to be valid in regions close to the free end of the structure. Although the formulation yields a physically inadmissible solution in the crack-tip region less than $10^{-6} - 10^{-4}$ of the crack length as discussed by Malyshev and Salganik [6], this limitation is frequently deemed of no practical consequence as all infinitesimal elasticity solutions for this type of problem must break down so close to the crack tip, including those which circumvent the problem of crack-tip material interpenetration [9].

The analytical expressions for the tractions in the plane of lamination can be shown to reduce to the following set of singular integral equations coupled in the u and v components of the displacement discontinuities $\theta_u(x)$ and $\theta_v(x)$ across the crack interfaces when the crack faces are free of tractions:

$$
\begin{Bmatrix} -\sigma_{yy}(x) \\ -\sigma_{xy}(x) \end{Bmatrix} = \begin{bmatrix} 0 & B_{12} \\ -B_{12} & 0 \end{bmatrix} \begin{Bmatrix} \theta_v(x) \\ \theta_u(x) \end{Bmatrix} +
$$

$$
+ \frac{1}{\pi} \int_{-a}^{+a} \begin{bmatrix} B_{11} & 0 \\ 0 & B_{11} \end{bmatrix} \begin{Bmatrix} \theta_v(x') \\ \theta_u(x') \end{Bmatrix} \frac{dx'}{x' - x} +
$$

$$
+ \frac{1}{\pi} \int_{-\infty}^{+\infty} \begin{bmatrix} A_{11}(s, x, x') & A_{12}(s, x, x') \\ A_{21}(s, x, x') & A_{22}(s, x, x') \end{bmatrix} \begin{Bmatrix} \theta_v(x') \\ \theta_u(x') \end{Bmatrix} ds\, dx' -
$$

$$
- \frac{1}{\pi} \int_{-\infty}^{+\infty} \begin{bmatrix} F_{11}(s, x) & F_{12}(s, x) \\ F_{21}(s, x) & F_{22}(s, x) \end{bmatrix} \begin{Bmatrix} P_y^+ \\ P_y^- \end{Bmatrix} ds, \tag{12.6}
$$

where

$\theta_v(x), \theta_u(x) = 0$ in the region $|x| > a$, a being half the crack length,

s = Fourier-transform variable,

B_{11}, B_{12} = constant parameters describing the degree of material property mismatch of the upper and lower beam,

$A_{ij}(s, x, x'), F_{ij}(s, x)$ = regular kernels obtained from inversion of the Fourier-transformed structural matrix of the composite beam,

P_y^-, P_y^- = externally applied normal tractions per unit width.

It ought to be mentioned that the applicability of the traction-free condition in the above formulation is verified by experimental data which indicates that the traction components $\sigma_{xy}(x)$ and $\sigma_{yy}(x)$ vanish on the crack faces.

The solution of the above system of equations closely follows the general methodology outlined by Erogan [3] where a weighted residual technique is employed to reduce the singular integral equations to a set of algebraic equations in unknown coefficients of Jacobi polynomials used to approximate the displacement discontinuities $\theta_v(x)$ and $\theta_u(x)$. Determination of the unknown coefficients, which has been carried out using an 11-term expansion in Jacobi polynomials and 384 Gaussian integration points in the numerical evaluation of the integrals in the above formulation, consequently allows one to obtain the full field interfacial stresses given above.

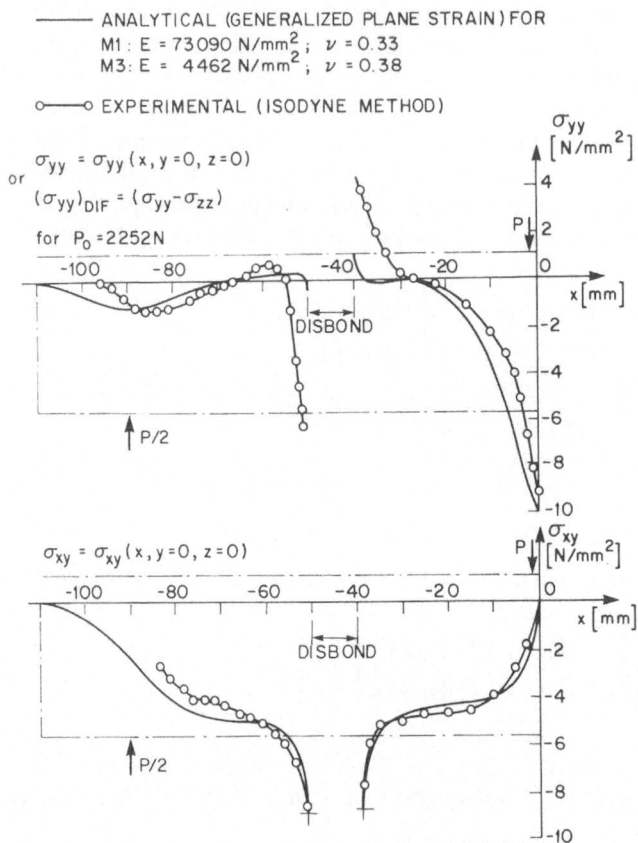

Fig. 12.8. Interlaminar traction distributions in the laminated beam with interfacial disbonds under three-point loading.

12.3.3 Experimental/analytical correlation and discussion

The predictions of the analytical model, equation (12.6), for the shear and normal traction components acting in the plane $z = 0$ along the interface containing the disbonds for the generalized plane strain case is included in Figure 12.8 for comparison with the experimentally generated results. It ought to be mentioned that the difference between the plane stress and generalized plane strain results for the interlaminar tractions as predicted by the analytical model is not very significant in this case and thus the plane stress results are not shown.

Several interesting features of the correlation study are noteworthy. The correlation between the experimentally obtained values of the shear traction component $\sigma_{xy}(x)$ and the analytical results is very good as the interlaminar crack is approached from the center of the beam, i.e., in the inner region. In the outer region the difference between experimental and analytical results increases somewhat with the distance from the crack tip, although the functional form of the two curves is very similar.

The comparison between the analytical results for $\sigma_{yy}(x)$ and the experimentally obtained differential stress values $(\sigma_{yy})_{diff} = \sigma_{yy} - \sigma_{zz}$ appears to indicate the presence of a significant through-the-thickness normal stress component in the polyester beam along the lamination plane in the vicinity of the interlaminar disbonds and the points of external load application. The difference between the externally applied normal tractions and the integral of the difference $\sigma_{yy} - \sigma_{zz}$ along the plane of lamination as determined from the isodyne recordings approaches 30%. On the other hand, this difference becomes very small in the plane located half way across the height of the polyester beam where the methods of transmission photoelasticity are applicable.

The fracture patterns of two failed bi-material beams are presented in Figure 12.9. The primary crack appears to have initated at the outer tip of the disbond at an angle of approximately 66 deg and then apparently changed direction and propagated dynamically at an angle of 59 deg with respect to the plane of lamination. The subsequent fracture process involved partial delamination of the beam components and clearly dynamic fracture across the height of the polyester beam not far from the inner crack tip of the other symmetrically located disbond. The chronological sequence of particular fracture and delamination processes was not determined. The overall fracture process was catastrophic for all the specimens tested with no apparently significant initial quasistatic crack growth along the plane of lamination in the outer region. All fractures occurred several, or more, minutes after application of the critical load.

It is interesting to note that the very rapid crack growth initiated at the outer tip where the normal traction component σ_{yy} is compressive as predicted both by the isodyne and analytical calculations. Furthermore, both the experimental data and the analytical model indicate that the shear traction component

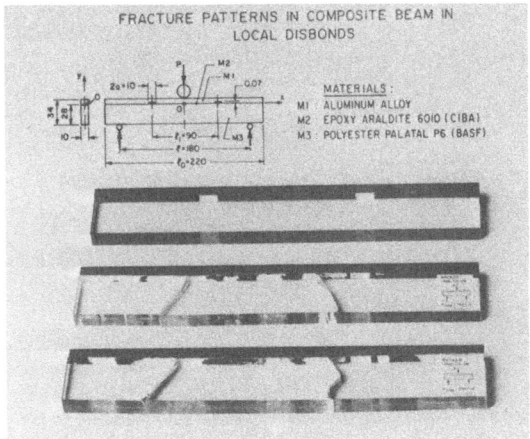

Fig. 12.9. Fracture patterns in laminated beams with interlaminar disbonds under three-point loading. Virgin beam (top) and two fractured beams (bottom). Primary (initial) fractures are on the left and secondary (dynamic) fractures are on the right.

increases faster near the external crack tip than the internal tip, although the analytical model predicts that this difference is smaller than obtained experimentally. On the other hand, the axial stress component σ_{xx} in the polyester beam along the plane of lamination is significant and has different sign in the vicinity of the outer and inner crack tips: it is tensile as the disbond is approached from the outer region and compressive as the disbond is approached from the inner region.

12.4 References

[1] Beuth, Jr., J. L. and Herakovich, C. T., "On Fracture of Fibrous Composites", Composites 86: Recent Advances in Japan and the United States, K. Kowata, S. Umekawa and A. Kobayashi eds., (Proc. Japan-US CCM-III), 267—272, Tokyo (1986).

[2] Buczek, M. B. and Herakovich, C. T., "A Normal Stress Criterion for Crack Extension Direction in Orthotropic Composite Materials", *J. Comp. Materials* **19**(6), 544—553 (1985).

[3] Erdogan, F., "Approximate Solutions of Systems of Singular Integral Equations", *SIAM J. Appl. Math.* **17**, 1041—1059 (1969).

[4] Erdogan, F. and Gupta, G., "Layered Composites with an Interface Flaw", *Int. J. Solids Structures* **7**, 1089—1107 (1971).

[5] Hashin, Z., "Analysis of Cracked Laminates: A Variational Approach", *Mechanics of Materials* **4**, 1—16 (1985).

[6] Malyshev, B. M. and Salganik, R. L., "The Strength of Adhesive Joints Using the Theory of Fracture", *Int. J. Fracture Mech.* **1**, 114—128 (1965).

[7] Sih, G., "Mechanics and Physics of Energy Density Theory", *Theoret. Appl. Fracture Mech.* **4**, 157—173 (1985).

[8] Vasil'ev, V. V., Dudchenko, A. A. and Elpat'evskii, A. N., "Analysis of the Tensile Deformation of Glass-Reinforced Plastics", *Mekhanika Polimerov* **1**, 144—147 (1970).

[9] Wang, S. S. and Choi, I., "Boundary Layer Effects in Composite Laminates: Part II — Free-Edge Stress Solutions and Basic Characteristics", *J. Appl. Mech.* **49**, 549—560 (1982).

[10] Zak, A. R. and Williams, M. L., "Crack Point Stress Singularities at a Bi-Material Interface" *J. Appl. Mech.* **30**, 142—143 (1963).

Author index

Subject index

ENGINEERING APPLICATION OF FRACTURE MECHANICS

Editor-in-Chief: George C. Sih

1. G. C. Sih and L. Faria (eds.), Fracture mechanics methodology: Evaluation of structural components integrity. 1984.
   ISBN 90-247-2941-6.
2. E. E. Gdoutos, Problems of mixed mode crack propagation. 1984.
   ISBN 90-247-3055-4.
3. A. Carpinteri and A. R. Ingraffea (eds.), Fracture mechanisms of concrete: Material characterization and testing. 1984.
   ISBN 90-247-2959-9.
4. G. C. Sih and A. DiTommaso (eds.), Fracture mechanics of concrete: Structural application and numerical calculation. 1984.
   ISBN 90-247-2960-2.
5. A. Carpinteri, Mechanical damage and crack growth in concrete: Plastic collapse to brittle fracture. 1986.
   ISBN 90-247-3233-6.
6. J. W. Provan (ed.), Probabilistic fracture mechanics and reliability. 1987.
   ISBN 90-247-3334-0.
7. A. A. Baker and R. Jones (eds.), Bonded repair of aircraft structures. 1987.
   ISBN 90-247-3606-4.
8. J. T. Pindera and M.-J. Pindera, Isodyne stress analysis. 1989.
   ISBN 0-7923-0269-9.